Handbook of Advanced Building Construction

Handbook of Advanced Building Construction

Contributors

A. W. M. van Schijndel et al.

AURIS
Reference

www.aurisreference.com

Handbook of Advanced Building Construction

Contributors: A. W. M. van Schijndel et al.

Published by Auris Reference Limited

www.aurisreference.com

United Kingdom

Handbook of Advanced Building Construction

ISBN: 978-1-78154-827-1

British Library Cataloguing in Publication Data
A CIP record for this book is available from the British Library

Printed in the United Kingdom

Exclusively distributed by CBS Publishers & Distributors Pvt. Ltd.

Sales & Distribution Rights only for India, Pakistan, Bangladesh, Sri Lanka, Nepal and Bhutan. This book is not to be sold outside these territories.

Contents

List of Abbreviations

AHP	Analytical Hierarchy Process
BIM	Building Information Modelling
BPM	Business Process Management
BPMS	Business Process Management System or Suite
BRANZ	Building Research Association of New Zealand
BREEAM	Building Research Establishment Environmental Assessment Method
BUs	Building Units
CBR	Case-based reasoning
CCA	Copper Chrome Arsenate
CDMA	Code-division multiple-access
CRM	Customer relationship management
CSI	Construction Standards Institute
DSS	Decision support system
EATT	Environmental Assessment Trade-off Tool
EDM	Electrical discharge machine
ESGs	Electrical strain gauges
FOSs	Fiber optic sensors
GBCs	Green Building Councils
GPS	Global positioning system
ICT	Information and Communication Technology
ILT	Inter-Laboratory Test
KPIs	Key performance indicators
LCA	Life Cycle Assessment
LCC	Life Cycle Cost
LDS	Laser displacement sensor
LEED	Leadership in Energy and Environmental Design
LF	Low-frequency
MCDM	Multi-criteria decision-making
MSDSS	Material Selection Decision Support System
NACA	National Advisory Committee for Aeronautics
NIST	National Institute Standards and Technology
PC	Personal computer
PPC	Percent Plan Completed
RC	Reinforced-concrete
RRT	Round Robin Test
SHM	Structural health monitoring

List of Contributors

A. W. M. van Schijndel
Unit Building Physics and Services, Department of the Built Environment, Eindhoven University of Technology, Eindhoven, Netherlands

Chiara Scrosati
Construction Technologies Institute, National Research Council of Italy, via Lombardia 49, 20098 San Giuliano Milanese (MI), Italy

Fabio Scamoni
Construction Technologies Institute, National Research Council of Italy, via Lombardia 49, 20098 San Giuliano Milanese (MI), Italy

María Dolores Andújar-Montoya
Department of Building and Urbanism, Polytechnic University College, University of Alicante, Carretera de San Vicente del Raspeig, s/n, 03690 Alicante, Spain

Virgilio Gilart-Iglesias
Department of Computer Science and Technologies, Polytechnic University College, University of Alicante, Carretera de San Vicente del Raspeig, s/n, 03690 Alicante, Spain

Andrés Montoyo
Department of Software and Computing Systems, Polytechnic University College, University of Alicante, Carretera de San Vicente del Raspeig, s/n, 03690 Alicante, Spain

Diego Marcos-Jorquera
Department of Computer Science and Technologies, Polytechnic University College, University of Alicante, Carretera de San Vicente del Raspeig, s/n, 03690 Alicante, Spain

Ian Skelton
School of Civil and Building Engineering, Loughborough University, LE11 3TU, UK

Peter Demian
School of Civil and Building Engineering, Loughborough University, LE11 3TU, UK

Jacqui Glass
School of Civil and Building Engineering, Loughborough University, LE11 3TU, UK

Dino Bouchlaghem
School of Architecture, Nottingham Trent University, NG1 4BU, UK

Chimay Anumba
Department of Architectural Engineering, Pennsylvania State University, University Park, State College, PA 16801, USA

Junli Yang
Department of Construction, School of Architecture and the Built Environment, University of Westminster, London, UK.

Ibuchim Cyril B. Ogunkah
Department of Construction, School of Architecture and the Built Environment, University of Westminster, London, UK.

Roger Birchmore
Department of Construction, Unitec Institute of Technology, Auckland, New Zealand

Andy Pivac
Department of Building Technology, Unitec Institute of Technology, Auckland, New Zealand

Robert Tait
Department of Building Technology, Unitec Institute of Technology, Auckland, New Zealand

Agnieszka Zalejska-Jonsson
Real Estate and Construction Management, School of Architecture and the Built Environment, KTH Royal Institute of Technology, Brinellvägen 1, Stockholm 10044, Sweden

Hans Lind
Real Estate and Construction Management, School of Architecture and the Built Environment, KTH Royal Institute of Technology, Brinellvägen 1, Stockholm 10044, Sweden

Staffan Hintze
Highway and Railway Engineering, School of Architecture and the Built Environment, KTH Royal Institute of Technology, Brinellvägen 23, Stockholm 10044, Sweden

Hyo Seon Park
Department of Architectural Engineering, Yonsei University, 134 Shinchon-dong, Seoul 110-732, Korea
Center for Structural Health Care Technology in Buildings, Yonsei University, 134 Shinchon-dong, Seoul 110-732, Korea

Sewook Son
Center for Structural Health Care Technology in Buildings, Yonsei University, 134 Shinchon-dong, Seoul 110-732, Korea

Se Woon Choi
Center for Structural Health Care Technology in Buildings, Yonsei University, 134 Shinchon-dong, Seoul 110-732, Korea

Yousok Kim
Center for Structural Health Care Technology in Buildings, Yonsei University, 134 Shinchon-dong, Seoul 110-732, Korea

Jing Ma
School of Civil Engineering & Architecture, Chongqing Jiaotong University, Chongqing, China

Preface

Construction is the process of constructing a building or infrastructure. Construction differs from manufacturing in that manufacturing typically involves mass production of similar items without a designated purchaser, while construction typically takes place on location for a known client. *Handbook of Advanced Building Construction* contains everything you need to know about the construction process. The objective of first chapter is to investigate inverse modeling techniques as a tool for the detection of moisture leakage locations in building constructions from inside surface moisture patterns. Second chapter shows the uncertainty of field measurements of a lightweight wall, a heavyweight floor, a façade with a single glazing window and a façade with double glazing window that are analyzed by a round robin test (RRT). A construction management framework for mass customization in traditional construction has been presented in third chapter. Fourth chapter describes the scientific advancement in applying aerodynamic theory, refined via modelling and testing, to a specific aspect of the building process of a tall building with potentially significant time and commercial benefits. Fifth chapter discusses the process of developing a decision-support system to support choices in low-cost green building materials. Sixth chapter outlines the early findings of a research project that moves research from desktop simulation to exploring the impact of a construction employing such a vapor check on unoccupied conditions in a real house. Seventh chapter examines how technologies used in energy-efficient residential building construction affect the available saleable floor area and how this impacts profitability of investment. In eighth chapter, a wireless laser range finder system has been employed to directly measure the deflection of structural members in an irregular building that is currently under construction. Influence analysis of a new building to the bridge pile foundation construction has been presented in ninth chapter. Last chapter outlines and draws conclusions about different aspects of the material efficiency of buildings and assesses the significance of different building materials on the material efficiency.

Chapter 1

EVALUATION OF INVERSE MODELING TECHNIQUES FOR PINPOINTING WATER LEAKAGES AT BUILDING CONSTRUCTIONS

A. W. M. van Schijndel

Unit Building Physics and Services, Department of the Built Environment, Eindhoven University of Technology, Eindhoven, Netherlands

ABSTRACT

The location and nature of the moisture leakages are sometimes difficult to detect. Moreover, the relation between observed inside surface moisture patterns and where the moisture enters the construction is often not clear. The objective of this paper is to investigate inverse modeling techniques as a tool for the detection of moisture leakage locations in building constructions from inside surface moisture patterns. It is concluded that although the presented methodology is promising, more research is needed to confirm its usability.

INTRODUCTION

Hunting Lodge St. Hubertus is one of the most prominent buildings from the beginning of the twentieth century and is noted in the top 100 list of Dutch monuments. The conservation of the building and its interior are of great importance. The Dutch Government Building Department, which takes care of the maintenance of the building, has expressed their concern about the observed damage due to high moisture levels by the rain that finds its way to the interior at places of inadequate detailing and therefore causes damage mainly near openings in the facade and on the inside of the facade below balconies. The main problem is that the location and nature of the moisture leakages are not easily detectable. We often don't know the relation between the observed inside surface moisture patterns and where the moisture enters the construction. The objective is to investigate inverse modeling techniques as a tool for the detection of moisture leakage locations in building constructions, i.e. we want to investigate the (in)possibilities of pinpointing moisture leakages from inside surface moisture patterns using inverse modeling techniques.

Summary of the Observed Moisture Problems at the Hunting Lodge St. Hubertus

Hunting Lodge St. Hubertus is located on the northern side of the Dutch National Park "De Hoge Veluwe". The Hunting Lodge is built as a guesthouse between 1916 and 1922, by Holland's most well-known architect from that time, H. P. Berlage. The building consists of a low-rise rectangular volume with wings that stretch out diagonally and with a characteristic high tower of over 30 meters height in the middle of the building (see Figure 1). A large pond is situated south-west of the building and the building is surrounded by forest in all other directions.

The damage that occurs in the tower was systematically inspected to enable a thorough assessment of the possible causes of the moisture problems by Briggen et al. (2009). The damage on the inside of the tower, and where possible also on the outside, is systematically inspected. The location and type of each moisture problem are documented in a table, illustrated with a picture of the damage. The moisture problems that manifest themselves in the tower of the Hunting Lodge can be divided in the following categories: efflorescence, cracking, soiling, moist spots, mechanical damage and biological growth.

A few pictures of the moisture damage that occurs in the tower are shown in Figure 2 (Briggen et al. [1]).

Figure 1: Hunting lodge St. Hubertus.

Regarding the location of the damage it can be concluded from the inspection that most damage occurs on the interior surface of the south-west facade of the tower. Since the prevailing wind direction in the Netherlands is south-west, which means that the south-west facade of the tower is subjected to wind-driven rain the most, there appears to be a connection between the rain load of the facade and the damage on the inside. There are no clear differences between the damage on lower or higher floors or between the damage on the middle and on the sides of the facade. Most damage occurs near openings in the facade and on the interior surface of the facade below balconies.

Figure 2: Observed moisture damage in the tower of the building: moist spots and efflorescence.

Methodology

The research method contained the following steps:

(Step 1) Measurements of the external climate, indoor climate and surface conditions (Section 2).

(Step 2) Computational model development and calibration of the parameters (Section 3).

(Step 3) From one of the prominent observed moisture spot, an inverse modeling technique was used to determine the moisture entrance for pinpointing the most likely water leakage (Section 4).

(Step 4) Discussion and conclusions of the approach (Section 5).

MEASUREMENTS

The data set is part of the measurement program at the Hunting Lodge St. Hubertus site, performed during 2006- 2007 by Briggen [1] . Details of this project can be found in Briggen et al. [2] [3] . One of problems seemed to be high moisture contents at the inside surface of the facade of the tower. The construction of this facade is shown in Figure 3.

The outside climate conditions were measured by a weather station within 50m from the building. The inside air temperature and relative humidity were measured using standard equipment (seeFigure 4). A representation of inside surface conditions was obtained by placing a small box (5 cm × 5 cm × 1 cm) against the wall and measures the air temperature and relative humidity inside. The estimation of the measurement error of this method is left over for future research.

The data consists of the measured time series of the indoor and outdoor climate as presented inFigure 5 and Figure 6.

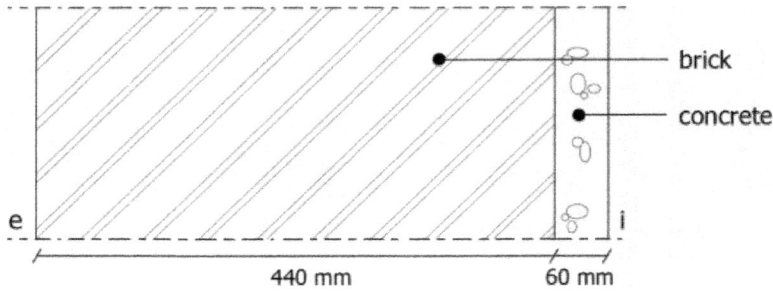

Figure 3: The building facade.

Figure 4: Measurement of the surface temperature and relative humidity using a box.

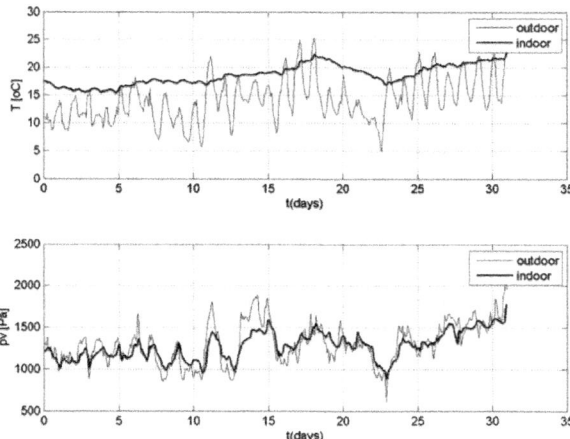

Figure 5: The measured air temperatures (top) and calculated vapour pressures (bottom, from measured T/RH).

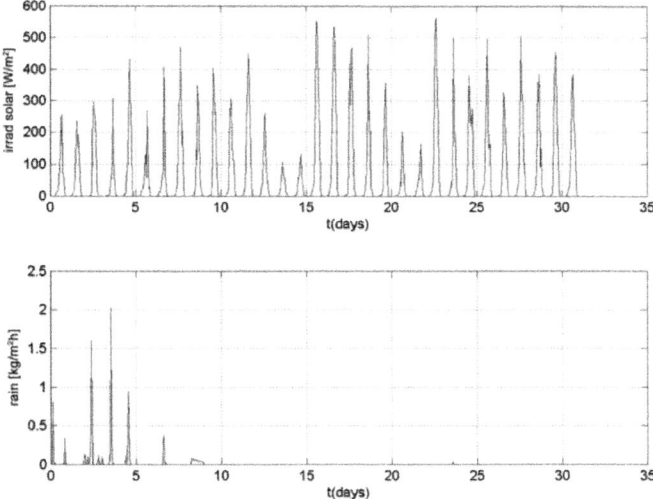

Figure 6: The measured solar irradiance (top) and rain intensity (bottom).

MODELING

The multiphysics modeling approach of van Schijndel [4] - [6] is used. The heat and moisture transport can be described by the following PDEs:

$$C_T \frac{\partial T}{\partial t} = \nabla \cdot \left(K_{11} \nabla T + K_{12} \nabla LPc \right)$$

$$(1)$$

$$C_{LPc} \frac{\partial LPc}{\partial t} = \nabla \cdot (K_{21} \nabla T + K_{22} \nabla LPc)$$

(2)

where t is time [s]; T is temperature [°C]; The reminder of the terms in the heat (1) and moisture Equation (2) are explained below:

$$LPc = {}^{10}\log(Pc)$$

(3)

$$C_T = \rho \cdot c$$

(4)

$$K_{11} = \lambda$$

(5)

$$K_{12} = -l_{lv} \cdot \delta_p \cdot \phi \cdot \frac{\partial Pc}{\partial LPc} \cdot \text{Psat} \cdot \frac{M_w}{\rho_a RT}$$

(6)

$$C_{LPc} = \frac{\partial w}{\partial Pc} \cdot \frac{\partial Pc}{\partial LPc}$$

(7)

$$K_{22} = -K \cdot \frac{\partial Pc}{\partial LPc} - \delta_p \cdot \phi \cdot \frac{\partial Pc}{\partial LPc} \cdot \text{Psat} \cdot \frac{M_w}{\rho_a RT}$$

(8)

$$K_{21} = \delta_p \cdot \phi \cdot \frac{\partial \text{Psat}}{\partial T}$$

(9)

where Pc is capillary pressure [Pa]; ρ is material density [kg/m³]; c is specific heat capacity [J/kg·K]; λ is thermal conductivity [W/mK]; l_{lv} is specific latent heat of evaporation [J/kg]; δ_p vapour permeability [s]; ϕ is relative humidity [-]; Psat is saturation pressure [Pa]; $M_w = 0.018$ [kg/mol]; $R = 8.314$ [J/mol·K]; ρ_a is air density [kg/m³]; w is moisture content [kg/m³]; K is liquid water permeability [s].

MatLab is used for the implementation of material and boundary properties. These functions are used to convert measurable material properties such as K, ϕ, δ_p and λ which are dependent on the moisture content into PDE coefficients which are dependent on the LPc and T. This is schematically shown in Figure 7.

The material database of DELPHIN [7] is used to provide material properties for the first guess. For brick, the Brick material properties of

DELPHIN are used with constant $\rho = 1700$; $c = 840$; $\lambda = 0.85$ and variable moisture properties using the tables. For concrete, the Lime plaster properties of DELPHIN $(\rho = 1800; c = 840; \lambda = 1.05)$ are used in the same way. From these data, the PDE coefficients were determined together with the boundary conditions implemented using the COMSOL [8] model of Section 3. Figure 8 and Figure 9 show the results.

Figure 8 shows that the simulated inside surface temperature is already quite close to the measured one.

The simulated relative humidity at the inside surface of Figure 9 seems to be less close to the measured one compared to the previous figure. This gives also rise to the just mentioned questions. For each material and at each point the vapour pressure can be calculated using a similar corresponding function.

DETERMINATION OF MOISTURE SOURCE CHARACTERISTICS

In this section we try to reproduce the following observed moisture spots (see Figure 10).

The modeling approach of the previous section was used. The mesh of (simplified) geometry is presented in Figure 11.

The first step of the inverse modeling procedure is to switch one or more boundary conditions from dry into wet and then investigate it's effect on the inside surface moisture print. For example,Figure 13 shows the simulated profile at the inner surface by switching the location provided inFigure 12 from dry into wet.

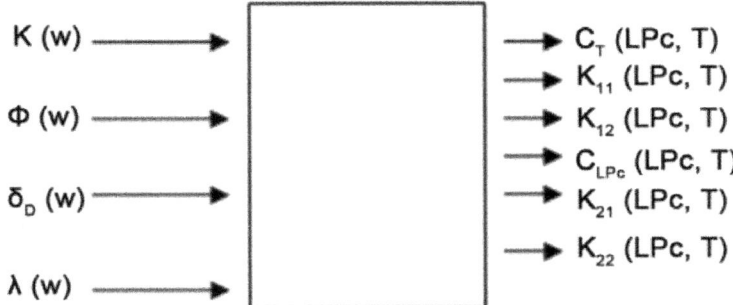

Figure 7: The conversion from measurable material properties into PDE coefficients.

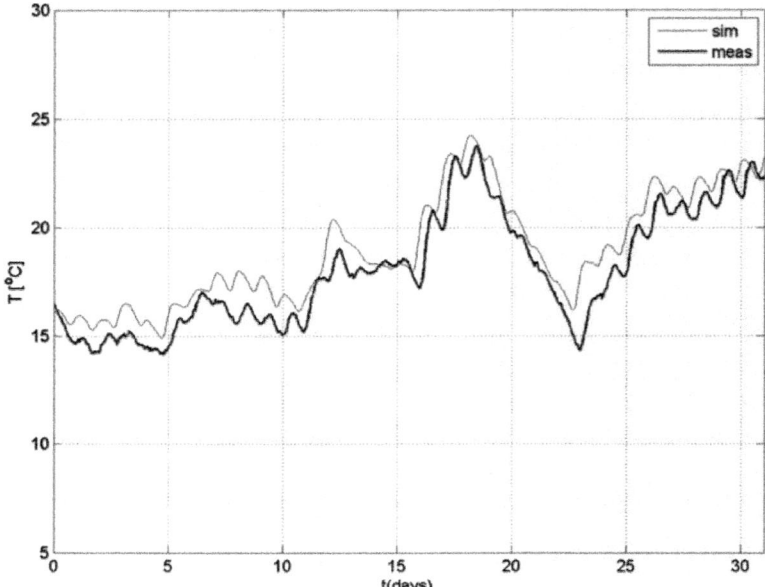

Figure 8: The measured and simulated inside surface temperature.

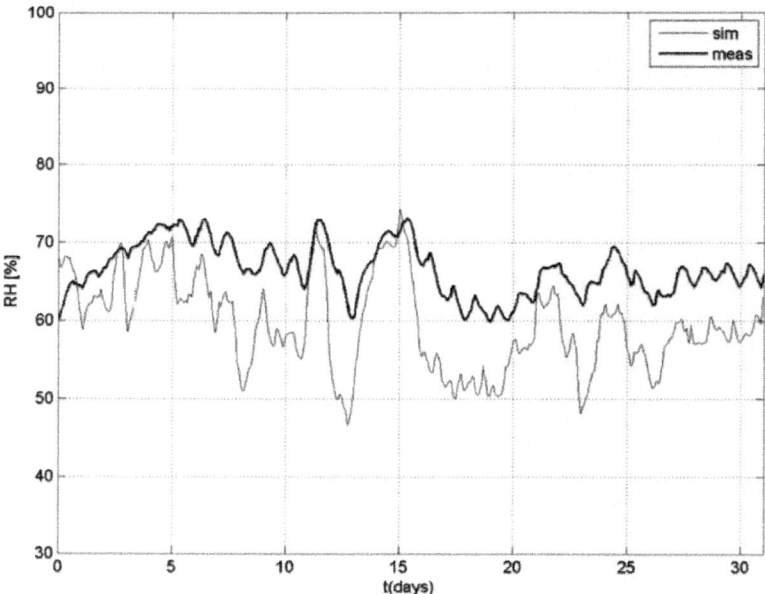

Figure 9: The measured and simulated relative humidity at the surface.

Figure 10: Observed moisture spots at the inner surface near the windows.

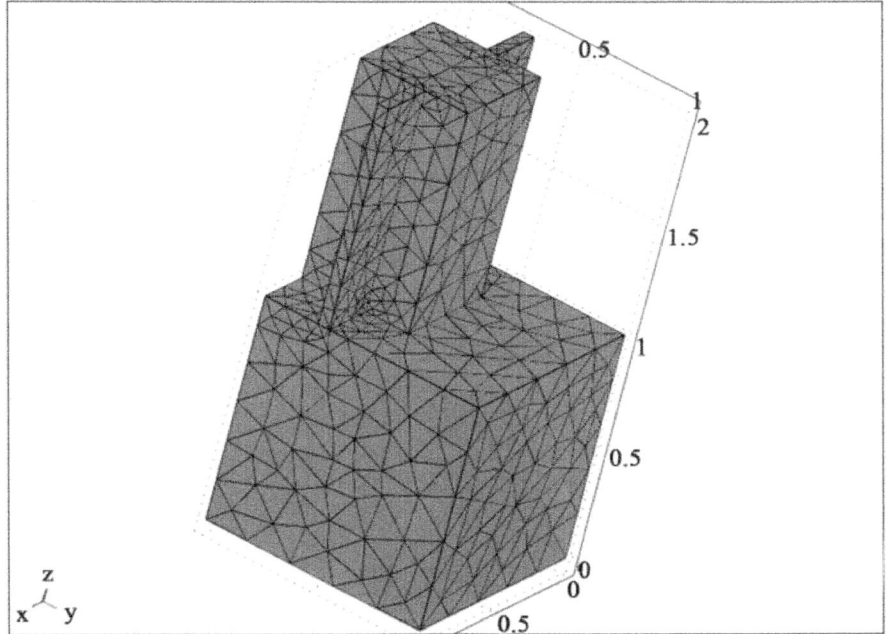

Figure 11: The mesh.

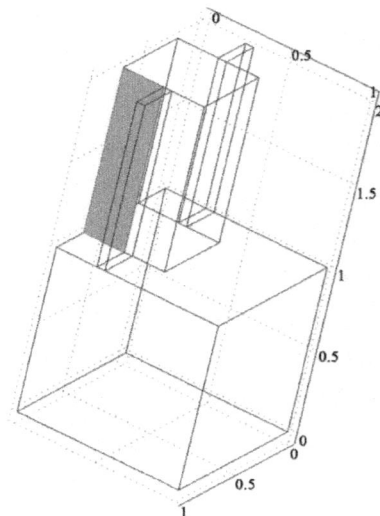

Figure 12: The location of the wet surface condition.

There is no match between the simulated profile at the inner surface of Figure 13 with the observed profile of Figure 10. Therefore it is concluded that the location of Figure 12 is not a possible candidate that causes the observed moisture spots.

In Appendix an overview is provided of nine more moisture profiles caused by corresponding possible wet locations. From these results, the best candidate for the moisture leakage location seems to be at the bottom of the window.

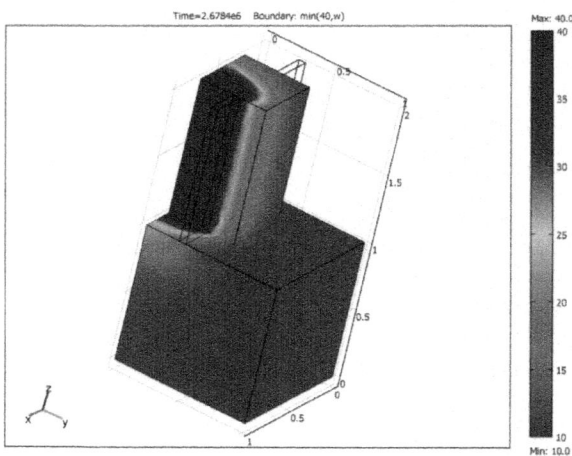

Figure 13: The steady state moisture pattern at the surfaces.

DISCUSSION AND CONCLUSIONS

This paper investigates the (in) possibilities of pinpointing moisture leakages from inside surface moisture patterns using inverse modeling techniques.

Limitations

1) In the presented method, the material properties were obtained from a database (DELPHIN [7]). In order to get very accurate simulation results, the material properties should be measured. This is however quite a problem because, often it is not allowed to take samples from monumental constructions. So the accuracy of the simulation results is limited. More research is needed to investigate exactly what the effect is of this uncertainty in the material properties.

2) The inverse modeling technique tries to match simulated moisture surface profiles with the observed moisture spots. It assumes a single "natural" cause, for example a leak between the window and brick caused by cracks. If a moisture spot is caused by some repair action, for example a complete replacement of the windows, this is not taken into account.

3) The current inverse modeling technique is still rather basic by manipulating the boundary conditions by hand. A more sophisticated method, where the boundary conditions are manipulated by a computer algorithm is under investigation.

Future Research

Future research include the testing of the method with laboratory experiments and a thoroughly evaluation of the approach as instrument for pinpointing the location of leakages. By selecting materials with well-known properties limitation (1) can be fixed. Also limitation (2) can be fixed because of the high level of control at the lab.

It is concluded that although the presented methodology is promising, more research is needed to confirm its usability.

APPENDIX

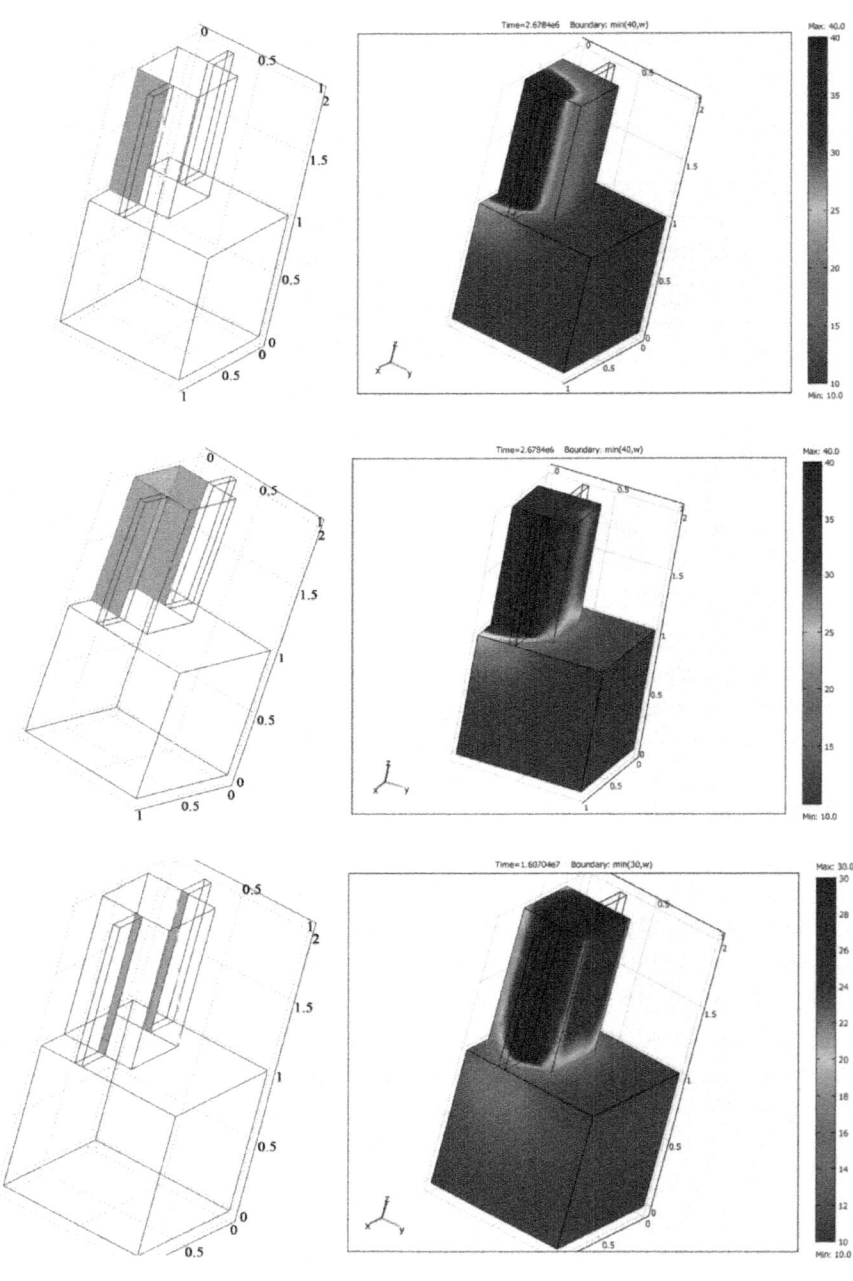

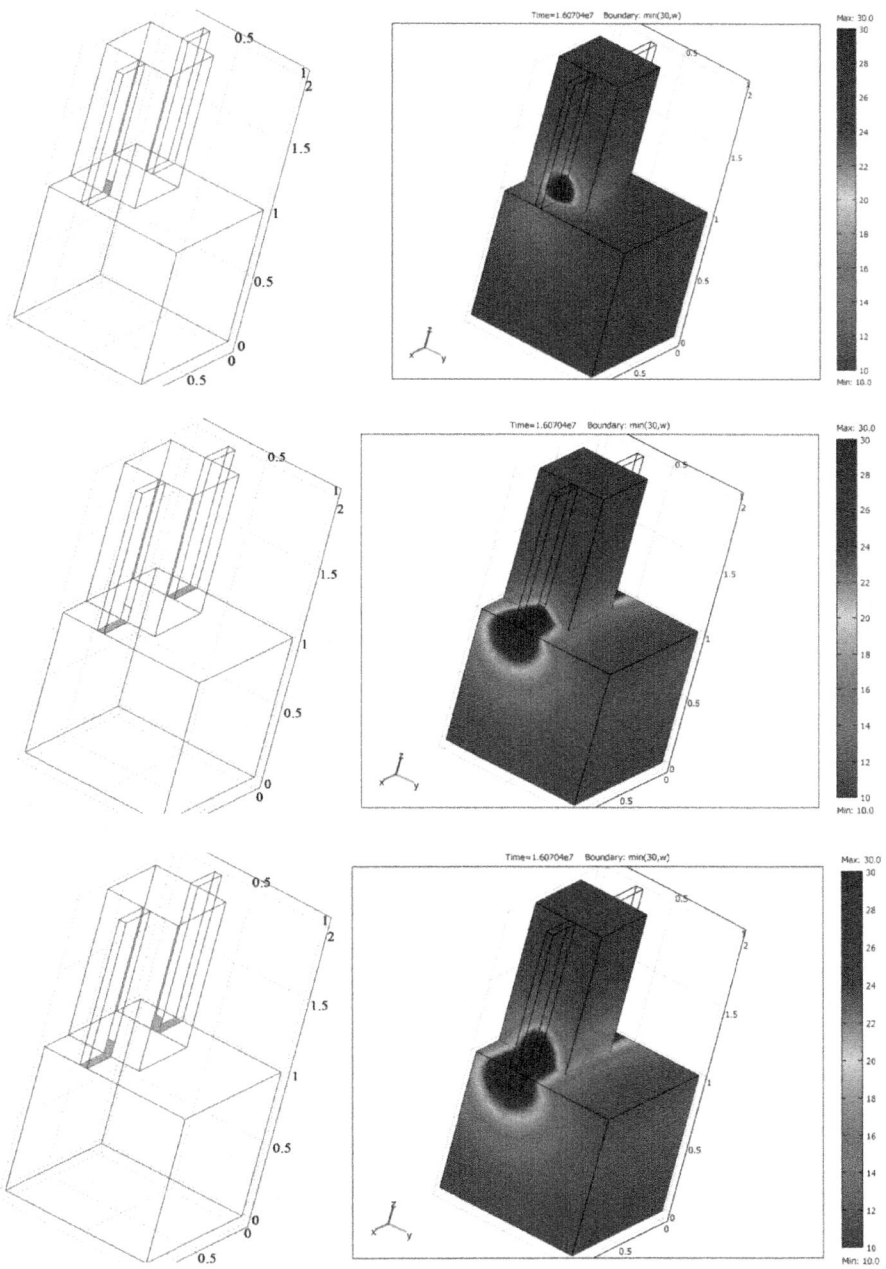

REFERENCES

1. Briggen, P.M. (2007) A Study on the Moisture Problems at Hunting Lodge St. Hubertus. M.Sc. Thesis, Technische Universiteit Eindhoven, Eindhoven, 120 p. (In Dutch)

2. Briggen, P.M., Blocken, B.J.E., van Schijndel, A.W.M. and Schellen, H.L. (2009) Wind-Driven Rain and Related Moisture Problems at the Tower of a Monumental Building. 4th International Building Physics Conference, 15-18 June 2009, Istanbul, 709-716.

3. Briggen, P.M., Blocken, B.J.E. and Schellen, H.L. (2009) Wind-Driven Rain on the Facade of a Monumental Tower: Numerical Simulation, Full-Scale Validation and Sensitivity Analysis. Building and Environment, 44, 1675-1690.

4. van Schijndel, A.W.M. (2007) Integrated Heat Air and Moisture Modeling and Simulation. Ph.D. Thesis, Technische Universiteit Eindhoven, Eindhoven, 200 p.

5. van Schijndel, A.W.M. (2009) Integrated Modeling of Dynamic Heat, Air and Moisture Processes in Buildings and Systems Using SimuLink and COMSOL. Building Simulation: An International Journal, 2, 143-155.

6. van Schijndel, A.W.M. (2009) The Exploration of an Inverse Problem Technique to Obtain Material Properties of a Building Construction. 4th International Building Physics Conference, 15-18 June 2009, Istanbul, 91-98.

7. DELPHIN (2014) http://de.wikipedia.org/wiki/Delphin_(Software)

8. COMSOL (2014) http://en.wikipedia.org/wiki/COMSOL

Chapter 2

MANAGING MEASUREMENT UNCERTAINTY IN BUILDING ACOUSTICS

Chiara Scrosati and Fabio Scamoni

Construction Technologies Institute, National Research Council of Italy, via Lombardia 49, 20098 San Giuliano Milanese (MI), Italy

ABSTRACT

In general, uncertainties should preferably be determined following the principles laid down in ISO/IEC Guide 98-3, the Guide to the expression of uncertainty in measurement (GUM:1995). According to current knowledge, it seems impossible to formulate these models for the different quantities in building acoustics. Therefore, the concepts of repeatability and reproducibility are necessary to determine the uncertainty of building acoustics measurements. This study shows the uncertainty of field measurements of a lightweight wall, a heavyweight floor, a façade with a single glazing window and a façade with double glazing window that were analyzed by a Round Robin Test (RRT), conducted in a full-scale experimental building at ITC-CNR (Construction Technologies Institute of the National Research Council of Italy). The single number quantities and their uncertainties were evaluated in both narrow and enlarged range and it was shown that including or excluding the low frequencies leads to very significant differences, except in the case of the sound insulation of façades with single glazing window. The results obtained in these RRTs were compared with other results from literature, which confirm the increase of the uncertainty of single number quantities due to the low frequencies extension. Having stated the measurement uncertainty for a single measurement, in building acoustics, it is also very important to deal with sampling for the purposes of classification of buildings or building units. Therefore, this study also shows an application of the sampling included in the Italian Standard on the acoustic classification of building units on a serial type building consisting of 47 building units. It was found that the greatest variability is observed in the façade and it depends on both the great variability of window's typologies and on workmanship. Finally, it is suggested how to manage the uncertainty in building acoustics, both for one single measurement and a campaign of measurements to determine the acoustic classification of buildings or building units.

INTRODUCTION

This paper is a revised and expanded version of the paper "Uncertainty in Building Acoustics" [1] presented at the 22nd International Congress on Sound and Vibration ICSV22.

When reporting the result of the measurement of a physical quantity, it is compulsory that some quantitative indications of the quality of the result be given so that those who use it can assess its reliability. Without such indications, measurement results cannot be compared, either with one another or with reference values given in a specification or standard. It is therefore necessary, in order to characterize the quality of the result of a measurement, to evaluate and to express its uncertainty. Generally, it is widely recognized that, when all of the known or suspected components of error have been evaluated and the appropriate corrections have been applied, an uncertainty about the correctness of the stated result still remains; that is, a doubt about how well the result of the measurement represents the value of the quantity being measured.

The word "uncertainty" means doubt, and thus in its broadest sense "uncertainty of measurement" means doubt about the validity of the result of a measurement. The formal definition of the term "uncertainty of measurement" developed in the Guide to the expression of uncertainty in measurement (GUM) [2] is as follows. Uncertainty (of measurement): parameter, associated with the result of a measurement, that characterizes the dispersion of the values that could reasonably be attributed to the measurand.

This definition of uncertainty of measurement is an operational definition that focuses on the measurement result and its evaluated uncertainty. However, it is not inconsistent with other concepts of uncertainty of measurement, such as a measure of the possible error in the estimated value of the measurand as provided by the result of a measurement; or an estimate characterizing the range of values within which the true value of a measurand lies. Although these two traditional concepts are ideally valid, they focus on unknowable quantities: the "error" of the result of a measurement and the "true value" of the measurand (in contrast to its estimated value), respectively.

THE UNCERTAINTY IN TERMS OF REPEATABILITY, RE-PRODUCIBILITY AND *IN SITU* STANDARD DEVIATION

Tests performed on samples made of materials presumed to be the same, in identical conditions, generally do not give the same results. This condition is due to inevitable errors (systematic and random) in test procedures, caused by the difficulties in controlling the several factors that influence the test. To determine the accuracy of a measurement method, both accuracy and precision should be considered; in particular, the latter indicates the correlation between the test results.

Precision is a general term for the variability between repeated tests. Two measures of precision, termed repeatability and reproducibility, have proved necessary and, for many practical cases, sufficient for describing the variability of a test method. Repeatability refers to tests performed on the same test object with the same method under conditions that are as constant as possible, with the tests performed during a short interval of time, in one laboratory by one operator using the same equipment. On the other hand, reproducibility refers to tests performed on identical test items with the same method, in widely varying conditions, in different laboratories with different operators and different equipment. Thus, repeatability and reproducibility are two extremes, the first measuring the minimum and the latter the maximum variability in results.

The building acoustic quantities include airborne sound insulation of internal partitions, airborne sound insulation of façades, impact sound insulation of floors and sound pressure level from service equipment in buildings. The quantities that have to be measured and their measurement methods, for all aspect involved, are described in the international standard series EN ISO 10140 [3] for laboratory measurements and in the international standard series ISO 16283 [4] for field measurements. The accuracy of these measurement method depends on several factors that influence the test, such as acoustic instrumentation, acoustic method (microphones and sources position), context (regular rooms or semi-open space, of any size), constructive details of the building (that could have effect on acoustic measures) and workmanship, and, concerning sound levels, influence of instrumentation working conditions (repeat configuration). Detailed information for each of these factors is hardly available. Both random and systematic errors affect the acoustic measurements results. The random effects can be determined by repeated independent measurements in essentially identical conditions. The systematic effects, however, are not easy to determine, but, as a general rule, they can be determined thanks to comparative measurements to be executed in different test facilities (for laboratory measurements) or carried out by different laboratories (for field measurements), and the knowledge of the random errors in those conditions. Therefore, it is necessary to refer to the concepts of repeatability and reproducibility, which provide a simple means for the expression of the precision of a test method and of the measurements performed according to the test method.

The best methodology to study the repeatability and reproducibility of building acoustic measurements is to carry out an Inter-Laboratory Test (ILT), or a Round Robin Test (RRT), tests consisting of independent measurements executed several times by different operators. Due to the particular nature of the sample in building acoustics, in addition to repeatability and reproducibility standard deviations, another standard deviation is defined, the *in situ* standard deviation (defined, for the first time, in ISO 12999-1 [5]), which could be useful to estimate. The *in situ* standard deviation is a particular kind of reproducibility standard deviation that is measured in the same location on the same object. In fact, in the case of RRT field measurements, when different operators, with their own equipment, perform measurements on a particular building element, both the location and the object under test are the same. Therefore, location is the only difference between reproducibility and *in situ* standard deviation: for the *in situ* standard deviation, the location is exactly the same as is the test object, while in the case of reproducibility standard deviation the locations are different and the test object can be either the same test object or identical test objects tested in the different locations. The *in situ* standard deviation, therefore, corresponds to a reproducibility standard deviation of the same object in the same location.

Round Robin Test

Generally, cooperative tests (ILT or RRT) assess the uncertainty of measurement methods using a reference value. One of the main aspects of these tests is the determination of this reference and its uncertainty. A reliable, low-uncertainty reference value is required in order to minimize the uncertainty of a cooperative test. Due to the typology of the sample test in acoustic measurements, a reference value does not exist; therefore an estimated value is used. The best measuring reference is the mean value. A RRT of sound insulation field measurements of building elements was carried out as part of a research sponsored by the Lombardy Region [6,7,8]; this study was based on the cooperation of three different bodies: a research body, ITC-CNR (Construction Technologies Institute of the National Research Council of Italy); a university laboratory, DISAT (Department of Earth and Environmental Sciences of the University of Milano-Bicocca); and a control organization, ARPA-Lombardy (Regional Agency for environmental protection) and it was coordinated by ITC-CNR. In the first approach to the problem [6], the analysis was centered on the single number values of the Italian regulation [9] and on the narrow frequency range (from 100 to 3150 Hz). In later studies [7,8], the analysis considered all the possible descriptors of the different European national legislations and was extended to the enlarged frequencies range (from 50 to 5000 Hz). Another study on the uncertainty of façade sound insulation [10] was carried out at the initiative of the Building Acoustics Group (GAE) of the Italian Acoustic Association (AIA). This study was focused on the low frequencies (from 50 to 80 Hz), in particular on the comparison between the procedure stated in ISO 140-5 [11] and the new low frequency procedure stated in ISO 16283 [4]. The main results of these studies are summarized in the following section.

Airborne Sound Insulation

Notwithstanding the importance of the uncertainty of the measurement method in building acoustics, the uncertainty of field measurements was not comprehensively investigated. There are only few examples in the literature [12,13] compared to those of laboratory tests [14,15,16,17,18]. The studies regarding laboratory tests conclude that the main influences are caused by the laboratory geometry and materials, the flanking transmissions, the type of border material, and the different test opening dimensions [15,16].

Nine teams coordinated by ITC-CNR were involved in the study about the uncertainty of airborne sound insulation [7]; each of them has replicated the tests five times, including the reverberation time.

No deviations occurred from the test procedure laid down in ISO 140-4 [19] but, repeating the measurements several times, the parameters left open in the measurement procedure were represented as best as possible. In particular, the set of microphone positions and source positions were selected anew, more or less randomly, for each repeated measurement. The measurands were a floor without floating floor (surface mass of 550 kg/m^2 and surface of about 19 m^2) and a lightweight wooden partition wall (surface mass of 30 kg/m^2 and surface of about 8.5 m^2). Considering the goal of European harmonization of acoustic parameters [20], the differences between the various descriptors (R', D_n and D_{nT}) were analyzed in terms of average, maximum and minimum values, and in terms of standard deviation of repeatability and reproducibility (*in situ* standard deviation, referring to ISO 12999-1 [5], where the reproducibility standard deviation of the same element is measured in the same location).

Figure 1 shows the standard deviations of repeatability s_r and *in situ* reproducibility standard deviation s_{situ} of all analyzed quantities. The descriptors extension at low frequencies (from 50 to 80 Hz) (LF) was also analyzed. From the graphs of Figure 1, it is evident that the uncertainty at LF is much greater than the uncertainty in the narrow frequencies range from 100 to

5000 Hz. From the comparison of the RRT s_{situ} values with the values of the ISO 12999-1 [5] for situations A (s_R) and B (s_{situ}) (see Figure 1), it was found that the values of situation B underestimate the uncertainty of *in situ* measurements in particular at low-medium frequencies. Moreover, the values of s_{situ} [7] obtained are higher also than the s_R values, in particular for the floor at low-medium frequencies from 80 to 200 Hz, and for the wall from 160 to 250 Hz.

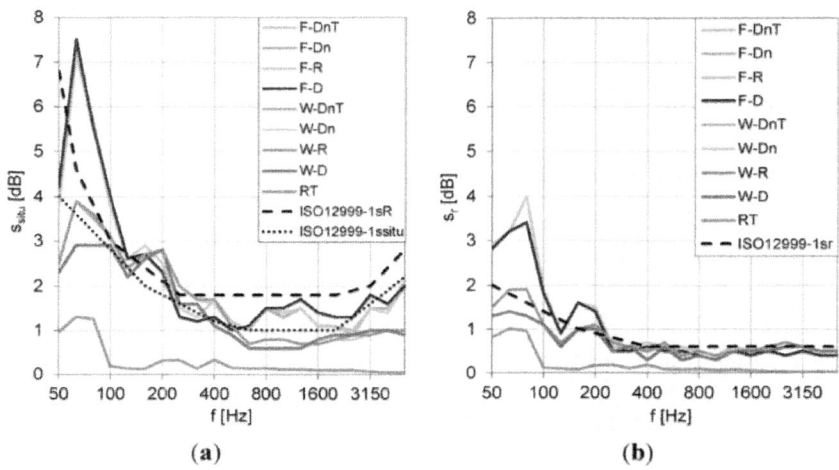

(a)   (b)

Figure 1: s_{situ} (a) and s_r (b) of floor (F) and wall (W) of R', D_n, D_{nT}, D and RT [7], with the comparison with the reproducibility, *in situ* (a) and repeatability (b) standard deviation of ISO 12999-1 [5].

The results of SNQ calculations are shown in Table 1. Two different ways to determine the SNQs have been considered for the above-mentioned study [7]. The former is to determine SNQ according to ISO 717-1 [21] by shifting the reference curve (value in the range from 100 to 3150) in steps of 1 dB toward the measured curve, until the mean unfavorable deviation is as large as possible but not more than 32 dB; all the laboratories involved in the RRT have followed this procedure. The latter is to determine SNQ plus the spectrum adaptation terms C and C_{tr} according to ISO 717-1 [21] both in the narrow frequency range from 100 to 3150 Hz, and in the enlarged frequency range from 50 to 5000 Hz; in both cases rounded to integer and with 1 decimal place (subscript 01), using Equation (1) [21]. The SNQs plus the spectrum adaptation terms were determined using a 0.1 dB resolution, following from the work of Wittstock [22], to obtain more accurate data for the analysis of standard deviation than the 1 dB resolution.

$$X_{Aj} = -10 \lg \sum_i 10^{(L_{ij} - X_i)/10} = X_w + C_j [dB]$$

(1)

where j is the index of the spectrum No. 1 to calculate C or No. 2 to calculate C_{tr} according to ISO 717-1 [21]; i is the index of frequencies; L_{ij} is the level indicated in ISO 717-1 [21] at frequency i for spectrum j; X_i is one of the quantities considered, R_i, D_{ni} or D_{nTi}; at frequency i for the spectrum j; X_w is the single number; and C_j is the spectrum adaptation term C or C_{tr} if calculated with spectrum No. 1 or No. 2, respectively.

Table 1: s_r and s_{situ} of SNQs of floor (F) and wall (W) in narrow (100–3150 Hz) and enlarged (50–5000 Hz) range [7]

		Narrow Range 100–3150 Hz					Enlarged Range 50–5000 Hz	
		X	X + C	X + C_{tr}	X_{01} + C	X_{01} + C_{tr}	X_{01} + C	X_{01} + C_{tr}
s_{situ}	F-D_{nT}	1.3	1.3	1.5	1.3	1.5	1.4	2.8
	F-D_n	1.2	1.2	1.5	1.3	1.4	1.4	2.8
	F-R'	1.2	1.2	1.5	1.3	1.5	1.4	2.7
	W-D_{nT}	0.7	0.9	1.2	0.9	1.2	0.8	1.4
	W-D_n	0.9	0.9	1.3	0.8	1.2	0.8	1.4
	W-R'	0.8	0.9	1.3	0.9	1.2	0.8	1.4
s_r	F-D_{nT}	0.7	0.6	0.6	0.5	0.7	0.6	1.3
	F-D_n	0.5	0.5	0.7	0.5	0.7	0.6	1.3
	F-R'	0.5	0.6	0.9	0.5	0.7	0.6	1.3
	W-D_{nT}	0.2	0.2	0.3	0.2	0.2	0.2	0.3
	W-D_n	0.3	0.3	0.3	0.2	0.2	0.2	0.4
	W-R'	0.2	0.2	0.4	0.2	0.2	0.2	0.4

The internal partitions considered in this RRT were a lightweight wall and a heavy floor. It was demonstrated that the uncertainties of lightweight samples are lower than the uncertainties of heavy types of construction; therefore it will be important for datasets of different constructions to be considered separately. A similar difference between the uncertainty of heavy and lightweight test samples was shown by Dijckmans and Vermeir [23] who made a numerical investigation of the repeatability and reproducibility of laboratory sound insulation measurements by investigating both the pressure method and the intensity method. Dijckmans and Vermeir [23] found that for large, heavy test elements, like concrete walls, the reproducibility in the lowest frequency bands is not improved by using the intensity method, while, for double plasterboard walls, the theoretical uncertainty is decreased by 1 dB by using the intensity method.

The results of Table 1 show that the one-third-octave band uncertainty at LF slightly affects the SNQs in the enlarged range plus C spectrum adaptation term but greatly affects (almost double than the narrow range standard deviation) the SNQs in the enlarged range plus C_{tr} spectrum adaptation term. This is mainly due to the fact that the spectrum adaptation term C_{tr} considers predominantly the low-medium frequencies noise components.

In their recent study on the correlations and implications of SNQ for rating airborne sound insulation in the frequency range 50 Hz to 5 kHz, Garg and Maij [24] showed that $R_{traffic}$ (as defined in ISO CD 16717-1 [25] and corresponding to $R_w + C_{tr50-5000}$) is highly sensitive to low frequency sound insulation as compared to the current SNQ and R_{living} (as defined in ISO CD 16717-1 [25] and corresponding to $R_w + C_{50-5000}$). Finally, the measurement uncertainty in the low frequency range (due to the presence of the normal modes of vibration, that imply that at the first three one-third-octave bands the measured levels can be strongly influenced by the measurement position) is too high to justify the decision to perform field measurements down to low frequencies, and therefore the scientific evidence for including the low frequency range should be significantly improved. Moreover, the fact that the higher uncertainty at LF is not well represented in the SNQs uncertainty confirms that further studies are needed to better understand all the implications of the inclusions of LF in the SNQs, from both a physical point of view and from a legislation point of view. Garg and Maij [24] found interconversion equations applicable for sandwich gypsum constructions and roof constructions. They stressed the fact that testing of sound transmission loss characteristics in the extended frequency range of 50 Hz to 5 kHz also implies the need to reformulate the sound regulation requirements in buildings including the low frequency spectrum adaptation terms.

Some recent studies [26,27,28,29] on the uncertainty of SNQs extended to the low frequencies range show an increase in the SNQs uncertainty due to the LF extension, confirming the results found in this RRT. Mahn and Pearse [26] studied the effect on uncertainty of expanding the frequency range included in the calculation of the single number ratings, using laboratory measurements of 200 lightweight walls as data. They found that the uncertainty of the single number ratings is highly dependent on the shape of the sound reduction index curve. The uncertainty obtained for R_{living} ($R_w + C$ in the enlarged frequency range) was greater than that of the traditional weighted sound reduction index for 98% of the 200 lightweight building elements included in the evaluation.

Hongisto et al. [27] focused their study on the two most important SNQs proposed by ISO CD 16717-1 [25]; that is, $R_{traffic}$($R_w + C_{tr}$ in the enlarged frequency range) and R_{living} ($R_w + C$ in the enlarged frequency range), and how their reproducibility values differ from the reproducibility values of their counterparts $R_w + C_{tr}$ and R_w. They found that the reproducibility values of the proposed single-number quantities (50–5000 Hz; R_{living}, $R_{traffic}$) are larger than the reproducibility values of the present SNQs (100–3150 Hz; R_w, $R_w + C_{tr}$)

with sound insulation measurements made with the pressure method; with the sound intensity method, the reproducibility values increased very little.

Machimbarrena *et al.* [28] presented an alternative procedure, aiming at evaluating the need of performing individual uncertainty calculations and the effect of extending the frequency range used to calculate sound insulation single number quantities. For this purpose they performed calculation in a set of 2081 field airborne sound insulation measurements on 22 different types of separating walls partitions of *in situ* airborne sound insulation measurements. The results of Machimbarrena *et al.* [26] show that the frequency range used for the evaluation affects the uncertainty of the single number quantity. In almost all the cases shown in their paper, the uncertainty is increased when the frequency range is extended.

António and Mateus [29] studied the influence of low frequency bands on airborne and impact sound insulation single numbers for typical Portuguese buildings. They found that the uncertainty is higher for the $D_{nT,w} + C_{tr}$ descriptor than for $D_{nT,w} + C$, confirming what was found in this RRT. They also found that when the low frequency bands are included in the calculation, the uncertainty of the descriptor increases on average and this increase is more evident when the adaptation term is for a spectrum of traffic noise.

Façade Sound Insulation

The uncertainty of field measurements, in particular façade sound insulation, has not been comprehensively investigated. There is only one example in the literature of a Round Robin Test conducted on a window of a façade [12].

In the study about the uncertainty of façade sound insulation [8], the measurand was a prefabricated concrete façade with a 4 mm single glazing wood-aluminum frame window with a MDF (Medium Density Fiberboard) shutter box. The façade is situated at first floor level. Nine teams coordinated by ITC-CNR were involved in this study; each of them has replicated the tests five times, including the reverberation time. One laboratory showed a significant presence of stragglers and outliers. After a statistical examination of this result, the laboratory was excluded. In fact, it turned out that the random effect estimated for laboratory was, in absolute value, the highest value [8]: the Grubbs test [30,31] for one outlier identified the laboratory as the first outlier. Therefore here are the eight reported laboratories results.

In this study, the highest values of s_r and s_{situ} were found at the frequencies of 50, 63 and 80 Hz. That paper [8] also underlined that the uncertainties in $D_{ls,2m,nT}$ are heavily contaminated by the inappropriateness of the reverberation time correction at low-frequencies and a comparison between the uncertainties of the standardized level difference $D_{ls,2m,nT}$ and the level difference $D_{ls,2m}$ shows the magnitude of the reverberation time at low frequencies (see Figure 2). This influence is noticeable in particular at 63 Hz and at 80 Hz, while at 50 Hz the uncertainties of $D_{ls,2m,nT}$ and $D_{ls,2m}$ are coincident.

The variations between laboratories at low frequencies are still very high even if the reverberation time correction is not included in the calculation (*i.e.*, just considering $D_{ls,2m}$), which implies that for the sound pressure level measurements the low frequencies also have a high uncertainty. The s_{situ} and s_r behavior of $D_{ls,2m}$ is similar to the behavior of the uncertainties of ISO12999-1 [5], which increase steadily and rapidly below 100 Hz. Thus the trend of the standard deviation curve at low frequencies of *in situ* reproducibility and repeatability standard deviation calculated from the RRT study is attributable to the reverberation time measurements.

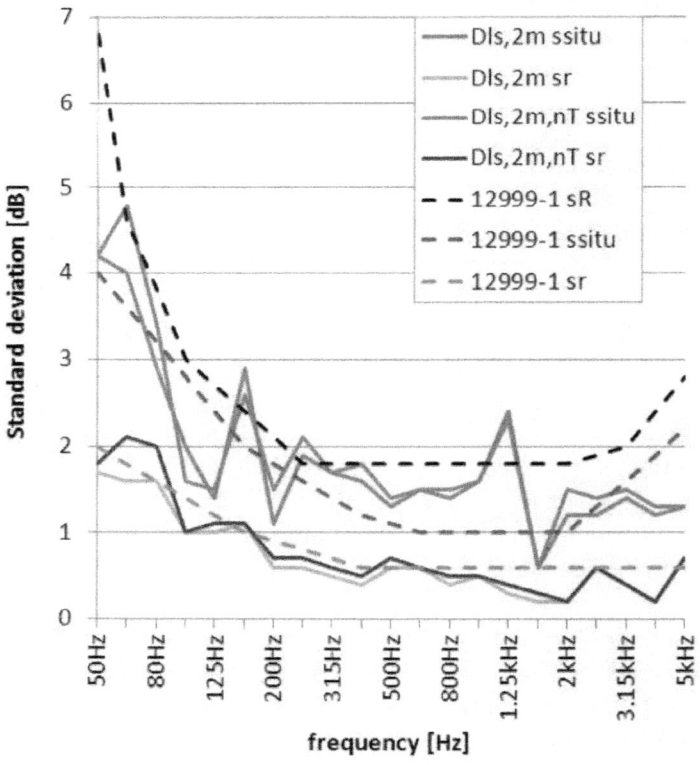

Figure 2: Comparison between the *in situ* and repeatability standard deviation of $D_{ls,2m,nT}$ and $D_{ls,2m}$ [8] and the reproducibility, *in situ* and repeatability standard deviation of ISO 12999-1 [5].

In Table 2 are shown the SNQs uncertainties, in terms of repeatability and *in situ* standard deviations. The SNQs were determined according to ISO 717-1 [21] shifting the reference curve both in steps of 1 dB and 0.1 dB (subscript 01), toward the measured curve, until the mean unfavorable deviation is as large as possible, but not more than 32 dB; all the laboratories involved in the RRT have followed this procedure. The shift in increments of 0.1 dB was evaluated because the 2013 update of the ISO 717-1 [21] provides for increments of 0.1 dB for the expression of uncertainty. The SNQs plus spectrum adaptation terms C and C_{tr} according to ISO 717-1 [21] in the extended range (from 50 to 5000 Hz), both at integer and with one decimal place (subscript 01) were calculated using Equation (1).

Table 2: s_{situ} and s_r of SNQs, calculated as one of the levels *j* of RRT [8]

Frequency Range	SNQs	s_{situ}	s_r
narrow range 100–3150 Hz	$D_{ls,2m,nT,w}$	0.8	0.3
	$D_{ls,2m,nT,w} + C$	1.0	0.4
	$D_{ls,2m,nT,w} + C_{tr}$	1.1	0.3
	$D_{ls,2m,nT,w01}$	0.9	0.3
	$D_{ls,2m,nT,w01} + C$	1.0	0.2
	$D_{ls,2m,nT,w01} + C_{tr}$	1.1	0.3
enlarged range 50–5000 Hz	$D_{ls,2m,nT,w01} + C$	0.9	0.2
	$D_{ls,2m,nT,w01} + C_{tr}$	1.1	0.3

In the study about the airborne sound insulation [7], it was found that the extension at low frequencies range increases the uncertainty of the SNQs. In the case of the façade, calculating the SNQs uncertainty handling the SNQs values as a level of the RRT itself (see Table 2), no significant differences are observed whether including or excluding the low frequencies. In this case, the low frequency uncertainty is not well reflected in the SNQs uncertainty. Considering the extension to low frequencies, the suitability of the reference spectra for rating airborne sound insulation should be validated.

On this topic, Masovic *et al.* [32] made a study on the suitability of ISO CD 16717-1 [25] reference spectra for rating airborne sound insulation. The ISO CD 16717-1 [25] spectra living and traffic correspond to the reference spectra C (50–5000 Hz) and C_{tr} (50–5000 Hz) of ISO 717-1 [21], respectively. Masovic *et al.* [32] demonstrated, with an extensive noise monitoring in a number of dwellings recordings of 38 potentially disturbing activities, that the reference spectrum for living noise (L_{living}), should be redefined to better match the typical spectrum of noise in dwellings because it seems to be rather high at lower frequencies, especially below 100 Hz. Moreover, in the case of noise generated by sources of music with strong bass content the reference spectrum for traffic noise ($L_{traffic}$) seems to be more appropriate above 100 Hz than L_{living}. This could suggest one of the reasons why the low frequencies uncertainty is not adequately reflected by the SNQs uncertainty extended to low frequencies and should be considered deeper before deciding to perform measurements down to LF range.

Therefore, considering this kind of façade (prefabricated concrete façade with a single glazing window and with a shutter box) including the low frequencies range in the façade sound insulation measurements, brings no obvious advantage, but rather the disadvantage of complicating and lengthening the measurement. In literature, there are some studies (e.g., Rindel [33] and Park and Bradley [34]) on the annoyance of noise from neighborhood at low frequencies that stress the importance of investigating the LF noise; nevertheless, at present time, effective protection systems against low frequency noise are still an open challenge both for researchers and components manufacturers, as underlined by Prato and Schiavi [35]. Hongisto et al. [27] suggested that scientifically valid socio-acoustic evidence for the need to include the frequency range 50–80 Hz should be significantly improved before deciding that the low frequency measurements are included in the calculation of the SNQs. Last but not least, if LF measurements are aimed at the protection against LF noise, the fact that the high uncertainty of the one-third octave LF band affects the reliability of the performance of the test element implies that the potential effectiveness of the protection system against low frequency noise is not quantifiable.

A prefabricated concrete façade with a PVC frame with double glazing 4/12/4 window was tested in the further RRT study concerning façade sound insulation uncertainty [10], focused on the new low frequencies measurement procedure stated in ISO/DIS 16283-3 [36], that will soon replace the standard ISO 140-5 [11]. Ten teams, coordinated by ITC-CNR were involved in this RRT, each of them operating with its own equipment and replicates the tests 5 times, including the new low frequencies procedure (explained below) and the reverberation time measurements. All teams performed measurements following the global loudspeaker method, which yields the level difference of a façade in a given place with respect to a position 2 m in front of the façade. All teams positioned the outside microphone 2 m in front of the façade, and the loudspeaker on the ground, with the angle of sound incidence equal to 45° ± 5°; as positioned directly in front of the façade by some teams, and in a lateral position by other teams. The statistical analysis of the data provides a three-step procedure for the identification of stragglers and outliers. Following this procedure, two teams were identified as outliers and excluded because they showed a significant presence of stragglers and outliers starting from 500 Hz to 3150 Hz [10]. The comparison of standard deviation values, repeatability and *in situ* standard deviation, from RRT (calculated for both $D_{ls,2m,nT}$ and $D_{ls,2m}$) and from ISO 12999-1 [5] are plotted in Figure 3.

Regarding the low frequency range (from 50 to 80 Hz), the reasons for the high values of s_r and s_{situ} can be sought in the presence of the normal modes of vibration, in fact at the first three one-third- octave bands (50, 63 and 80 Hz), the measured levels can be strongly influenced by the measurement position.

At low frequencies, the s_{situ} and s_r behavior of both $D_{ls,2m,nT}$ and $D_{ls,2m}$ is not similar to the behavior of the uncertainties of ISO 12999-1 [5], in terms of reproducibility s_R and *in situ* standard deviation, which increase steadily and rapidly below 100 Hz, as it can be seen in graphs of Figure 3. Contrary to what was found in the previous RRT [8], this difference is not attributable to the reverberation time measurements. This different behavior could be attributable to the differences of the façade test samples: the façade of the previous RRT [8] is a prefabricated concrete façade with a 4 mm single glazing wood-aluminum frame window with a MDF shutter box; the façade of the second study is a prefabricated concrete façade with a PVC frame with double glazing 4/12/4 window. Also the loudspeaker position could be relevant and its influence is under investigation.

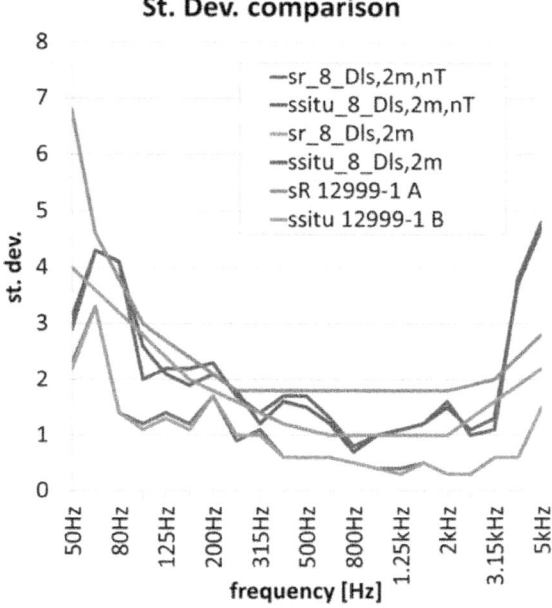

Figure 3: Comparison of standard deviation values from RRT (calculated for both $D_{ls,2m,nT}$ and $D_{ls,2m}$) and from ISO 12999-1 [10].

With respect to the high frequency range, in particular at 4000 and 5000 Hz, the RRT and ISO 12999-1 [5] standard deviations values show the same behavior, *i.e.*, an increase with frequency, but the RRT s_{situ} values are higher than the ISO 12999-1 [5] values. Moreover the RRT s_{situ} values are higher than the low frequency s_{situ} values of both RRT and ISO 12999-1 [5]. This is probably due to the different positions of the loudspeaker with respect to the façade [10] and it is still under investigation. In the previous RRT [8], where all the teams involved placed the loudspeaker in the same position (directly in front of the façade), the high frequency uncertainty was lower, in particular lower than ISO 12999-1 [5] values and much lower than the low frequencies uncertainty.

In the first RRT on façade sound insulation [8] a behavior similar to the behavior found by Lang [12] in the Austrian RRT was observed, where the RRT values exceed the values of the ISO 140-2 [37] (the standard on acoustics measurement uncertainty available at the time of Lang's RRT) in the range of mass-spring-mass resonance frequency and in the range of the coincidence frequency of the double glazing. Lang suggests that such behavior may be caused by the difficulty of arranging the loudspeaker at an angle of incidence of 45°.

The first RRT [8] faced no difficulty with the arrangement of the loudspeaker at an angle of incidence of 45°. Such behavior is thus exclusively attributable to the nature (*i.e.*, critical frequencies) of the measurand itself. However, the uncertainty dependence from the loudspeaker position could be found at high frequencies as shown in the second RRT [10] and, as already said, it must be more deeply investigated. Berardi *et al.* [38] and Berardi [39] considered the position of the loudspeaker as a variable, but its influence on the high frequencies was not comprehensively evaluated.

In this RRT [10] all the participating laboratories repeated the measurements with the low-frequency procedure included in the upcoming standard ISO 16283-3 (ISO/DIS 16283-3 [36]). In his recent paper Hopkins [40] gives the background to the revision of the ISO 140 standards relating to field measurement of airborne, impact and façade sound insulation that form the new ISO 16283 series. The low-frequency procedure was first studied and proposed by Hopkins and Turner [41] in a work about the airborne sound insulation between rooms. For each of the 50, 63 and 80 Hz bands, they proposed that the average low frequency sound pressure level in the room, L_{LF}, be calculated from $L_{ISO140-4}$ (the average sound pressure level in a room measured according to the normative guidance in ISO 140-4) and L_{corner} (the corner sound pressure level measured according to the normative guidance in ISO 16283-1) according to:

$$L_{LF} = 10 \lg \left[\frac{2\left(10^{0.1 L_{ISO140-4}}\right) + 10^{0.1 L_{corner}}}{3} \right] [dB]$$

(2)

The low-frequency (LF) procedure is mandatory in case of room volume lower than 25 m³. As the volume of the receiving room in this RRT is 40 m³, it was possible to compare the results of the two procedures: the LF procedure and the default procedure. The results of this comparison, for the LF range are shown in Table 3. The results refer both to 8 and to 10 teams, as the two outlier teams that are excluded from the calculation of standard deviation for the all frequencies considered (from 50 to 5000 Hz), can be included in the evaluation of the LF standard deviation because these teams showed a significant presence of stragglers and outliers starting from 500 Hz to 3150 Hz.

Table 3: Low frequency s_r and s_{situ} values for the two measurement methods (default and LF) for both 8 and 10 teams [10]

Standard	50 Hz		63 Hz		80 Hz	
Deviations	Default	LF	Default	LF	Default	LF
s_r_10	2.7 dB	2.5 dB	3.1 dB	4.5 dB	1.4 dB	2.3 dB
s_{situ}_10	3.1 dB	3.1 dB	4.8 dB	5.5 dB	4.0 dB	4.1 dB
s_r_8	2.3 dB	2.3 dB	3.3 dB	5.0 dB	1.4 dB	2.5 dB
s_{situ}_8	2.9 dB	3.2 dB	4.3 dB	5.2 dB	4.1 dB	4.2 dB

With the low-frequency procedure there is an increase of the uncertainty, particularly noticeable at 63 Hz: the repeatability standard deviation increases by about 1.5 dB while the *in situ* standard deviation increases by about 1 dB. The results shown in Table 3 indicate that the low-frequency measurement procedure does increase the uncertainty. This cannot be attributed to the operators whose experience is well proven; this aspect is still under investigation.

To deal with the measurement issue in the low frequency domain, Prato and Schiavi [35] and Prato *et al.* [42] suggest the modal approach. At frequencies below 100 Hz, the acoustic field is non-diffuse, as it is characterized by large fluctuations of sound pressure levels in space and frequency domains. Because of the inhomogeneity of the acoustic field, Prato *et al.* [42] suggest to move from a statistical approach typical of diffuse sound field (average sound energy) to a discrete one, focused at highest noise and annoyance points, *i.e.*, the points of highest sound pressure level in space (corners) and frequency (resonance modes): the so-called modal approach.

In this RRT [10], it was found that the differences between including and excluding low frequencies are a little higher for SNQ plus C_{tr} when using standard measurement procedure and are very high for SNQ plus C_{tr} when using the LF measurement procedure, as shown by comparing Table 4 (SNQs without LF) and Table 5 (SNQs with LF), contrary to what was found in the previous RRT [8] that showed that the differences between including or not the low frequencies were practically negligible.

Table 4: Standard uncertainties of SNQs without low frequencies for the 8 teams [10]

Descriptor (SNQs)	s_r (dB)	s_{situ} (dB)
$D_{ls,2m,nT,w}$	0.4	0.7
$D_{ls,2m,nT,w} + C_{(100-3150)}$	0.6	0.8
$D_{ls,2m,nT,w} + C_{tr(100-3150)}$	0.8	1.0
$D_{ls,2m,nT,w01}$	0.3	0.7
$D_{ls,2m,nT,w01} + C_{(100-3150)}$	0.5	0.8
$D_{ls,2m,nT,w01} + C_{tr(100-3150)}$	0.7	1.0
$D_{ls,2m,nT,w01} + C_{(100-5000)}$	0.6	1.2
$D_{ls,2m,nT,w01} + C_{tr(100-5000)}$	0.7	1.0

Table 5: Standard uncertainties of SNQs with low frequencies for the 8 teams [10]

Descriptor (SNQs)	s_r (dB)		s_{situ} (dB)	
	Default	LF	Default	LF
$D_{ls,2m,nT,w01} + C_{(50-3150)}$	0.5	0.6	0.8	1.0
$D_{ls,2m,nT,w01} + C_{tr(50-3150)}$	0.8	1.9	1.0	2.1
$D_{ls,2m,nT,w01} + C_{(50-5000)}$	0.6	0.6	1.2	1.3
$D_{ls,2m,nT,w01} + C_{tr(50-5000)}$	0.8	1.9	1.0	2.1

This different behavior could be attributable to the differences of the façade test samples: the façade of the previous RRT [8] is a prefabricated concrete façade with a 4 mm single glazing wood-aluminum frame window with a MDF shutter box; the façade of the second study is a prefabricated concrete façade with a PVC frame with double glazing 4/12/4 window.

In fact, from the experience derived from many measurements of façade sound insulation [43,44], the lower the insulation of a window, the lower the spectrum adaptation term C_{tr} and *vice versa*, the higher the window insulation, the higher C_{tr}. For this reason, in the case of the previous RRT [8] (a façade with low insulation window) the difference between $D_{ls,2m,nT,w}$ and $D_{ls,2m,nT,w} + C_{tr}$ averages, was not a large one, only 1.5 dB, while in the case of the present study (a façade with higher insulation window), the difference between the average values of $D_{ls,2m,nT,w}$ and of $D_{ls,2m,nT,w} + C_{tr,50-5000}$ is 5.3 dB for default measurements and 6.8 dB for the low-frequency method.

HOW TO MANAGE THE COOPERATIVE TESTS UNCERTAINTY

As stated in the introduction, current knowledge in building acoustics suggests that the best methodology to study the measurements uncertainty is to carry out an Inter-Laboratory Test or a Round Robin Test. Therefore the results of ILTs and RRTs are very important to know the uncertainty magnitude that is reasonably expected for a measurement result. However, even if an ILT or RRT gives the uncertainty of a measurement method, the uncertainty magnitude depends also on the measurand. An example of the dependence on the method can be drawn from the results of uncertainty of façade sound insulation measurements discussed in Section 2.1.2, where the high frequency uncertainty depends on the loudspeaker position (which is still under study). On the other hand, the uncertainty magnitude also depends on the test sample, as showed in Section 2.1.1 concerning the sound insulation of internal partitions where it was found that the uncertainties of lightweight samples are lower than the uncertainties of heavy types of construction. The dependence on the measurand, in particular for including or not the LF in SNQs, was also found in the case of façade sound insulation uncertainty (see Section 2.1.2) where the comparison of the two RRTs results highlighted that that the differences are attributable to the windows, on which the C_{tr} coefficient depends: a single glazing window and a double 4/12/4 glazing window.

ISO 12999-1 [5] gives the medium uncertainty on all the ILTs and RRTs considered (and available at the time when the standard draft was being written) in that standard, for airborne sound insulation, without distinction of the type of measurand. At the current level of knowledge and due to the number of cooperative tests available, this seems to be the only way to give an idea of the uncertainty magnitude. The fact that the values of ISO 12999-1 [5] are the best estimates for the uncertainty of sound insulation measurements that can be obtained today, was also underlined by Wittstock [45] in his paper that describes how the average uncertainty values standardized in ISO 12999-1 [5] were derived. Therefore, it is important to keep that standard constantly updated in order to increase the number of available data on which the average uncertainty values could be calculated. This specific standard is inaccurate as far as the façade sound insulation is concerned, because its uncertainty is considered equal to the airborne sound insulation uncertainty; indeed, the façade sound insulation measurement method is extremely different from the airborne sound insulation measurement method for party walls and floors. *A priori*, the reproducibility standard deviation is higher than the *in situ* standard deviation because of, as far as reproducibility is concerned, the geometry of the rooms and wall can change, while this is not the case for the *in situ* standard deviation as defined in Section 2. Because the geometry (*i.e.*, modal behavior) has a large influence at low frequencies, s_R is larger than s_{situ} (*cf.* Table 6). The use of s_{situ} is thus only appropriate when the geometry is the same. In the case of façade sound insulation, however, there are no literature data that referred to RRT of the same object in different situations and it will be appropriate in the future that ISO 12999-1 [5] include this difference (*i.e.*, reproducibility and *in situ* standard deviation for façade sound insulation), considering the following: the measurement method of façade sound insulation is extremely different from the laboratory measurement of airborne sound insulation; the uncertainty at high frequencies (which exceed, in the case of the second façade RRT [10], the s_R values of ISO 12999-1 [5] as shown in Figure 3) is mainly dependent on the loudspeaker position (as supposed in the case of the second RRT of Façade [10], as said before), and the RRT [12] values exceed the values of the ISO 140-2 [37] in the range of mass-spring-mass resonance frequency and in the range of the coincidence frequency of the double glazing. At the present state of knowledge, the reproducibility standard deviation values included in ISO 12999-1 [5] seem to be the only available uncertainty that could be used also in the case of façade sound insulation, keeping in mind that the façade sound insulation measurement method is very different.

Table 6: Standard uncertainties for single-number values in accordance with ISO 717-1, as per ISO 12999-1 [5]

Descriptor	s_R dB	s_{situ} dB	s_r dB
R_w, R'_w, D_{nw}, $D_{nT,w}$	1.2	0.9	0.4
$(R_w, R'_w, D_{nw}, D_{nT,w}) + C_{100-3150}$	1.3	0.9	0.5
$(R_w, R'_w, D_{nw}, D_{nT,w}) + C_{100-5000}$	1.3	0.9	0.5
$(R_w, R'_w, D_{nw}, D_{nT,w}) + C_{50-3150}$	1.3	1.0	0.7
$(R_w, R'_w, D_{nw}, D_{nT,w}) + C_{50-5000}$	1.3	1.1	0.7
$(R_w, R'_w, D_{nw}, D_{nT,w}) + C_{tr,100-3150}$	1.5	1.1	0.7
$(R_w, R'_w, D_{nw}, D_{nT,w}) + C_{tr,100-5000}$	1.5	1.1	0.7
$(R_w, R'_w, D_{nw}, D_{nT,w}) + C_{tr,50-3150}$	1.5	1.3	1.0
$(R_w, R'_w, D_{nw}, D_{nT,w}) + C_{tr,50-5000}$	1.5	1.0	1.0

Therefore, in the case of a single measurement, the uncertainty that should be associated to this measurement is the reproducibility standard deviation given in ISO 12999-1 [5] multiplied by the appropriate coverage factor to obtain the expanded uncertainty. Now, considering what was stated in the introduction, when reporting the result of the measurement of a physical quantity, it is compulsory that some quantitative indications of the quality of the result be given. Such an indication should be independent on the final use of the results (verification of a requirement or determination of predicted values), and shall be stated as follows, as provided by GUM [2] and ISO 12999-1 [5]:

$$Y = y \pm U \qquad (3)$$

where Y is the measurand; y is the best estimate (obtained through the measurement) of the value attributable to the measurand; and U is the expanded uncertainty, calculated for a given confidence level for the two-sided test, defined as the product of the measurement uncertainty u (which is the reproducibility standard deviation s_R) with a coverage factor k.

Therefore, for example, for a single measurement of the airborne sound insulation of a partition floor R'_w (C;C_{tr}) = 53 (−1;−4), considering the values given in ISO 12999-1 (see Table 6), the airborne sound insulation of this partition wall shall be given to one decimal place (R'_w = 52.6; C = −1.0; C_{tr} = −4.1) to state also its uncertainty and should be designated as [5]:

$$R'_w = (52.6 \pm 2.4)dB(k = 1.96, two - sided) \qquad (4)$$

$$R'_w + C = (51.6 \pm 2.6)dB(k = 1.96, two - sided) \qquad (5)$$

$$R'_w + C_{tr} = (48.5 \pm 2.9)dB(k = 1.96, two - sided) \qquad (6)$$

where k = 1.96 corresponds to a confidence level of 95% for a two-sided test.

On the other hand, when a measurement is made in order to verify a requirement, the expanded uncertainty that should be given with the result, should be calculated using a coverage factor for one-sided test, as laid down in ISO 12999-1 [5]. Then the expanded uncertainty should be added to or subtracted from the measurement result to check whether that measurement result is smaller or larger than the requirement, respectively.

The Italian standard on the acoustic classification of building units UNI 11367 [46] first considers the measurement uncertainty from RRTs as a basis for the expanded uncertainty U. When a national regulation has to be met, the choice of the confidence level is very important. The Italian standard on the acoustic classification [46] has faced for the first time the problem related to the confidence level. In the case of measurement uncertainty, the standard recommends to use a coverage factor k for one-sided test equal to 1, which corresponds to an 84% probability; for buildings performances, in fact, in order to meet the limit, it is not realistic to use a 95% or 90% confidence level, which is normally used in other contexts. As the update of the ISO 717-1 [21] allows applying the weighting procedure by 0.1 dB steps for the expression of measurement uncertainty, it could now possible also be to use, in building acoustics, a coverage factor k for one-sided test equal to 1.65 corresponding to a 95% confidence level.

Generally, when measurements are made to verify the acoustic requirements of buildings, one single measurement might not be enough to this end, and therefore more measurements and more results for the same requirement are necessary. In this case, the measurement uncertainty is combined in a certain way with the uncertainty due to the number of tests performed.

SAMPLING

There are two different types of surveys that can be used to analyze the acoustic requirements of building units, or buildings: a census (the entire population is taken into account) or a sample survey (only a part of the elements that make up the population are considered). For building acoustics, a sample survey is the best solution in terms of cost and time. To make meaningful comparisons with both national regulations and acoustic classification, it is therefore necessary to determine the type and amount of the measurements. In order to make any sample survey on certain features (acoustical) of a finite population, it is essential to formulate a strategy of selection, which is closely connected with the purposes, the cost and the execution time of the survey. In addition, the sample obtained from it, is the only valuable information that could be used for the interpretation of the results.

Among the different sampling strategies currently available, the two main ones used in building acoustics, for the time being, are the following: the stratified sampling as adopted by UNI 11367 [46] (see next section) and a sampling procedure taking into account a certain percentage associated with a selection criterion as adopted by UNI 11444 [47] and proposed by ISO/ WD 19488 [48]. Only the former strategy (stratified sampling) includes the sampling uncertainty. The strategy of UNI 11444 [47] consists in the selection of a minimum number of Building Units (BUs): not less than 10% of the total amount of BUs composing the building system and not less than 2 BUs, if the total amount is 4, and not less than 3 BUs for building systems up to 30 BUs. These BUs must be the most critical BUs from an acoustic point of view. The selection of the most critical BUs must take into account all the critical acoustic features of the building elements of the BU. The selection criteria for each type of acoustic performance (façade sound insulation, sound insulation of horizontal and vertical partitions, impact sound insulation and equipment noise level) are stated in standard UNI 11444 [47]. This standard does not include the sampling uncertainty but, for each measurement, it includes the measurement uncertainty as stated in UNI 11367 [46].

The standard proposal ISO/WD 19488 [48] considers, as a general principle, that, when verifying the acoustic class of a unit, a sufficient number of measurements of each relevant acoustic characteristic must be performed in order for the result to represent the unit. It also suggests that care should be taken to include the critical site/rooms, e.g., partitions with critical flanking constructions. At the current stage, the proposal includes neither the sampling uncertainty nor the measurement uncertainty, but it considers that compliance is granted if the average results comply with the class limits and no individual result deviates unfavorably by more than 2 dB. Moreover, if classification for different dwellings, rooms or acoustic characteristics varies, the classification assigned is the minimum class obtained.

Considering the pros and cons of these two sampling strategies, the first thing to keep in mind is the scope of the measurements; *i.e.*, to determine a class within the acoustic classification or to verify the legal requirements. In the former case, it is obvious that a value as close as possible to the value of all the elements is suitable. In the latter case, the scope is to identify the worst acoustic performances and to verify if also the critical site/rooms is/are in compliance with the legal requirements.

The stratified sampling strategy allows increasing the efficiency of a sampling plan, without increasing the sample size. With this strategy it is possible to obtain the best representative value of a class to be attributed to the entire building system, as if the entire population were taken into account. Another pro is the stratified sampling uncertainty related to the final result that gives a confidence level, which is important both for the owners and the builders. The con of this strategy is that it requires a large number of measurements (a minimum of three measurements for each homogeneous group).

A strategy that takes into account a certain percentage of the population, including all the critical site/rooms, could not be representative of all situations but would give the worst results and therefore, if this result complies with the legal requirements, the whole building complies with them. On the other hand, not all the critical site/rooms may have been taken into account and therefore the confidence level and the sampling uncertainty to be associated with the results is not known. Moreover, the sampling strategy proposed in ISO/WD 19488 has the obvious drawback that it cannot guarantee to have spotted all critical situations: for example a workmanship failure that cannot be detected by visual inspection can be identified only after the measurements. Thus, a sampling criterion based on generic rules cannot find it. However, the con of this strategy is that in general the number of measurements is limited.

Stratified Sampling

The stratified sampling is the most direct procedure that allows increasing the efficiency of a sampling plan, since it allows reducing the order of magnitude of the sampling error without increasing the sample size.

Stratification is made possible by means of additional information about one or more characters of the population, which is about the structure of the population itself. This allows, based on informed choice, dividing the population into a number of layers as homogeneous as possible, as meaning that within each layer, the considered character has a lower variability. A simple random sample is extracted from each layer; therefore there are as many simple samples as there are layers. These samples are independent of each other and can have different sample sizes. The stratification, due to the way it is implemented, allows obtaining an improvement in the estimates for the same sample size, or to contain the sample size at the same level of efficiency [49].

Considering the above mentioned advantages offered by the stratified sampling, this latter is the solution adopted by UNI 11367 [46] in the case of classification of serial type buildings.

The part of the Italian standard on the classification of buildings and building units that refers to the stratified sampling procedure can be applied in the case of a serial type building. The stratified sampling procedure is based on the concept of homogeneous group. The population of all the building elements that have to be measured for the acoustic classification has to be divided in the homogeneous groups that are defined in the Italian standard on classification. Referring to UNI 11367 [46], generally, a set of test items can be considered homogeneous and therefore subject to a possible sampling (in reference to a specific requirement), if the following conditions are satisfied: item dimensions (with 20% tolerance); dimensions (with 20% tolerance with respect to the volumes) of both transmitting and receiving rooms where the test item is located; the same test methodology; stratigraphy, materials and surface mass; structural constraints (flanking transmissions); presence of equipment passing through the test item; installation techniques. In this section an example is given with reference to the paper presented by the authors at the 38th National Congress of the Acoustical Society of Italy in 2011 [50], concerning the acoustic classification of a building system of a total volume of about 15,000 m^3, consisting of two similar buildings, identified as body A1 and body A2, on three floors, with apartments on the ground floor, first and second floor and, in the body A1, a third floor attic. In total, the building system consists of 47 Building Units (BUs), distributed according to their type: six four-room apartments, eight three-room apartments, 25 two-room apartments and eight studios.

The building system was considered a serial type building system, based on the following considerations: it is possible to identify a typical floor (see Figure 4) in which the distribution of BUs is symmetrical with respect to the stairwells; the two-room apartment type is repeated 25 times; the rooms with the same intended use (bedrooms, living rooms, kitchens, *etc.*) have the same shape and size.

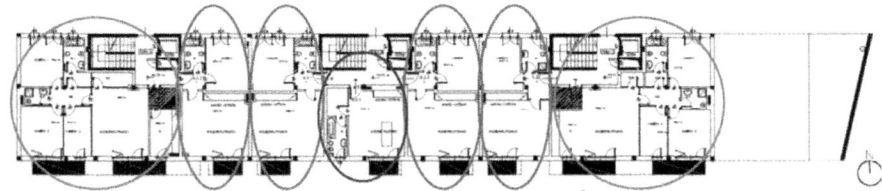

Figure 4: Typical floor of the building system considered: the BU typologies are highlighted, in green the four-room apartments, in red the two-room apartments and in blue the studio [50].

For the application of the stratified sampling procedure defined in UNI 11367 [46], it would have been sufficient to use a minimum number of items to be tested equal to at least 10% of the total number of elements of the homogeneous group and not less than three for each homogeneous group. However, in order to obtain the most useful data for a critical examination of the results, the number of items to be tested was higher than the minimum required. In particular, 84% of the vertical partitions were measured. For some requirements, the number of items to be tested of some homogeneous group was equal to two, which is less than the minimum required of three for reasons related to the impossibility to perform further tests (inaccessibility of the rooms). When more measurements than the minimum necessary number were made, in order to simulate the case in which only the minimum sampling number (3) of measurements was performed, the results were reconsidered on the basis of all the combinations without repetition of three elements actually measured and calculating the average. This was done in order to make the choice of the three elements under test as random as possible, and to evaluate the probability of obtaining a specific standard deviation, and a specific class according to the variability and to the randomness of selection of the three elements. The results, obtained on the basis of both the performed measurements and the statistical analysis of the sampling procedure, are as follows. First of all, a methodological indication: a review and a possible redefinition of the homogeneous groups retrospectively (*i.e.*, when the measurement are concluded), this may be useful to formulate an acoustic classification closer to the real situation. This indication comes from the fact that, in the case under study, the values of the impact sound insulation differ greatly and in a systematic manner between the two bodies; the results of body A2 show a worse performance than those of body A1. This difference is due to the installation that, in one case (body A1), was evidently very well done. The influence of workmanship was studied by Craik and Steel [51] who found that workmanship can account for a variation of approximately 2 dB in airborne sound insulation. Within this distinction, the variability of the impact sound insulation values is on the average when compared with the airborne sound insulation of vertical and horizontal partitions.

From the analysis of all the combinations without repetition of the measurements, two possible classifications are found: in one of them, the percentage of BUs of class III (43%) becomes smaller compared to that of class IV, while in the other, the percentage of BUs of class III is rather prevalent (64%); in particular, this is due to the requirement of airborne sound insulation of internal partitions R'_w. Actually, the values relating to the airborne sound insulation R'_w, for vertical partitions, are in the vast majority of cases very close to the lower limit value (50 dB) of class III; therefore, in the random choice, there is a higher probability that the choice falls on these values straddling the two classes, with the result of moving the larger percentage of BUs from class to class. This analysis makes it clear that it is necessary to adopt, at design level, more conservative design solutions.

Table 7 shows the average, minimum, maximum and standard deviation of the measured performances for each type of technical element; in particular, the variability of the data is described by the standard deviation and it increases with the increase of the latter. The values shown in Table 7 are the net values, as defined in UNI 11367 [46], *i.e.*, the results of a measurement corrected with the measurement uncertainty.

The greatest variability is found for façades; for the building system under classification this is caused mainly by the typical variability of façades, dependent on many types of window frames and the presence of balconies, irrespective of a proper installation and, to a lower extent, also to workmanship.

The lower variability is observed in internal partitions, and in particular floors (horizontal partition), with respect to the sound insulation requirement. The variability of the impact sound insulation is comparable with that of the other requirements. Moreover, the variability of the impact sound insulation for the two bodies separately is comparable, confirming the systematic difference found in body A2.

Table 7: Performances variability of technical elements [50]

Technical Element	Façade	Vertical Partitions	Horizontal Partitions	Horizontal Partitions A1	Horizontal Partitions A2
Quantity used in law requirements	$D_{2mnT,w}$	R'_w	R'_w	L'_n	L'_n
number of test elements	35	36	21	9	10
average	39.6	50.1	52.6	57.6	65.5
standard deviation	3.7	1.87	1.4	2.2	2.6
minimum	30	47	50	54	62
maximum	45	54	55	61	70

Stratified Sampling Uncertainty

When a sample survey is used to define the classification of building or BUs, it is necessary to consider the uncertainty associated with the sampling procedure. Moreover, considering that each single measurement result that contributes to the value attributed to a certain requirement has its own measurement uncertainty, it becomes necessary to combine these two uncertainties in a certain way.

In the case of UNI 11367 [46] the representative value of a homogeneous group (Equations (9) and (10)), *i.e.*, the arithmetic mean value of the group with the sampling uncertainty with a one-sided coverage factor, already includes the measurement uncertainty. In fact, the arithmetic mean values X_{he} and Y_{he} are calculated from the net values (*i.e.*, the results of the measurement corrected with the measurement uncertainty) of the homogeneous group itself [46,52] as indicated in UNI 11367 [46] as follows:

$$X_{he} = \frac{\sum_{c=1}^{C_h} X_{hc}}{C_h}$$

(7)

$$Y_{he} = \frac{\sum_{c=1}^{C_h} Y_{hc}}{C_h}$$

(8)

where X_{hc} is the net value of a sample of a specific requirement (façade sound insulation or airborne sound insulation of the internal partition), Y_{hc} is the net value of a sample of a specific requirement (impact sound insulation or sound pressure level for service equipment), and C_h is the number of samples within a homogeneous group.

The "representative value" X_h and Y_h of each homogeneous group is then obtained as follows [46]:

$$X_h = X_{he} - U_{sh} \tag{9}$$

$$Y_h = Y_{he} + U_{sh} \tag{10}$$

where U_{sh} is the sampling uncertainty equal to the sampling standard deviation s_{sh} times the coverage factor k:

$$U_{sh} = s_{sh} \cdot k \tag{11}$$

where s_{sh} is the sampling standard deviation, determined with Equations (12) and (13):

$$s_{shX} = \sqrt{\frac{\sum_{c=1}^{C_h} (X_{he} - X_{hc})^2}{C_h - 1} \frac{(M_h - C_h)}{(M_h - 1)}} \tag{12}$$

$$s_{shY} = \sqrt{\frac{\sum_{c=1}^{C_h} (Y_{he} - Y_{hc})^2}{C_h - 1} \frac{(M_h - C_h)}{(M_h - 1)}} \tag{13}$$

where s_{shX} is the standard deviation referred to the façade sound insulation or to the airborne sound insulation of internal partitions, s_{shY} is the standard deviation referred to the impact sound insulation or to the sound pressure level for service equipment and M_h is the number of all the measurable technical elements within a homogeneous group.

CONCLUSIONS

This study showed that the measurement uncertainty in building acoustics is very high, in particular if the measurements are extended at low frequencies. Therefore, it is extremely important to define the way to manage measurement uncertainty in building acoustics, depending on the different situations.

In the case of a single measurement, the uncertainty that should be associated with this measurement is the reproducibility standard deviation given in ISO 12999-1 [5] multiplied by the appropriate coverage factor to obtain the expanded uncertainty. Such an indication should be independent of the final use of the results, and shall be stated as the expanded uncertainty with a 95% confidence level for a two-sided test. When the single measurement is made in order to verify a requirement, the expanded uncertainty that should be given with the result, should be calculated using a coverage factor for one-sided test and the confidence level should be set to 95%.

When measurements are made to verify the acoustic requirements or the acoustic classification of building units, or buildings, one single measurement might not be enough to this end, and therefore more measurements and more results for the same requirement are necessary. For building acoustics, a sample survey is the best solution in terms of cost and time. There are two main types of sampling strategies used in building acoustics, for the time being: the stratified sampling as stated in UNI 11367 [46] and a sampling procedure taking into account a certain percentage associated with a selection criterion as adopted UNI 11444 [47] and proposed in ISO/WD 19488 [48]. In the former case, the measurement uncertainty is combined with the sampling uncertainty to obtain a reliable classification for BUs, while in the latter case, the sampling uncertainty is not taken into account because the strategy selection includes all the critical acoustic situations. In UNI 11444 [47], the measurement uncertainty is included for each measurement result, as stated in UNI 11367 [46]. In ISO/WD 19488 [48], if classification for different dwellings, rooms or acoustic characteristics varies, the classification assigned is the minimum class obtained, and therefore all the other BUs complied with that class.

In any case, a measurement, whether single or part of a set of measurements for the sampling, should always be associated with its measurement uncertainty. In the case of sampling, either the sampling uncertainty is considered, obtaining a value representative of all the situations considered, or selection criteria of the most critical cases is taken into account, obtaining a value that is not representative of all the situations but is a precautionary value.

ACKNOWLEDGMENTS

We gratefully acknowledge Paolo Cardillo for revising the English text of the manuscript.

AUTHOR CONTRIBUTIONS

Chiara Scrosati designed and conceived the study, acquired data, analyzed and interpreted data, supervised the RRTs and drafted the manuscript.

Fabio Scamoni revised the manuscript critically for important intellectual content.

REFERENCES AND NOTES

1. Scrosati, C.; Scamoni, F. Measurement uncertainty in building acoustics Invited and peer reviewed paper. In Proceedings of the 22nd International Congress on Sound and Vibration (ICSV22), Florence, Italy, 12–16 July 2015.

2. *ISO/IEC Guide 98-3:2008 The Guide to the Expression of Uncertainty in Measurement (GUM:1995)*; ISO: Genève, Switzerland, 2008.

3. *ISO 10140 Acoustics—Laboratory Measurement of Sound Insulation of Building Elements*; ISO: Genève, Switzerland, 2010.

4. *ISO 16283 Acoustics—Field Measurement of Sound Insulation in Buildings and of Building Elements*; ISO: Genève, Switzerland, 2014.

5. *ISO 12999-1:2014 Acoustics—Determination and Application of Measurement Uncertainties in Building Acoustics—Part 1: Sound Insulation*; ISO: Genève, Switzerland, 2014.

6. Scamoni, F.; Scrosati, C.; Mussin, M.; Galbusera, E.; Bassanino, M.; Zambon, G.; Radaelli, S. Repeatability and reproducibility of field measurements in buildings. In Proceedings of the Euronoise 2009, Edinburgh, Scotland, 26–28 October 2009.

7. Scrosati, C.; Scamoni, F.; Bassanino, M.; Mussin, M.; Zambon, G. Uncertainty analysis by a round robin test of field measurements of sound insulation in buildings: Single numbers and low frequency bands evaluation—Airborne sound insulation. *Noise Control Eng. J.* 2013, *61*, 291–306.

8. Scrosati, C.; Scamoni, F.; Zambon, G. Uncertainty of façade sound insulation in buildings by a round robin test. *Appl. Acoust.* 2015, *96*, 27–38. [Google Scholar]

9. Decree of the President of the Council of Ministers D.P.C.M. Determinazione Dei Requisiti Acustici Passivi Degli Edifici (Determination of Building Passive Acoustic Requirements). In *G.U. (Official Journal) General Series n.297*; D.P.C.M.: Rome, Italy, 1997.

10. Scrosati, C.; Scamoni, F.; Asdrubali, F.; D'Alessandro, F.; Moretti, E.; Astolfi, A.; Barbaresi, L.; Cellai, G.; Secchi, S.; di Bella, A.; *et al.* Uncertainty of façade sound insulation measurements obtained by a round robin test: The influence of the low frequencies extension, Invited paper. In Proceedings of the 22nd International Congress on Sound and Vibration (ICSV22), Florence, Italy, 12–16 July 2015.

11. *ISO 140-5:1998 Acoustics—Measurement of Sound Insulation in Buildings and of Building Elements—Part 5: Field Measurements of Airborne Sound Insulation of Façade Elements and Façades*; ISO: Genève, Switzerland, 1998.

12. Lang, J. A round robin on sound insulation in building. *Appl. Acoust.* 1997, *52*, 225–238.

13. Simmons, C. Uncertainty of measured and calculated sound insulation in buildings—Results of a round robin test.*Noise Control Eng. J.* 2007, *55*, 67–75.

14. Farina, A.; Fausti, P.; Pompoli, R.; Scamoni, F. Intercomparison of laboratory measurements of airborne sound insulation of partitions. In Proceedings of the 1996 International Congress on Noise Control Engineering, (Internoise' 96), Liverpool, UK, 30 July–2 August 1996.

15. Fausti, P.; Pompoli, R.; Smith, R.S. An inter-comparison of laboratory measurements of airborne sound insulation of lightweight plasterboard walls. *J. Build. Acoust.* 1999, *6*, 127–140.

16. Smith, R.S.; Pompoli, R.; Fausti, P. An Investigation into the reproducibility values of the european inter-laboratory test for lightweight walls. *J. Build. Acoust.* 1999, *6*, 187–210.

17. Schmitz, A.; Meier, A.; Raabe, G. Inter-laboratory test of sound insulation measurements on heavy walls: Part I—Preliminary test. *J. Build. Acoust.* 1999, *6*, 159–169.

18. Meier, A.; Schmitz, A.; Raabe, G. Inter-laboratory test of sound insulation measurements on heavy walls: Part II—Results of main test. *J. Build. Acoust.* 1999, *6*, 171–186.

19. *ISO 140-4:1998 Acoustics—Measurement of Sound Insulation in Buildings and of Building Elements—Part 4: Field Measurements of Airborne Sound Insulation between Rooms*; ISO: Genève, Switzerland, 1998.

20. COST Action TU0901 Integrating and Harmonizing Sound Insulation Aspects in Sustainable Urban Housing Constructions. 2009–2013. Available online: http://www.costtu0901.eu/ (accessed on 15 September 2015).

21. *ISO 717-1:2013 Acoustics—Rating of Sound Insulation in Buildings and of Building Elements—Part 1: Airborne Sound Insulation*; ISO: Genève, Switzerland, 2013.

22. Wittstock, V. On the uncertainty of single-number quantities for rating airborne sound insulation. *Acta Acust. United Acust.* 2007, *93*, 375–386.

23. Dijckmans, A.; Vermeir, G. Numerical investigation of the repeatability and reproducibility of laboratory sound insulation measurements. *Acta Acust. United Acust.* 2013, *99*, 421–432.

24. Garg, N.; Maji, S. On analyzing the correlations and implications of single-number quantities for rating airborne sound insulation in the frequency range 50 Hz to 5 kHz. *Build. Acoust.* 2015, *22*, 29–44.

25. *ISO CD 16717-1:2013 Acoustics—Evaluation of Sound Insulation Spectra by Single-Number Values. Part 1: Airborne Sound Insulation*; ISO: Genève, Switzerland, 2013.

26. Mahn, J.; Pearse, J. The uncertainty of the proposed single number ratings for airborne sound insulation. *J Build. Acoust.* 2012, *19*, 145–172.

27. Hongisto, V.; Keränen, J.; Kylliäinen, M.; Mahn, J. Reproducibility of the present and the proposed single-number quantities of airborne sound insulation. *Acta Acust. United Acust.* 2012, *98*, 811–819.

28. Machimbarrena, M.; Monteiro, C.R.A.; Pedersoli, S.; Johansson, R.; Smith, S. Uncertainty determination of *in situ*airborne sound insulation measurements. *Appl. Acoust.* 2015, *89*, 199–210.

29. António, J.; Mateus, D. Influence of low frequency bands on airborne and impact sound insulation single numbers for typical Portuguese buildings. *Appl. Acoust.* 2015, *89*, 141–151.

30. Grubbs, F.E. Procedures for detecting outlying observation in samples. *Technometrics* 1969, *11*, 1–21.

31. Grubbs, F.E.; Beck, G. Extension of sample sizes and percentage points for significants tests of outlying observations.*Technometrics* 1972, *14*, 847–854.

32. Masovic, D.B.; Sumarac Pavlovic, D.S.; Mijic, M.M. On the suitability of ISO 16717-1 reference spectra for rating airborne sound insulation. *J. Acoust. Soc. Am.* 2013, *134*, EL420–EL425.

33. Rindel, J.H. On the influence of low frequencies on the annoyance of noise from neighbours. In Proceedings of the InterNoise 2003, Seogwipo, Korea, 25–28 August 2003.

34. Park, H.K.; Bradley, J.S. Evaluating standard airborne sound insulation measures in terms of annoyance, loudness, and audibility ratings. *J. Acoust. Soc. Am.* 2009, *126*, 208–219.

35. Prato, A.; Schiavi, A. Sound insulation of building elements at low frequency: A modal approach. In Proceedings of the Energy Procedia—6th International Building Physics Conference, IBPC 2015, Turin, Italy, 14–17 June 2015.

36. *ISO/DIS 16283-3:2014 Acoustics—Field Measurement of Sound Insulation in Buildings and of Building Elements—Part 3: Façade Sound Insulation*; ISO: Genève, Switzerland, 2014.

37. *ISO 140-2:1991. Acoustics—Measurement of Sound Insulation in Buildings and of Building Elements—Part 2: Determination, Verification and Application of Precision Data*; ISO: Genève, Switzerland, 1991.

38. Berardi, U.; Cirillo, E.; Martellotta, F. Interference effects in field measurements of airborne sound insulation of building façades. *Noise Control Eng. J.* 2011, *59*, 165–176.

39. Berardi, U. The position of the instruments for the sound insulation measurement of building façades: From ISO 140-5 to ISO 16283-3. *Noise Control Eng. J.* 2013, *61*, 70–80.

40. Hopkins, C. Revision of international standards on field measurements of airborne, impact and facade sound insulation to form the ISO 16283 series. *Build. Environ.* 2015, *92*, 703–712.

41. Hopkins, C.; Turner, P. Field measurement of airborne sound insulation between rooms with non-diffuse sound fields at low frequencies. *Appl. Acoust.* 2005, *66*, 1339–1382.

42. Prato, A.; Ruatta, A.; Schiavi, A. Transmission of impact noise at low frequency: A modal approach for impact sound insulation measurement (50–100 Hz). In Proceedings of the 22nd International Congress on Sound and Vibration (ICSV22), Florence, Italy, 12–16 July 2015.

43. Masovic, D.; Miskinis, K.; Oguc, M.; Scamoni, F.; Scrosati, C. Analysis of façade sound insulation field measurements—Influence of acoustic and non-acoustic parameters. In Proceedings of the INTER-NOISE 2013, Innsbruck, Austria, 15–18 September 2013.

44. Masovic, D.; Miskinis, K.; Oguc, M.; Scamoni, F.; Scrosati, C. Analysis of façade sound insulation field measurements—Comparison of different performance descriptors and influence of low frequencies extension. In Proceedings of the INTER-NOISE 2013, Innsbruck, Austria, 15–18 September 2013.

45. Wittstock, V. Determination of measurement uncertainties in building acoustics by interlaboratory tests. Part 1: Airborne sound insulation. *Acta Acust. United Acust.* 2015, *101*, 88–98.

46. *UNI 11367:2010 Building Acoustics—Acoustic Classification of Building Units—Evaluation Procedure and In Situ Measurements*; UNI: Milan, Italy, 2010.

47. *UNI 11444:2012 Building Acoustics—Acoustic Classification of Building Units—Guidelines for the Selection of Building Units in not Serial Building Systems*; UNI: Milan, Italy, 2012.

48. *ISO/WD 19488:2015 (Rev. 20 July 2015) Acoustics—Acoustic Classification Scheme for Dwellings*; ISO: Genève, Switzerland, 2012.

49. Scrosati, C.; Pontarollo, C.M.; Scamoni, F.; di Bella, A.; Elia, G. Procedure di verifica delle prestazioni acustiche di edifici: Analisi tramite campionamento, (Acoustic performances of buildings: Sampling procedure analysis). In Proceedings of the 37th AIA National Congress, Siracusa, Italy, 26–28 May 2010.

50. Scamoni, F.; Scrosati, C.; Cera, S. Classificazione acustica in un edificio residenziale a tipologia seriale: Analisi statistica della procedura di campionamento (Acoustic classification of a residential building of serial type: Statistical analysis of sampling procedure). In Proceedings of the 38th AIA National Congress, Rimini, Italy, 8–10 June 2011.

51. Craik, R.J.M.; Steel, J.A. The effect of workmanship on sound transmission through buildings: Part 1. Airborne sound. *Appl. Acoust.* 1989, *27*, 57–63.

52. Di Bella, A.; Fausti, P.; Scamoni, F.; Secchi, S. Italian experience on acoustic classification of buildings. In Proceedings of the Internoise 2012, New York, NY, USA, 19–22 August 2012.

Chapter 3

A CONSTRUCTION MANAGEMENT FRAMEWORK FOR MASS CUSTOMISATION IN TRADITIONAL CONSTRUCTION

María Dolores Andújar-Montoya[1], Virgilio Gilart-Iglesias[2], Andrés Montoyo[3] and Diego Marcos-Jorquera[2]

[1]Department of Building and Urbanism, Polytechnic University College, University of Alicante, Carretera de San Vicente del Raspeig, s/n, 03690 Alicante, Spain

[2]Department of Computer Science and Technologies, Polytechnic University College, University of Alicante, Carretera de San Vicente del Raspeig, s/n, 03690 Alicante, Spain

[3]Department of Software and Computing Systems, Polytechnic University College, University of Alicante, Carretera de San Vicente del Raspeig, s/n, 03690 Alicante, Spain

ABSTRACT

A Mass Customisation model is discussed as a competitive positioning strategy in the marketplace adding value to the customer's end-use. It includes the user as part of the construction process responding to the customer's demands and wishes. To the present day, almost all proposals for Mass Customisation have been focused on the design phase and single family houses. The reality is that the processes carried out in the work execution are so inefficient that the costs of the Mass Customisation models are assumed by the customer and they do not offer solutions that support the change management. Furthermore, this inefficiency often makes Mass Customisation unfeasible in terms of deadlines and site management. Therefore, the present proposal focuses on achieving the paradigm of Mass Customisation in the traditional residential construction complementary to the existing proposals in the design phase. All this through the proposal of a framework for the integral management in the work execution, which will address change management introduced by the users offering an efficient and productive model that reduces costs in the process. This model will focus on the synergy between different strategies, techniques and technologies currently used in the construction management (such as Lean Construction or Six Sigma), together with, other strategies and technologies that have proven to be valid solutions in other fields (such as Business Process Management, Service Oriented Architecture, *etc.*).

INTRODUCTION: MASS CUSTOMISATION IN HOUSING MARKET

Mass Customisation is emerging as an inevitable strategy to ensure competitiveness in a customer-oriented market [1], this means to satisfy more closely the individual wants and needs of the customers, by providing a wide variety of options and individual customisation at prices comparable to standard goods [2]. So, mass customization is the capability to integrate these varieties derived from the individual customer's needs and the efficiency of standard mass production [3,4]. Specifically, in the Spanish housing sector the need of adopting competitive strategies oriented to the prospective buyer is more pronounced due to the current economic slowdown, the sharp drop in activity and the oversupply in dwelling that difficulty the home sales. The strong adjustment of the sector, the high number of dwellings in stock, the largest social requests from the future users and the increased normative requirements, especially in terms of sustainability, safety and quality in buildings, address companies to the imminent need of being reinforced to compete in the market [5]. Facing this new development, enterprises need to be positioned in the market through a differentiation with competitors by implementing new ways to add value to their offers [6]. Through the offer of Mass Customisation in the property development, value-added services are created replying to the need of some potential customers who are changing the way they used to buy, integrating customer into the process. So a new agile and flexible approach is needed not only to satisfy the initial customers' requirements, but also to provide a greater ability to adapt to their needs as they evolve [7] with the same cost as in mass production [8].

However, the barriers to achieving this approach in the field of building and construction are high. At the same time that enterprises are forced to react to the growing individualization of demand, the increased pressure of competitors and the financial conditions of the market dictate the need of shorter construction time, reduction in total costs and higher production quality. Companies must adopt strategies that include both cost efficiency and a closer relationship to the customer's needs [9]. This need of efficiency in building costs is one of the principal barriers to Mass Customisation, as it leads to increased material costs, higher manufacturing costs, lower on time deliveries, increased level of needed inventory for every variant of Customisation and reduction in product quality [1]. These increases in costs cannot be absorbed by customers who usually expect individualized products at the same price as they would pay for mass-produced items [10].

Besides the increase in cost, Mass Customisation accompanies problems related to term, due to the fact that identical houses cannot be produced. Construction firms require more time to carry out the site works for not being totally familiarity with the plans, owing the variability of them from some property to another [11].

Furthermore, Ph.D. Noguchi describes in "ZEMCH Research Initiatives: Mass Customisation and Sustainability" [12] the additional obstacles related to customisation in property developments, such as lack of pilot housing, more complexity in communication, greater chance of material delays, and disadvantages in negotiations with subcontractors.

Another obstacle attributed to the housing customisation is the delay in the buyer decision making, a fact that can cause delays in the planning of works [13]. Also the risk of doubtful customers and changes in opinion derived from the array of choices.

In addition, this flexibility and variability increase the complexity of the management and the execution of the whole process given that the diversity of options may exceed the capacity of the organization to manage works efficiently.

Currently, construction companies do not have information systems to address efficiently the management of Mass Customisation in site. This obstacle is compounded by the inefficiency in productivity indexes that has been dragging the industry in the last years, in comparison with other sectors of the Spanish economy.

Consequently of the foregoing and considering the current situation of low profit margins, the main barrier to adopt Mass Customisation in housing is the lack of a structured system to help the site management efficiently, minimizing the extra costs, the deviations in schedules, the materials delays and the uncertainty in the planning. Therefore, the short-term challenge for the industry is to increase efficiency by reducing these barriers at the same time the communication and negotiations are improved bringing simplicity to the management, and providing added value to the product.

This requires the adoption of a new management model that makes feasible Mass Customisation in property developments, allowing the satisfaction of the customer requirements, thus accentuating the competitive advantage of firms.

At the present time there are proposals with large repercussion to reinforce company competitiveness, and then overcome the main barriers to adopt the Mass Customisation. They are mainly centred on solving the inefficiency in cost and time of the construction works, while they focus on meeting the customers' needs. To develop and emphasize the competitive advantage of a corporation some construction companies are starting to carry out relevant business management strategies to strength their position in the market. One of the most emerging approach to improve organizational effectiveness is the application of Lean Philosophy [14] to construction, which is consider as a part of a "cost leadership [15]" competitive strategy [16] that maximize customer value while minimizing waste. Also another upward trend to increase productivity is the implementation of Information and Communication Technology (ICT) as a competitive approach that is increasingly being used in the construction industry [17]. Several options of specific software for the works are available to increase productivity in the construction process (e.g., software for quality control, planning, project based accounting, *etc.*). Also software tools of enterprise performance systems (ERP) or customer relationship management (CRM), among others, are implemented by many companies.

Therefore to exploit the whole potential of each specific proposal it should be approached in a holistic solution [18], avoiding the technological waste with the "point solutions" for each stage of the construction process and looking for a well-suited solution to construction [19,20]. Accordingly, to get a holistic solution in the site execution, the main novelty lies in using for the first time, Business Process Management (BPM) in construction, in particular to provide an integrated management model by using BPM as a backbone solution. BPM is an evolving trend in management science whose effectiveness has been commonly demonstrated in other fields such as the industry sector, banking, retail, government, and health care. The importance of BPM is confirmed by the existence of specialized international journals (e.g., the Business Process Management Journal), conferences (e.g., the Business Process Management Conference) and courses institutionalized at several universities [21]. This is a powerful instrument to gain competitive advantage through a holistic process-oriented view [22] based on the ICT to automate or support processes and their lifecycle. This holistic view can involve a lot of management disciplines and techniques (such as lean, Six Sigma, lean, ERP, Service Oriented Architecture SOA, and other enablers [23]) and embraces parts of management, like Change management, Information Technology (IT) management, Project management and deals with suppliers, customers, employees, *etc*.... [22]. BPM is a comprehensive approach to process improvement and it is focused on integrating, aligning, managing and measuring all the business processes of an organization. It is an inclusive approach of other approaches, it includes the application of all the other available methods (Process Excellence, Performance Improvement, Six Sigma, Business Process Reengineering, Lean, Business Process Engineering, Customer Expectation Management) [24]. BPM is the synthesis and extension of several technologies and techniques into a unified whole, a new foundation upon which to build sustainable competitive advantage [25]. Therefore our proposal lies in an integrated solution that embraces all the earlier proposals that allow Mass Customisation, while provides continuous improvement, customer satisfaction and constructive sustainability.

The rest of the paper is structured in the following way: In Section 2 a review and analysis about the state of art of the research proposals and related works is done; in Section 3 the research methodology is explained;, in Section 4 the problems and weaknesses associated with the current process are described; in Section 5 partial solutionsare presented and the integrated management model proposed; in Section 6 an use case implementation and validation is defined, and finally inSection 7, the main contributions and main conclusions of the research and the way forward for future research are presented.

THE STATE OF THE ART IN MASS CUSTOMISATION APPLIED TO CONSTRUCTION INDUSTRY

Nowadays, there are several proposals that suggest the implementation of Mass Customisation inside the construction industry. One of the most cited authors in the field, Ph.D. Masa Noguchi [26], highlights the importance of Mass Customisation in the industrialized housing as a marketing strategy to solve customer dissatisfaction inside the Asian real estate market. Noguchi focuses on resolving the customer dissatisfaction through various customisation options. By offering customers the opportunity to express their specific needs with a support system for decision making in customizing homes. Based on online information technologies, Noguchi's approach provides the necessary support, with the available solutions and the estimated costs of the choices. This option is only available for prefabricated homes, so what about the traditional construction? In this way, the author [11] also introduces the implementation of Mass Customisation in traditional construction projects, but this time the newness consists in the application of a communication system that allows users to take part in the customisation of their homes.

In the same way, other authors [27] suggest that customers are involved in their housing customisation through a support system for decision-making according to their preferences using a hybrid approach, that combines case-based reasoning (CBR) and genetic algorithm (GA). Through this system the possibility of customer dissatisfaction with the final product is reduced, solving the problems caused by inexperience for decision making cost-quality of the chosen option, or by discrepancies between expectations and final outcome of the house.

Also other researches [28] improve the customer participation by encouraging the exchange of information through an interactive models for customers that include BIM (Building Information Modelling) technology. With this approach they incentivize customer satisfaction while gaining the benefits of using BIM (Building Information Modelling) technology, as getting a better integration of data, fewer mistakes and inconsistencies in the projects. In addition to this, through the BIM system can be extracted 2D drawings and 3D models from any point of view, reducing time, work and mistakes that come from project modifications.

Although all these solutions improve customer satisfaction, giving them the opportunity to participate in the design process by customizing the house according to one's preferences, they provide a partial solution because customers are only involved in the design phase but they are not in the construction phase. So in none of them is proposed an integrated solution to the problems of efficiency in traditional construction.

Related to this another paper [29] presents an improvement, through the implementation of a production model that includes three sets, a design system, a construction system and a computer system, based on the codification of prefabricated building systems. The integration is achieved through a tool to visualize solutions and the automatic generation of the required information in the production phase. They present an integral system based on technology that speeds up the construction process while providing customer satisfaction, without increased cost. Nevertheless they offer a partial solution because it is only oriented to a prefabrication business model and actually not in traditional construction. In addition, it is only available for a modular design. Also the building system outlined in the paper follows a procedure not a customer oriented process. It is not process oriented and there is a delinking between process and technology.

As well, some authors [30] propose the use of technology to overcome the disadvantages arising from the Mass Customisation, such us increased manufacturing cost, fewer on-time deliveries, low supplier delivery performance, and increased order response time, increased material costs and reduction in product quality. To reduce these shortcomings and the bad influence on the construction costs they use information technology (IT) to give speed to the process, accessibility and exchange of information. By the use of a system called FIS—Finishing Information System, they improve the productivity in housing construction, minimizing the additional manpower due to the use of Mass Customisation, improving the efficiency of communication between providers and consumers, even if the process participants are geographically separated.

Although the proposal demonstrates the adequacy of Information Technology (IT), it is thus important to note that it is not focused on process management or the continuous improvement processes, neither are they measured. In addition to this there aren't defined actors and the functionalities are not established.

Despite all the literature reviewed for Mass Customisation applied to construction, the mostly analysed solutions focus their interest on the design phase. Thus, they do not do in the site execution phase. Also hardly any of them are centred on a technological solution. And the ones that take advantage of ICT's do not offer an integral management system to solve the problems that impede to achieve Mass Customisation in the field, such as the lack of business strategies that allow better efficiency in operations and people involved in them using technology holistically.

RESEARCH METHODOLOGY

Based on the analysis performed in the previous sections, it has been identified the need to propose a comprehensive model for the execution of the works oriented towards mass customisation. It allows achieving a strategic position in the market and satisfying customer needs by adding value to the product. At the same time, the necessary requirements to achieve this model were identified, obtained from observation, experience and analogy to other areas where similar problems have occurred. The synergy between the different management strategies, that are identified, together with ICTs will be the key factor to achieve these requirements.

To carry out the proposal it has been followed a research methodology based on business process management [31,32,33]. Process management is a strategy for structuring a complex process into a sequence of tasks understood as actions that transform inputs into some other output elements. This transformation must be aligned with previously defined goals, which are considered as strategic, to meet some needs or gaps identified in the environment. In this way, the defined process will achieve the object model of this work in a systematic way, selecting the most appropriate techniques and tools to meet the objectives and thus solving the problems identified. In the proposed methodology, each of the tasks identified represents a stage of the investigation, and these will have associated one or more scientific methods as described below.

The proposed research methodology is based on Eriksson-Penker formal notation [34]. It is an extension of UML for the representation of business processes that is characterized for being very intuitive and easily understood by all the stakeholders involved in the process. This is a notation that facilitates the understanding of the problem and its subsequent analysis to identify a solution according to the objectives.

Specifically, from the starting hypothesis Figure 1 shows the main process carried out in the research, where the input element (<<*input*>>) represents current models of mass construction. This <<*input*>> must be transformed through this design process into a building model oriented towards mass customization (<<*output*>>), meeting the requirements identified above and which now represent the strategic objectives that will guide the research process (<<goals>>).

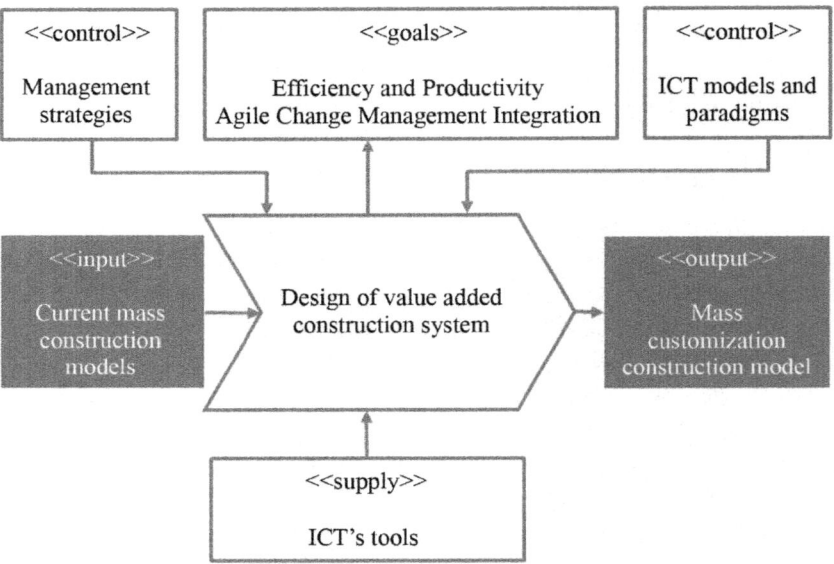

Figure 1: Modelling of value added design construction system process through Eriksson-Penker notation.

To perform the transformation to achieve the objectives established, must be identified the controllers (<<Control>>) and facilitators (<<supply>>) needed to guide this transformation. Particularly, these elements represent strategies, paradigms, techniques and technologies that will be integrated into our proposal. To facilitate its identification from a better comprehension of the problem, processes can be divided into sub processes or tasks. In our case a first division has been required as illustrated in Figure 2.

The resulting sub processes are, in first place, the sub process "identification of the construction site problems", focused on locate those general problems of mass construction current models that impede the achievement of the goals of the main process. In second place the sub process "partial solutions analysis", centered on the identification and analysis of solutions to the problems identified in the previous sub process. And finally, the sub process "integration of solutions" focused on obtaining an integral solution to the problem.

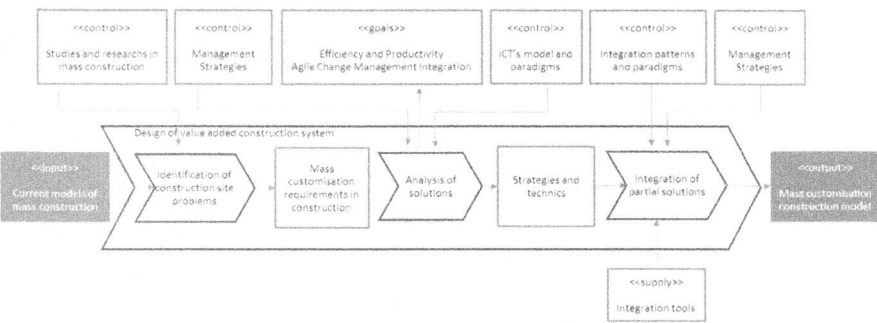

Figure 2: Modelling of value added design construction system sub processes through Eriksson-Penker notation.

In the following sections and using this methodology, will be developed each mentioned sub process, justifying the achievement of the general and particular goals and describing the actions performed to obtain the proposed model.

THE ANALYSIS OF THE CONSTRUCTION SITE PROBLEMS

As we have specified before, one of the main barriers for adopting Mass Customisation as a competitive strategy in the sector is the increase in uncertainty, complexity, cost and time. These inconveniences are further compounded by the already low standards of efficiency and productivity that the sector is dragging in recent years. Compared with the traditional manufacturing industry, the construction sector has a 30% of less productivity levels [35] even though it sets up as a key sector within the framework of the Spanish economy. It reached at its peak the 11% of Spanish Gross Domestic Product GDP [36].

Identifying the Causes of Such Problems

These shortcomings in productivity and efficiency levels produce a significant increase in construction costs, due mainly to some causes that have their origin in the business organization such us obsolete management models, lack of business integration, improvisation in strategic decisions or search for short-term benefit. Some of them are a result of the real estate boom experienced by companies during the housing bubble [35]. Therefore, to be successful, the construction companies need to supplement their current short term approaches with more long term strategies [37,38] and move from traditional models management to advanced models management, based on the inter-and intra-enterprise integration [39] that respond to the new situation of the sector. Through better business integration and organizational structured models it is possible to take full advantage of functional synergies and reduce costs. With a systematic organization structure that takes into account the capabilities of stakeholders and the environment, avoiding making decisions by relying exclusively on intuition or experience.

Besides the business causes that promote low levels of productivity, mostly of the problems come from people who participate in the process. The construction works are characterized by the concurrence of variety of firms and participants. These problems appear mainly in two kind of agents: Contractors-subcontractors and workmanship [40]. On the one hand, they talk about the necessity of eliminating the adversity and competition atmosphere, that it is usually generated because of the war costs generated between subcontractors and their low involvement in the project. On the other hand, they point the low level of specialization and experience of the sector's manpower. Also they stipulate that the successful key is to become the adversity tendency between different parties, in adaptations and collaborations for mutual success, and to implement security and training oriented programs for all the workers. Related to the lack of formation also it can be included as human causes the unstructured traditional works prone to improvisation, without a systematized procedure [41]. Besides to this, they state the human causes that promote low productivity as lack of involvement, poor worker performance, low motivation, absenteeism, turnover and changes in the work teams, insufficient space to do their jobs efficiently, and the presence of obstacles in workspaces. There are also quite frequent incidents during the course of the construction works, such as project changes during execution, design errors and re-works made by the repetition of activities badly performed [42].

As all they have suggested, the human causes have a huge impact in productivity. Furthermore this situation worsens because of the heavy dependence of the building sector on human resources rather than adopting advanced or automated technology [43].

Although technology is one of the greatest allies of the construction companies to improve the competitiveness of their business, the specific software for business management does not have much presence in the sector. Indeed, the large majority of the construction micro-enterprises (97.9%) do not have ERP software tools [44]. Furthermore, the ones that use them show up a technological waste because the ERP package can hardly meet the needs of the construction industry [45] because they come from another sector. Due to its manufacturing origin, none of the construction management modules of commercial ERP packages provided by software suppliers have been found suitable for construction firms [46].

Then also, the main technological trend in the sector is the use of management software focused only on specific tasks such as project planning and monitoring, cost control, risk management, scheduling, *etc.* [47]. Also the whole process is divided into many isolated stages such as design, tender, construction, and maintenance, so the software applications used in each stage has led to the creation of "islands of automation", *i.e.*, incompatible hardware and software, because of the lack of attention given to the integration of these applications [48].

This fact favours the tightness of information, impeding information being transferred from one stage to another [49]. Besides this and related to the incompatibility problem mentioned earlier, there is a technological gap because of the lack of a standardised platform for information exchange that can facilitate the flow of information between incompatible hardware and software [50]. This technological gap promotes the existence of errors due to the inadequate, incomplete and outdated information that usually lead to delays and extra costs during the execution of a construction project [51]. About these extra time and cost of operations, in [52] is emphasised the current lack of discipline in planning and execution tasks, due to the flow variation that increases operating cycle times. And in this sense, it makes worse the coordination of interdependent disciplines or crafts with the consequent unneeded inventory or lack of materials that produce incompletions tasks, and again, errors requiring work to be redone and re-handly. As well there are other factors that also cause missed deadlines and increased cost because of the uncertainty in the building construction, and the shortage of tools to address this uncertainty and these changes in a flexible manner. Change is inherent in construction works [53] due to some peculiarities such as complexity and uncertainty [54] resulting from one-of-a-kind nature of projects, the production site, and the temporary multi-organization [55].

The Classification of the Causes

Following the Leavitt [56] model classification the main factors involved in an organization and their interaction are distinguished. After that, it has been made a classification of the causes in order to help to identify the solutions. Then, all the previous causes that promote uncertainty, complexity, and deviations in cost and time, have been encompassed in four mainly groups of factors:

Causes that have its origin in the structure or business organization, causes that have its origin in people who participates in the hole construction process, causes that have its origin in the applied technology and causes that have its origin in the tasks of the process.

Furthermore these causes are aggravated on account of the lack of integration and interaction, hampering the improvement of competitiveness in construction. This interdependence is shown in some researches [51,57] where is stablished the four interacting components that make up an organization (task, structure, people and technology) and how the change in one them inevitably affect the other three. This synergy implies that major improvements are made by house building companies by focusing on all the interrelated actions areas.

Therefore to reach the paradigm of Mass Customisation in building is necessary a comprehensive model that solves the requirements defined in Figure 3. To solve these causes is need competitive essential actions such as a change of mind, a shift toward process orientation and improved communications [51]. By encouraging better communication intra (inside organization level) and inter (customers-subcontractors and suppliers) companies with cohesive relationships based on trust, mutual commitment, understanding of each other's individual expectations and an open book culture (open exchange of information), organizations remove problems about information and the bottleneck that can lead to delays and extra costs. Also some key drivers of change to solve problems of cost and time are: Committed leadership, a focus on the customer, integrated processes and teams, a quality driven agenda and commitment to people. This last one also includes decent site conditions, care for the health and safety of the work force, including commitment to training for all participants in the process, involving everyone in sustained improvement and learning [58].

In this regard, some of the current new paradigms or technological solutions to achieve construction efficiency are only sub optimizing individual factors instead of optimizing the entire system [59]. Then, the expected benefits are not achieved because there is not a strategic vision in the organization or technology do not form an integral part of the organization business strategy [60]. Then, the appropriate way of implementing changes and innovation is enhancing the overall organizational performance [61], emphasizing a holistic model that includes innovation in business strategy, organization of work, technology, and people. So to take full advantage of functional synergies, changes cannot be done in isolation, it is necessary to develop a holistic system where the factors, structure, tasks, people and technology, are linked (Figure 4).

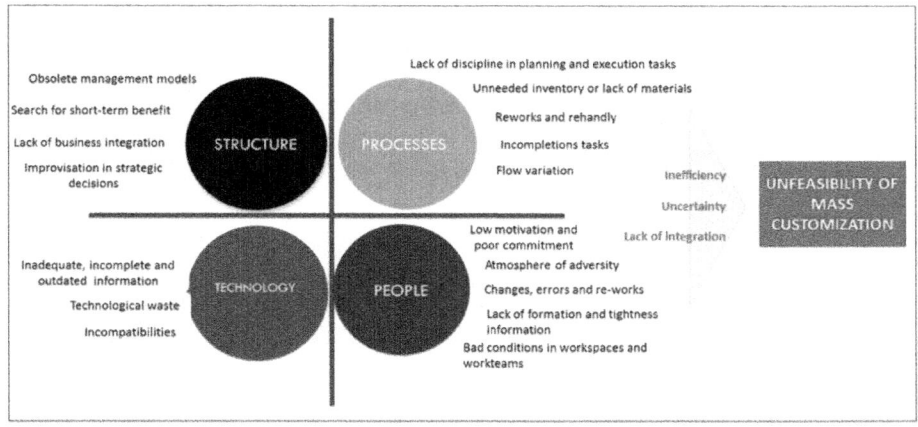

Figure 3: The Classification of Causes.

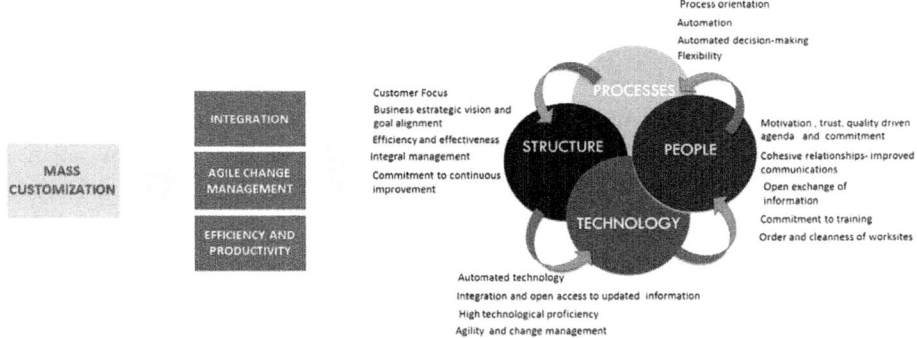

Figure 4: Requirements for Mass Customisation Feasibility.

AN INTEGRAL MANAGEMENT MODEL FOR THE WORKS EXECUTION ORIENTED TOWARDS MASS CUSTOMISATION

Within the construction field, there are different novel proposals that are focused on some of the objectives identified above. However, it comes to independent and partial solutions that do not offer an integral solution, which contains wholly the above objectives (Table 1). Therefore, a general and integral model it is proposed to achieve new paradigms, like Mass Customisation, in construction. The proposal is based on the integration of the above solutions, which offer partial solutions to the considered problem, together with other proposals that have proven successful in other sectors to solve the rest of problems that impede achieving Mass Customisation.

Table 1: Objectives Achieved with the Existing Techniques

Existing techniques	Maximize customer value	Standardization. Reduce variability	Minimize waste	Continuous improvement	Trust and commitment	Improved communications	Open exchange of information	Order and cleanness of worksites	Automated technology	Open access to updated information	Automated decision-making	Integration (not just technology)	Compatibility and high technological proficiency	Agility and change management	Flexibility
Lean Construction	■	■	■	■	■	■	■	■							
Six Sigma	■	■		■	■		■			■					
ICT's										■	■	■			
BPM	■	■	■	■		■	■			■	■	■	■	■	■

One of the emerging mainstream approaches in the construction industry is the application of the Lean Construction philosophy, the new form of production management that comes from the manufacturing industry [14].

The aim of Lean Construction is to reconceptualise construction as flows to maximize customer value while minimizing waste, understanding waste [62] as any activity which absorbs resources but creates no value, such as mistakes which require rectification, processing steps which are not actually needed, movements of employees and transport of goods from one place to another without any purpose, groups of people in a downstream activity standing around waiting because an upstream activity has not delivered on time. Thus, Lean helps the firm to define customer value focusing on activities that add value, through standardizing the types of waste and offering best practices to remove waste. And therefore achieving, among other things, to reduce inventory and waiting times [63]. In addition, as it is established by Professor L. Koskela [54,55] other main focus of Lean Construction is to increase output value through systematic consideration of customer requirements, to reduce the cycle time, to simplify by minimizing the number of steps, parts and linkages, to increase output flexibility, to increase process transparency, to focus control on the complete process, to build continuous improvement into the process, to balance flow improvement with conversion improvement and benchmark, and to reduce variability. Understanding as variability the construction uncertainties that produce late delivery of material and equipment, design errors, change orders, equipment breakdowns, tool malfunctions, improper crew utilization, labour strikes, environmental effects, accidents, and physical demands of work [64], to eliminate these variations and create workflow in a process is widely recognized the use of the LSS method [65]. Six Sigma is a statistical-based methodology that provides a structured framework to organize and implement strategic process improvement initiatives to attain reductions in process variability [66]. Within the construction industry, Six Sigma has been applied independently to improve the overall performance by reducing process variability in current construction operations [67]. Moreover it has been implemented as a quality initiative in traditional construction [68], increasing customer satisfaction and profitability by improving the quality of products [69]. Also it has been used for sustainability and quality improvement in prefabricated composite structures, applying Six Sigma to increase quality management while reducing the consumption of energy during construction, pollution, noise pollutions and waste [70]. Similarly, to reduce and eliminate waste in prefabricated residential construction, Lean principles and Six Sigma has been used by working in unison as an integrated model [71]. This combination of Lean tools and Six Sigma has been also applied in traditional construction, where each methodology complements the other, to eliminate waste and variations and create workflow [72,73]. Lean does not possess the tools to reduce variation and bring a process under statistical control, and in the same way, despite the fact Six Sigma offers a lot of gains alone the process would be slow and its costs will be too high. More specifically, lean benefit Six Sigma in the identification of waste providing the powerful value stream map tool that highlight waste and delays, while improve process speed or cycle time [74].

Despite the common factors and the synergy benefit, Lean mainly revolves about the relation between people and process but it does not on technology [75], in a similar way the use of IT in Six Sigma projects is often carry out with specific statistical software [76]. Therefore, despite the benefits obtained in the use of LSS, it is not resolved the gap in terms of integration with the other factors identified above technology, structure, people and structure.

Furthermore, the introduction of these paradigms requires a great effort of change in thought and culture, in most cases made up for a change agent to facilitate its immersion [30]. There are other issues that do not cover the integration and agility to manage change. ICT misguided (BPR) can be counterproductive for achieving integration and agility (bottom-up approach).

Moreover, ICT's are presented as an indispensable facilitators that can help to achieve mass customisation in construction, by addressing some of the goals that have not been resolved by earlier paradigms such as the process automation, the technology integration, the support in decision making, *etc.* [77].

In other traditional sectors, there are now some proposals to implement strategies such as mass customisation [31], by using a set of paradigms and technologies oriented to solve some of the main obstacles that hamper the application in the construction industry.

These paradigms and technologies are focused on getting flexible systems that allow adapting quickly the technology to the business processes of an organization, by proposing fully interoperable open systems (such as Service Oriented Architecture SOA, Web Services and Restful or Enterprise Service Bus) [78,79,80].

In addition, they consider the reuse of legacy systems like the ones used today in construction, the implementation of which has involved high cost and should not be discarded if not transformed into elements with the necessary functionality to support the processes of the organization.

Also the use of intelligent techniques based on semantic reasoning have brought great benefits to achieve mass customisation, enabling the automation of certain management steps like modelling (such as ontologies and associated techniques) [33].

BPM as the Backbone

To meet the stated goals, it is necessary to join these views or partial solutions because when the integration is done properly, the benefits provided individually will be increased [71,72,81]. For this it is necessary to achieve the integration of all the factors and resources involved in the execution of the work. And also to be able to reach an agile change management, allowing the immediate alignment between the strategic objectives, the defined business processes to achieve those objectives and the technologies that support the execution of the processes. In this sense as the main novelty to cover these gaps, it is proposed the strategy Business Process Management (BPM) [82] to support the model which not only allows achieving those goals, but also enhancing the benefits obtained by the other proposals such as efficiency, productivity and agility (Figure 5).

BPM is a process-oriented business strategy focused on continuous improvement which includes IT as a fundamental element, unlike other strategies, filling the gap between business needs and IT capabilities [83], managing the complexity of the diverse inherited technology portfolio, creating transparency in the business environment and creating the agile link between business strategy and its execution [84]. All this is achieved in form of a continuous process management lifecycle [85], consisting of discovery, design, implementation, deployment, execution, interaction, control monitoring, analysis and optimization phases.

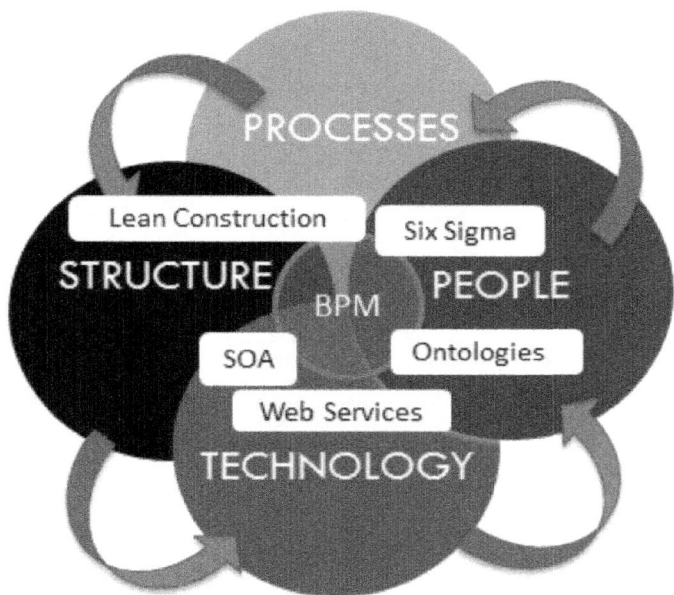

Figure 5: BPM as the Backbone of the Four Interacting Factors in Organizations.

Associated to the BPM movement has emerged a software solution focused on process management that supports the principles and the BPM lifecycle model, allowing the integration of people, systems and suppliers, the automation and the processes collaboration. They are called the BPMS (Business Process Management System or Suite) [86]. The technological components linked to this system are shown in Figure 6 [25,82] and they are what give its key features and benefits: Monitoring and control the information of the key performance indicators process, to reduce costs by automating and improving processes in real time, to adapt processes to changes in an agile way, immediate deployment of process-centric applications, to carry out solutions that face up to the needs resulting from continuous change, and use the investments in existing IT.

Taking as reference the BPM paradigm, whose life cycle will be supported by a BPMS system, the benefits of the solutions presented individually will increase [66,71,81]. The presented model is structured according to the top-down strategy by providing two levels, the level of management and the level of resources respectively, which will be connected through the BPM system (Figure 4). This approach is important because to achieve the proposed objectives it is necessary a complete understanding of the processes before being automated.

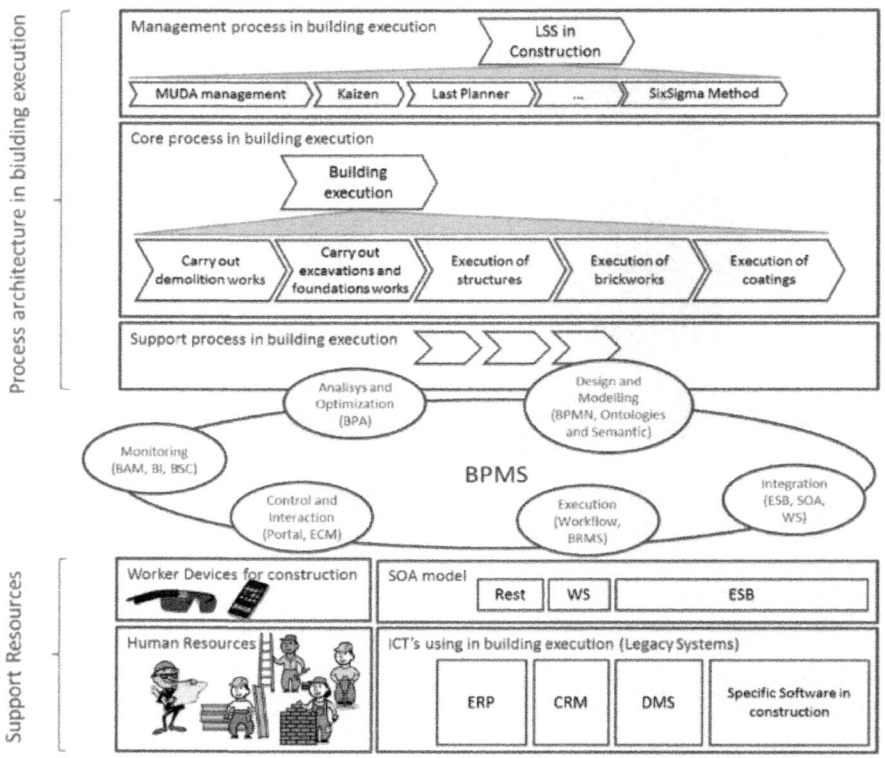

Figure 6: A Construction Management Framework for Mass Customisation in Traditional Building.

At the management level it is placed the work execution process architecture including process flow, the objectives to be achieved, the key performance indicators (KPIs) associated with the objectives of achievement measures, and the different stakeholders.

The BPMS includes tools for modelling and to include the elements defined in the architecture. The processes are modelled using BPMN standard notation, specifying in each process or task how to calculate KPIs and where to get the data, and associating the actors that should or may perform the different tasks to the BPMN model. BPMN is the standard notation for process modelling and that at any BPMS has a direct translation to the format that supports the processes execution, workflows owners or based on standards such as BPEL, BPEL4People and WS-HumanTask. Also BPMN is an understandable and intuitive notation based on flow charts, with a high degree of expressiveness that allows the process participants to understand how they should perform the tasks. The BPMS will support the processes included in this architecture that have been structured into three types, following any of the recommendations from some of the main authors [23,25,87] processes management, core and support processes.

A key aspect of the proposed model is that it has been included within the layer of management processes the ones related to Lean Construction philosophies, Six Sigma in construction or LSS in construction that are focused on the processes efficiency of the organization. Specifically, are implemented as processes and procedures the different techniques and principles associated with these disciplines, such as Kaizen, Last planner system, statistical techniques or any other analysis technique for continuous improvement.

One of the problems identified in the application of Lean and Six Sigma in the construction environment is the difficulty to assimilate the implementation of these philosophies, due to their high cultural dependence [30]. The automation of these principles and techniques with tools related to BPM will help to facilitate the inclusion. BPMS includes process execution engines and incorporates task managers that show the work to be done by each actor according to the running of the process flow, storing the relevant information and disposing of it in real time anytime and anywhere. Not only will this help to accelerate the process but also to know what to do, how, when, and how long (standardization of Lean) to each participant, both in terms of production processes and to apply continuous improvement techniques for the government of such processes.

For example, the site manager will be guided by the Last Planner process to manage daily in an agile manner the project execution, assigning tasks every day and even changing that assignment, appearing instantly to every member of the affected team the process to perform, how and when. If a construction's worker has allocated the paving execution of the houses on the first floor and each customer have chosen a different kind of pavement, in each case, it would appear the correct custom process that the worker should follow for put it in the more efficient manner, as well as the estimated time (tacktime in Lean). Finally, when he would have finished the task, would conclude the work indicating the completion of the task, and then, the time spent on it and the materials used would be recorded. This could lead to a request for material purchase depending on stock and needs, or a process improvement analysis if the time spent were far above the tacktime. Even, any worker could throw a proposal for improvement initiating the Kaizen process to analyse the proposal and carry it out.

In the layer of core processes y support processes will be located the associated production processes of the value chain in the building execution and the processes that support the achievement of them, respectively. These processes will be designed according to the principles and techniques of Lean Six Sigma and BPM as it is shown in some studies [24,81] applied to the construction peculiarities. For example, standard processes based on the main guidelines and recommendations in the building execution field must be created to achieve greater efficiency.

However, in contrast to what happened with the BPR paradigm, it is not about getting fixed and inflexible processes but rather the proposed model will include the variability among its principles, allowing in an agile way to adapt the processes to the specific objectives of the organization and its continuous change through composition techniques included in BPM and related technologies like SOA and Web Services [78,79,80].

A further key and innovative aspect of the proposed model is the use of ontologies and semantic reasoners to streamline the generation of process modelling for the work execution. This issue directly affects the customisation efficiency, now that depending on customers choices, the appropriate processes to build the custom home will be generated connecting sub processes as if it were a puzzle. These processes will guide the work of other staff to avoid errors. These processes will be executed over the process engine or BPMS workflow, but to execute certain tasks or sub processes they must interact with external information systems that were presented as functional isolated islands within an organization, performing functions inside the process of the work execution. These systems are located in the resource layer of the model (Figure 6) and they will be exposed following the principles of SOA paradigm using Web Services technologies or ESB infrastructures.

In this way is removed the stiffness of these systems and their functionalities, providing aspects such as reusability, interoperability and alignment with business processes that increase the efficiency of the model. And thus, to take advantage of the existing resources which have been a high cost for the organization. The BMPs include an integration module that allows these technologies to interoperate perfectly. In fact, although BPM and SOA have emerged as two independent approaches, today we have tested the benefit of the joint use of both philosophies. Another essential element in the model is the staff, which is directly involved in the process of building execution.

BPMS provides several technologies that can be integrated into these processes. Firstly, technologies such as portals or ECM will allow different roles to perform several tasks associated with the process managing or human tasks that are a part of the process itself, as the buyer of the material. For it, it is proposed in the model the use of appropriate devices for the site features, where the use of devices such as mobile devices or google glass can be great allies.

Finally, the BPMS includes a model that offers different technologies allowing the process performance measuring in real-time, and the variability of the results through the KPI's defined. For the first case, this is where may be used Lean Construction techniques that would imply an improvement of the process. These indicators can be analysed manually by the person responsible, and this imply to throw one of the continuous improvement processes or may generate the execution of other processes to correct automatically possible deviations, as the application of material required. Another example would indicate when certain tasks are being diverted from planned tacktime.

In the second case, the indicators would be generated using statistical methods Six Sigma to locate deviations in the result of processes, such as e.g., the execution of the interior woodwork. If abnormalities were detected, process would be launched immediately to analyse the causes and propose possible solutions. This would allow us to reduce the work performed during the realization of the processes of finished works.

USE CASE, IMPLEMENTATION AND VALIDATION

The theory model obtained previously presents an ambitious proposal in a framework that serves as a reference for the integral management for the execution of the works orientated to the achievement of mass customisation. The development of this model is to obtain a realistic proposal will be realized in different self-contained phases that by themselves contribute to achieving the efficient objectives and productivity and agile change management and integration. In concrete, in the present article it has been developed one use case of which the objective is the implementation of some representative techniques and principles defined in the model which permit reaching the strategic goals proposed. This relation is shown in Figure 7.

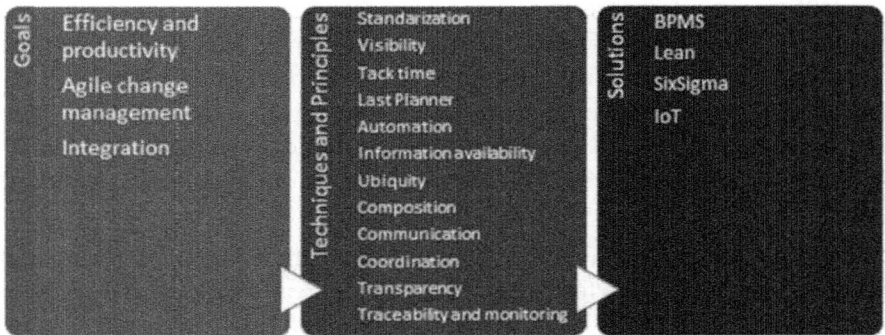

Figure 7: Relationship between objectives, principles and techniques and solutions.

In concrete, the use case includes two types of processes widely related, the process of weekly planning which implements the phase of the short-term planning of the Last Planner and which is located in the layer of management and the process of flooring, located in the layer core processes. Both processes have been chosen because of its repercussion in the mass customisation. The final customers could choose diverse types of floor and this includes that some customers could choose several types of floor for different rooms inside their houses. Also, this situation gets complicated and makes planning difficult because it requires a very high level of knowledge and control about the state of the execution of the works, concretely regarding the access of real time to this information because it exists too much variability and heterogeneity and human errors could increase considerably and therefore, increases the costs of execution and delays delivery deadlines laid out. On the other hand, it is possible that it is not known all of the proper procedures for each option of flooring or passing from one to another in a short period of time could derive in the inanition of some the steps.

Last Planner is a new technique of planning, used in Lean Construction that has broadly demonstrated its benefits in relation with efficiency and productivity. Nevertheless this is a complex process which implies a systematic methodology and a high level of coordination between the Construction Company and the different subcontractors. Also the information of planning carried out is realised in an extensive murals on the wall with the use of sticky's, which this information should be consulted in person. The election of the floor process has been because of different reasons. The flooring in a mass customisation model is a complex task due to the variability of the types of floor and therefore, to different procedures used for its execution.

The use of BPM as a management strategy to integrate other solutions has increased the benefits provided individually. In this sense, the use case was performed using the BPM open source Bonita BPM system (version 3.8). Firstly it has been done the modeling of the mentioned processes by using the standard modeling of business processes called BPMN (Figure 8 andFigure 9).

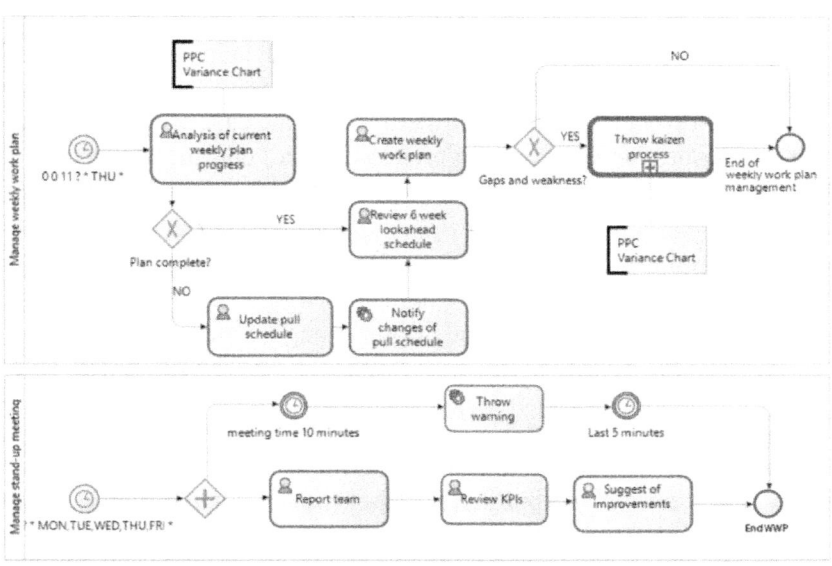

Figure 8: Weekly Work Plan BPMN model.

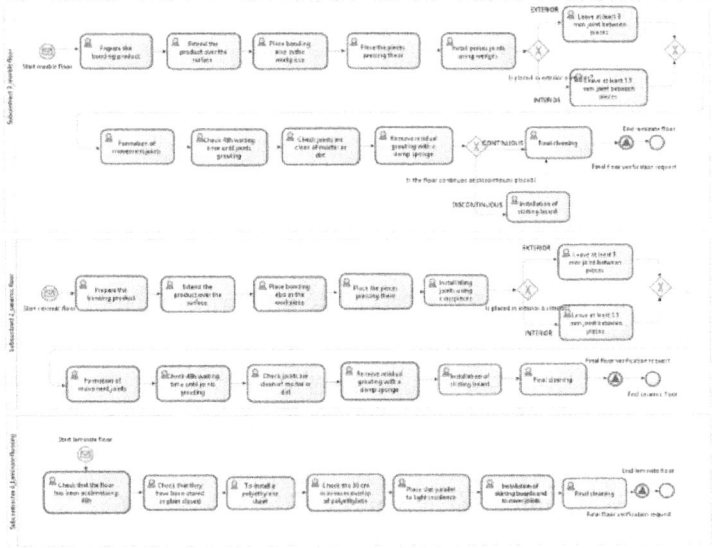

Figure 9: Flooring process BPMN model.

The process modeling has allowed us to achieve the following principles and their associated benefits:

- Standardization. It is a Lean principle that is increased with its realization by a standard modeling easily understandable by any role of the organization and based on a formal language that enables its automation. In both processes it will allow to stakeholders know what to do, how and by whom. Furthermore, in the case of the flooring, it has been done a redesign of the processes as a good practice based on the different techniques recommendations and the experience of several site managers.

- Another important aspect is the principle of processes composition in which BPMN is based. Modifications or link some processes with others is a very agile task that allows changing the behavior of a process in minutes. Any anomaly in the process could be rectified in a very short space of time, reducing the number of errors by the same problem. Furthermore, each subcontractor could include its own modeled processes, facilitating the integration and incorporation of its system with the one of the builder. Another Lean principle that is facilitated thanks to the use of BPM.

- In this first phase of modeling, another important Lean principle that has allowed implement Bonita BPM, is the tack time. The tack time is the time required to perform a task depending on the time available for the entire process. To determine this time is essential to control more accurately the progress of the work execution. In fact, as we shall see, automatically we will get information in real time about what tasks are not being done on time, which can help the site manager to identify problems early.

From the process modeling has been done the implementation of each of the tasks that compose the process, with particular emphasis on the needs of usability and accessibility in the built environment.

Firstly it has been developed a very flexible software tool that implements the Weekly Work Plan Process, till now performed manually. The entry point of this process is an event that is automatically launched every week, notifying everyone involved that must be met for the weekly schedule. This meeting could be virtual since they share a collaborative environment (Figure 10), involving considerable savings with respect to traditional methods. This planning is done starting from the tasks to be performed according to the 6 Weeks Look Ahead Schedule process. The environment implements a canvas and sticky's metaphor and includes a panel which allows using a simple drag and drop process to distribute tasks during the week, assign actors and specify the location where it must be carried out. At all times the user will have visual aids that reflect the state of the task.

Each task has an associated BPM process that establishes how to be executed and marks the execution time of the task from the tack time of each of the component tasks. Once planning is closed, the site manager can launch the process associated to each task, allowing the beginning of the execution of construction processes. In the tool these activities shall include a mark on the image that represents the task, enabling visualization at any time and from any location and device.

As shown in the Figure 10, for the same day of the week, Monday, it has been planned different types of floor to be executed. The execution of each process starts a series of events that would notify to the responsible of the tasks what to do together with the additional information for its management. These are human tasks, so to integrate the workers with the system devices have been used to facilitate interaction, offering a friendly, usable and simple interface. In this case we have done a prototype using platforms such us a smartwatch and a smartphone, as is shown in Figure 11. As the operator is carrying out the tasks, the processing sequence is advanced. The task management is very simple, is focused on reporting the task to be done, to execute the basic actions of it (start, stop, validation, *etc.…*).

Figure 10: Screenshot of Weekly Work Plan application embedded in Bonita BPM Plannig phase.

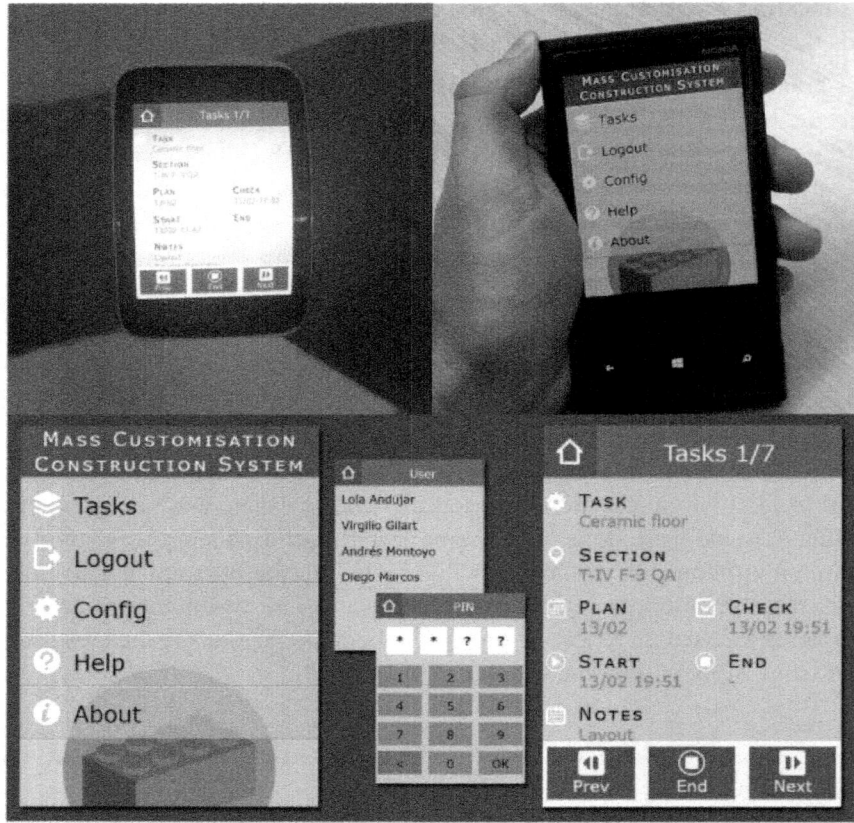

Figure 11: User agent for system interaction. Smartwatch and smartphone versions and interface screenshots.

Once a task is closed, the system notifies the next task to the appropriate worker. When the closure of a task involves the completion of a process, such action will be reflected in the application the Weekly Work Plan, showing the image of the task with a cross over it (Figure 12). Before starting a task, an operator can validate that the above task has been completely finished. This action involves, in the terminology of the weekly work plan, a double strikethrough on the visual representation of the task and updating the Percent Plan Completed (PPC) (Figure 12).

Figure 12: Screenshot of Weekly Work Plan application embedded in Bonita BPM. Traceability phase.

The implementation involves achieving the following principles, as is shown above in Figure 7, and its associated benefits to achieve the objectives of the proposal:

Automation. This is one of the BPM fundamentals, which allows the organization to automate the implementation, monitoring and control of its processes through the orchestration and integration of Processes, systems, people and information to create uniform, streamlined, and repeatable Business Processes. This automation is based on a process-centric approach where automation is performed on a redesign of the process based on best practices. Its automation implies an increase in performance in the process execution and therefore greater efficiency and productivity. Furthermore, the use of everyday devices based on IoT paradigm, in our case connected via Rest Services to the process facilitates the integration of human people in the same, allowing interaction and improving the efficiency and integration. Finally, it has been automated through events the notifications of tasks and actions to guide workers in the development of their activities. All this has led to improve the understanding and learning of the methodologies, techniques and philosophies implemented. Information available in real time and ubiquity. The proposal allows access to the state of the building process at any time, allowing each member involved to have complete information about it. Furthermore, such information can be consulted from any location and computer devices (pc's, laptops, tables, smartphones and wareables). This means saves on the time execution and therefore cost, also reductions of errors due to lack of information.

Communication, coordination and transparency. As was mentioned before, both the general framework of Bonita BPM and the applications developed on it, have been implemented in a collaborative work environment which together with the above principles of availability and ubiquity, enable synchronization and coordination of all stakeholders, which achieves the objectives of efficiency and productivity and integration. Traceability and monitoring. On one hand, the implementation of the Weekly Work Plan application allows for a visual feedback of the coarse-grained tasks. The automation of processes in Bonita BPM provides us full traceability and monitoring on the execution and the status of tasks, the time taken for completion, the person responsible who performed it and the task information that has been added. The level of detail provided is fine grained which will allow the site manager and work teams more precise control about each task. In this way it is possible to detect in real time the deviations of daily, and weekly planning and longer term planning the tack time defined (Figure 13). In fact, the system notifies the responsible any deviation on the tack time. All this information can be accessed from anywhere with different devices representing a complete knowledge about the execution of the work that influences in productivity, efficiency, and together with the composition, agile change management.

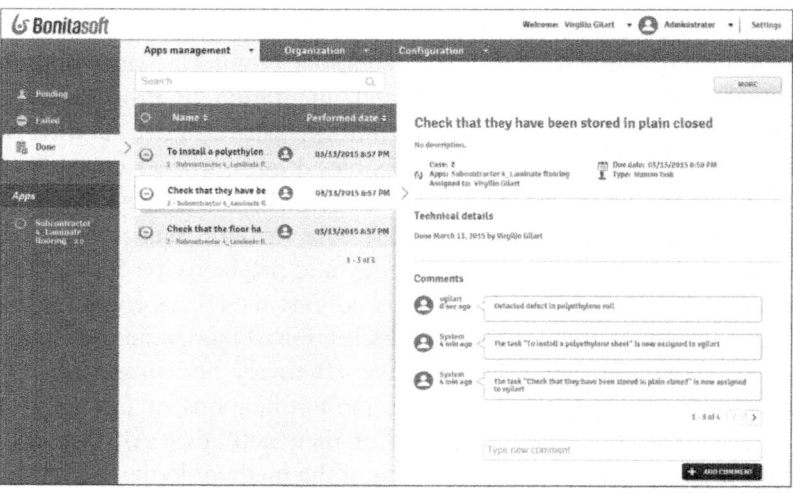

Figure 13: Bonita portal screenshot. Traceability and monitoring of flooring process.

CONCLUSIONS, CONTRIBUTIONS AND FUTURE WORK

In this paper we have presented a theoretical model for managing the execution of the work aimed at achieving carry out mass customization paradigm in traditional construction.

We have implemented a prototype model which has allowed showing its viability and has highlighted how the implementation of a representative set of techniques achieve the objectives of mass customization such as efficiency, productivity, agility and change management integration.

The main contributions of this work can be summarized in:

- It is the first time that BPM is applied to the field of construction in general, and the execution of the works, in particular. BPM allows achieving the principles of standardization, automation, agility, flexibility, integration. On the other hand, it is a generic framework that allows brings together and promotes the benefits of other strategies and to align management philosophies to align with information technology.

- It has been done an integral management model for mass customization in the execution of the work, particularly for traditional construction from the current management models.

- It has been carried out an integration of Lean and BPM paradigms; in particular, it has been automated using Bonita BPM a part of the Last Planner process (Weekly Work Plan). Until now there were some theoretical proposals offering the synergy benefits of these strategies, but none of them are focused on the specific area of construction.

- It has been incorporated everyday items based on the paradigm IoT to facilitate the integration of the construction actors in the process automation. Due to the characteristics of the work environment, have been selected devices such as wareables with transparency features, usability, flexibility and accessibility.

- A redesign and standardization of the flooring process has been presented through the BPMN modelling language.

Currently, we are working on the implementation of new processes and modules of the proposed model, including Six Sigma strategy, in order to include them in the present approach.

Over this implementation, as future lines of work, we are going to develop an ontology of the building execution standard core processes, allowing the dynamic and automatic generation of the full process from customer's choices. In addition, this study is considered the first step to derive a global model that integrates the entire value chain for the construction, from the design stage to the delivery of the property.

ACKNOWLEDGMENTS

This work was supported in part by the Vicerrectorado de Investigación, Desarrollo e Innovación, University of Alicante and by the Conselleria d'Educació, Cultura i Esport, Generalitat Valenciana.

AUTHOR CONTRIBUTIONS

Virgilio Gilart-Iglesias and Andrés Montoyo are the Directors of the research in which all authors have contributed. María Dolores Andújar-Montoya has made the conceptualization and the problem definition, and together with Virgilio Gilart-Iglesias and Andrés Montoyo all they have developed the proposed model. Likewise, the designing of the system and its validation has been carried out by Virgilio Gilart-Iglesias, Diego Marcos-Jorquera and Maria Dolores Andújar-Montoya. The four authors wrote the paper, revised, read and approved the final manuscript.

REFERENCES

1. Bernard, A.; Daaboul, J.; Laroche, F.; da Cunha, C. Mass customisation as a competitive factor for sustainability. In Enabling Manufacturing Competitiveness and Economic Sustainability, Proceedings of the 4th International Conference on Changeable, Agile, Reconfigurable and Virtual Production, Montreal, QC, Canada, , 2–5 October 2011; Springer: Montreal, QC, Canada, 2011; pp. 18–25.

2. Pine, B.J. *Mass Customization: The New Frontier in Business Competition*; Harvard Business School Press: Boston, MA, USA, 1999.

3. Tseng, M.M.; Hu, S.J. Mass customization. In *Cirp Encyclopedia of Production Engineering*; Laperrière, L., Reinhart, G., Eds.; Springer: Berlin, Germany, 2014; pp. 836–843.

4. Tseng, M.M.; Jiao, J.; Merchant, M.E. Design for mass customization. *CIRP Ann.-Manuf. Technol.* 1996, *45*, 153–156.

5. Confederación Nacional de la Construcción (CNC). *Informe de Evolución del Sector de la Construcción*; National Confederation of Construction: Madrid, Spain, 2011; pp. 17–46.

6. Gilmore, J.H.; Pine, B.J. *Authenticity: What Consumers Really Want*; Harvard Business School Press: Boston, MA, USA, 2007; Volume 1.

7. Barlow, J. From craft production to mass customisation? Customer focused approaches to house building. In From Craft Production to Mass Customisation? Customer Focused Approaches to House Building, Proceedings of IGLC 6th Annual Conference, Sao Paulo, Brazil, 13–15 August 1998.

8. Nambiar, A.N. Mass customization: Where do we go from here? In Mass Customization: Where do We Go from Here? Proceedings of the World Congress on Engineering WCE, London, UK, 1–3 July 2009; pp. 1–3.

9. Piller, F.T.; Moeslein, K. From economies of scale towards economies of customer integration: Value creation in mass customization based electronic commerce. In Proceedings of the 15th Bled Electronic Commerce Conference e-Reality: Constructing the e-Economy, Bled, Slovenia, 17–19 June 2002.

10. Radder, L.; Louw, L. Mass customization and mass production. *The TQM Mag.* 1999, *11*, 35–40.

11. Noguchi, M.; Hernàndez-Velasco, C.R. A 'mass custom design' approach to upgrading conventional housing development in Mexico. *Habitat Int.* 2005, *29*, 325–336.

12. Noguchi, M. The "mass custom design" approach to the delivery of quality affordable homes. Available online: http://www.masscustomhome.com/ (accessed on 8 November of 2014).

13. Cuperus, Y. Mass customization in housing, an open building/lean construction study. In Proceedings of Dense Living Urban Structures International Conference on Open Building, Hong Kong, China, 23–26 October 2003; pp. 1–13.

14. Aziz, R.F.; Hafez, S.M. Applying lean thinking in construction and performance improvement. *Alexandria Eng. J.* 2013, *52*, 679–695.

15. Porter, M.E. *Competitive Advantage of Nations: Creating and Sustaining Superior Performance*; Simon and Schuster: New York, NY, USA, 2011.

16. Almeida, J.C.; Salazar, G.F. Strategic issues in lean construction. In Proceedings of the 11th Annual Conference of the International Group for Lean Construction, Blacksburg, VA, USA, 22–24 July 2003; pp. 1–10.

17. Ahuja, V.; Yang, J.; Shankar, R. Strategic use of ict by construction organisations-requirements and implementation issues. *Bus. Process Manag. J.* 2006, *15*, 968–989.

18. Rausch, P.; Stumpf, M. Linking the operational, tactical and strategic levels by means of cpm: An example in the construction industry. In *Business Intelligence and Performance Management*; Rausch, P., Sheta, A.F., Ayesh, A., Eds.; Springer: London, UK, 2013; pp. 27–42.

19. Cork, R. Is construction ready for project based erp? Available online: http://download.ifsworld.com/home/if1/page_480/is_construction_ ready_for_project_based_erp_yet.html (accessed on 24 April 2015).

20. Chung, B.; Skibniewski, M.J.; Kwak, Y.H. Developing erp systems success model for the construction industry. *J. Construct. Eng. Manag.* 2009, *135*, 207–216.

21. Looy, A.V. *Business Process Maturity. A Comparative Study on A Sample of Business Process Maturity Models*; Springer International Publishing: Vaduz, Liechtenstein, 2014; Volume 1, p. 87.

22. Willaert, P.; van den Bergh, J.; Willems, J.; Deschoolmeester, D. The process-oriented organisation: A holistic view developing a framework for business process orientation maturity. In *Business Process Management*; Springer: Berlin, Germany, 2007; pp. 1–15.

23. Harmon, P. *Business Process Change: A Guide for Business Managers and Bpm and Six Sigma Professionals*; Elsevier/Morgan Kaufmann: Burlington, MA, USA, 2010.

24. Ohtonen, J. Combining all bpm methods–is that possible? Available online: http://www.bpmleader.com/2012/07/02/combining-all-bpm-methods-%E2%80%93-is-that-possible/ (accessed on 2 October 2014).

25. Smith, H.; Fingar, P. *Business Process Management (BPM): The Third Wave*; Meghan-Kiffer Press: Tampa, FL, USA, 2003.

26. Noguchi, M. The effect of the quality-oriented production approach on the delivery of prefabricated homes in Japan.*J. Hous. Built Environ.* 2003, *18*, 353–364.

27. Juan, Y.-K.; Shih, S.-G.; Perng, Y.-H. Decision support for housing customization: A hybrid approach using case-based reasoning and genetic algorithm. *Expert Syst. Appl.* 2006, *31*, 83–93.

28. Lee, S.; Ha, M. Customer interactive building information modeling for apartment unit design. *Autom. Construct.* 2013,*35*, 424–430.

29. Benros, D.; Duarte, J. An integrated system for providing mass customized housing. *Autom. Construct.* 2009, *18*, 310–320.

30. Shin, Y.; An, S.-H.; Cho, H.-H.; Kim, G.-H.; Kang, K.-I. Application of information technology for mass customization in the housing construction industry in Korea. *Autom. Construct.* 2008, *17*, 831–838.

31. Gilart-Iglesias, V. Metodología para la gestión integral de los procesos de producción: Modelado de la maquinaria industrial como un sistema de gestión de procesos de negocio. Ph.D. Thesis, Polytechnic University College, University of Alicante, Alicante, Spain, 2010.

32. Gilart-Iglesias, V.; Maciá-Pérez, F.; Marcos-Jorquera, D.; Mora-Gimeno, F.J. Industrial machines as a service: Modelling industrial machinery processes. In Proceedings of the 2007 5th IEEE International Conference on Industrial Informatics, Wien, Austria, 23–27 June 2007; pp. 737–742.

33. Ferrándiz Colmeiro, A.; Maciá Pérez, F.; Gilart Iglesias, V.; Berná Martínez, J.V.; Gea Martínez, J. Automatización del modelado y gestión de procesos de fabricación dirigida por semántica. In Proceedings of VI Jornadas para el Desarrollo de Grandes Aplicaciones de Red (JDARE'09), Alicante, Spain, 2009; pp. 181–198.

34. Penker, M.; Eriksson, H.-E. Business Modeling With UML: Business Patterns at Work. Available online: http://dsc.ufcg.edu.br/~sampaio/Livros/Wiley-Business-Modeling-with-UML-Business-Patterns-at-Work.pdf (accessed on 24 April 2015).

35. ISEA. La innovación, una prioridad para el sector de la construcción. Available online: http://www.iseamcc.net/news/revista/revista-isea-03.pdf (accessed on 13 October 2014).

36. Cuadrado-Roura, J. *El sector de la Construcción en España: Análisis, Perspectivas y Propuestas*; Cuadernos del Colegio Libre de Eméritos: Madrid, Spain, 2010. Available online: http://www.colegiodeemeritos.es/docs/repositorio/es_ES/Cuadernos_del_Colegio/el_sector_construccion_en_espana_%28final%29.pdf (accessed on 13 October 2014).

37. Bakar, A.H.A.; Tufail, M.A.; Yusof, M.N.; Pinang, W.M.; Virgiyanti, W. Implementation of strategic management practices in the malaysian construction industry. *Pak. J. Commer. Soc. Sci.* 2011, *5*, 140–154.

38. Price, A.; Newson, E. Strategic management: Consideration of paradoxes, processes, and associated concepts as applied to construction. *J. Manag. Eng.* 2003, *19*, 183–192.

39. Vicedo, J.C.; Pérez, A.H.; Esteban, F.C.L. Análisis de adaptación al sector de la construcción de estructuras organizativas y de producción existentes en el sector del automóvil. In Proceedings of V Congreso de Ingeniería de Organización, Valladolid-Burgos, Spain, 4–5 September 2003.

40. Martín, A.I.F.; Frías, V.G.; Romero, B.P. La cadena de suministro en proyectos de construcción. In Proceedings of the II International Conference on Industrial Engineering and Industrial Management, Burgos, Spain, 3–5 September 2008; pp. 1715–1724.

41. Arcudia Abad, C.E.; Solís Carcaño, R.G.; Baeza Pereyra, J.R. Determinación de los factores que afectan la productividad de la mano

de obra de la construcción. *Ingeniería* 2004, *8*, 145–154.

42. Giménez Palavicini, Z.; Suárez Isea, C. Diagnóstico de la gestión de la construcción e implementación de la constructabilidad en empresas de obras civiles. *Revista Ingeniería de Construcción* 2008, *23*, 4–17.

43. Chiang, Y.-H.; Tang, B.-S. "Submarines don't leak, why do buildings?" Building quality, technological impediment and organization of the building industry in hong kong. *Habitat Int.* 2003, *27*, 1–17.

44. Ontsi, F.Y.E. Análisis sectorial de implantación de las tic en la pyme española. Available online: http://www.ipyme.org/Publicaciones/ InformePyme2013.pdf (accessed on 17 October 2014).

45. Shi, J.J.; Halpin, D.W. Enterprise resource planning for construction business management. *J. Construct. Eng. Manag.*2003, *129*, 214–221.

46. Yang, J.-B.; Wu, C.-T.; Tsai, C.-H. Selection of an erp system for a construction firm in Taiwan: A case study. *Autom. Construct.* 2007, *16*, 787–796.

47. Forcada Matheu, N. Life Cycle Document Management System for Construction. Ph.D. thesis, Polytechnic University of Catalonia, University of Catalonia, Catalonia, Spain, 2005.

48. Cembellín, B.H. Herramientas informáticas para construir y fabricar. *Técnica Industrial* 2009, *283*, 36–39.

49. Alshawi, M.; Faraj, I. Integrated construction environments: Technology and implementation. *Constr. Innov.: Inf. Process Manag.* 2002, *2*, 33–51.

50. Alshawi, M.; Ingirige, B. Web-enabled project management: An emerging paradigm in construction. *Autom. Construct.*2003, *12*, 349–364.

51. Hong-Minh, S.; Barker, R.; Naim, M. Identifying supply chain solutions in the uk house building sector. *Eur. J. Purch. Supply Manag.* 2001, *7*, 49–59.

52. Ballard, G.; Howell, G. Implementing lean construction: Stabilizing work flow. Available online: http://www.leanconstruction.dk/media/18181/ Implementing_Lean_Construction__Stabilizing_Work_Flow_.pdf (accessed on 24 April 2015).

53. Smith, N.J.; Merna, T.; Jobling, P. *Managing risk in Construction Projects*; John Wiley & Sons: Oxford, UK, 2013.

54. Koskela, L. *Application of the New Production Philosophy to Construction*; Technical Report No. 72; Stanford University: Stanford, CA, USA, 1992.

55. Koskela, L. Lean production in construction. Available online: http:// www.iaarc.org/publications/fulltext/Lean_production_in_construction.

PDF (accessed on 24 April 2015).

56. Leavitt, H.J. Applied organizational change in industry: Structural, technical and human approaches. In *New Perspective in Organization Research*; Cooper, W.W., Leavitt, H.J., Shelly, M.W., II, Eds.; Wiley: New York, NY, USA, 1964; Volume 55, p. 71.

57. Orozco, F.A.; Serpell, A.F.; Molenaar, K.R.; Forcael, E. Modeling competitiveness factors and indexes for construction companies: Findings of Chile. *J. Construct. Eng. Manag.* 2011, *140*, 1–13.

58. Egan, S. J.; Williams, D. Rethinking construction: The report of the construction task force. Available online: http://constructingexcellence. org.uk/wp-content/uploads/2014/10/rethinking_construction_report.pdf (accessed on 28 September 2014).

59. Newman, M.; Zhao, Y. The process of enterprise resource planning implementation and business process re-engineering: Tales from two chinese small and medium-sized enterprises. *Inform. Syst. J.* 2008, *18*, 405–426.

60. Love, P.E.; Irani, Z. An exploratory study of information technology evaluation and benefits management practices of SMEs in the construction industry. *Inf. Manag.* 2004, *42*, 227–242.

61. Sexton, M.; Barrett, P.; Miozzo, M.; Wharton, A.; Leho, E.; Hughes, W. Innovation in small construction firms: Is it just a frame of mind? In Proceedings of 17th Annual ARCOM Conference, Salford, UK, 5–7 September 2001.

62. Womack, J.P.; Jones, D.T. *Lean Thinking: Banish Waste and Create Wealth in Your Corporation*; Simon and Schuster: New York, NY, USA, 2010.

63. Howell, G.A. What is lean construction-1999. In Proceedings of Seventh Annual Conference of the International Group for Lean Construction, IGLC-7, Berkeley, CA, USA, 26–28 July 1999; pp. 1–10.

64. Abdelhamid, T.S.; Everett, J.G. Physical demands of construction work: A source of workflow unreliability. In Proceedings of the 10th Conference of International Group for Lean Construction IGLC-10, Gramado, Brazil, 6–8 August 2002.

65. Oguz, C.; Kim, Y.-W.; Hutchison, J.; Han, S. Implementing lean six sigma: A case study in concrete panel production. In Proceedings of the 20th Annual Conference of the International Group for Lean Construction IGLC-20, San Diego, CA, USA, 18–20 July 2012.

66. Abdelhamid, T.S. Six Sigma in Lean Construction Systems: Opportunities

and Challenges. In Proceedings of the Eleventh Annual Conference of the International Group for Lean Construction IGLC-11, Blacksburg, VA, USA, 22–24 July 2003.

67. Han, S.H.; Chae, M.J.; Im, K.S.; Ryu, H.D. Six sigma-based approach to improve performance in construction operations. *J Manag Eng* 2008, *24*, 21–31.

68. Pheng, L.; Hui, M. Implementing and applying six sigma in construction. *J Construct Eng Manag* 2004, *130*, 482–489.

69. Tutesigensi, A.; Pleim, V. Title of Presentation. In In Why Small and Medium Construction Enterprises do not Employ Six Sigma, Proceedings of 24th Annual ARCOM Conference, Cardiff, UK, 1–3 September 2008; pp. 267–276.

70. Tchidi, M.F.; He, Z.; Li, Y.B. Process and quality improvement using six sigma in construction industry. *J. Civ. Eng. Manag.* 2012, *18*, 158–172.

71. Shan, Y.-H.; Li, Z.-F. Integration and application of lean principles and six sigma in residential construction.*Proceedings of 2012 3rd International Asia Conference on Industrial Engineering and Management Innovation*, Available online: http://link.springer.com/chapter/10.1007/978-3-642-33012-4_25 (accessed on 25 April 2015).

72. Banawi, A.; Bilec, M.M. A framework to improve construction praocesses: Integrating lean, green and six sigma. *Int. J. Constr. Manag.* 2014, *14*, 45–55.

73. Sunil, V.D.; Sharad, V.D. Minimising waste in construction by using lean six sigma principles. *Int IJCIET* 2013, *4*, 1–8.

74. George, M.L.; George, M. *Lean Six Sigma for Service*; McGraw-Hill: New York, NY, USA, 2003.

75. Visser, B. Lean principles in case management. Master's Thesis, Radboud Universiteit, Nijmegen, The Netherland, 2009.

76. Chiarini, A. Discussion and comparison about the common characteristics. In *From Total Quality Control to Lean Six Sigma*; Springer: Milan, Italy, 2012; pp. 47–51.

77. Chen, Q.; Reichard, G.; Beliveau, Y. Interface management—A facilitator of lean construction and agile project management. In Proceedings of the 15th Annual Conference of the IGLC 15, East Lansing, MI, USA, 18–20 July 2007; pp. 57–66.

78. Erl, T. *Service-Oriented Architecture: Concepts, Technology, and Design*; Pearson Education Crwafordsville: Indiana, IN, USA, 2005.

79. Erl, T. *Soa Design Patterns*; Pearson Education: Boston, MA, USA, 2008.

80. Marks, E.A.; Bell, M. *Service Oriented Architecture (SOA): A Planning and Implementation Guide for Business and Technology*; John Wiley & Sons: Hoboken, NJ, USA, 2006.

81. Morris, D.; Field, G. BPM, lean and six sigma better together the whole is greater than the sum of the parts. Available online: http://c.ymcdn.com/sites/www.abpmp.org/resource/resmgr/Docs/news_events_bpm_lean_6.pdf (accessed on 25 April 2015).

82. Jeston, J.; Nelis, J. *Business Process Management*; Elsevier Butterworth Heinemann: Burlington, MA, USA, 2014.

83. Woodley, T.; Gagnon, S. BPM and SOA: Synergies and challenges. In *Web Information Systems Engineering—Wise 2005. 6th International Conference on Web Information Systems Engineering*; Springer: New York, NY, USA, 2005; pp. 679–688.

84. Kirchmer, M. *High Performance through Process Excellence: High Performance through Process Excellence*; Springer-Verlag: Berlin, Germany, 2009.

85. Zur Muehlen, M.; Ho, D.T.-Y. Risk management in the bpm lifecycle. In *Business Process Management BPM 2005 Workshops*; Springer-Verlag: Berlin, Germany, 2005; pp. 454–466.

86. Chang, J.F. *Business Process Management Systems: Strategy and Implementation*; Auerbach Publications: Boca Raton, FL, USA, 2005.

87. Barros, O. Business processes architecture and design. *Bus. Process Trend* 2007, *1*, 1–28.

Chapter 4

LIFTING WING IN CONSTRUCTING TALL BUILDINGS—AERODYNAMIC TESTING

Ian Skelton[1], Peter Demian[1], Jacqui Glass[1], Dino Bouchlaghem[2], and Chimay Anumba[3]

[1]School of Civil and Building Engineering, Loughborough University, LE11 3TU, UK

[2]School of Architecture, Nottingham Trent University, NG1 4BU, UK

[3]Department of Architectural Engineering, Pennsylvania State University, University Park, State College, PA 16801, USA

ABSTRACT

This paper builds on previous research by the authors which determined the global state-of-the-art of constructing tall buildings by surveying the most active specialist tall building professionals around the globe. That research identified the effect of wind on tower cranes as a highly ranked, common critical issue in tall building construction. The research reported here presents a design for a "Lifting Wing," a uniquely designed shroud which potentially allows the lifting of building materials by a tower crane in higher and more unstable wind conditions, thereby reducing delay on the programmed critical path of a tall building. Wind tunnel tests were undertaken to compare the aerodynamic performance of a scale model of a typical "brick-shaped" construction load (replicating a load profile most commonly lifted via a tower crane) against the aerodynamic performance of the scale model of the Lifting Wing in a range of wind conditions. The data indicate that the Lifting Wing improves the aerodynamic performance by a factor of up to 50%.

INTRODUCTION

The primary concern in the engineering of tall buildings is the effect of the wind on the building's structure. Each uniquely shaped section of the world's tallest tower (Burj Dubai) prevents the wind from becoming organised and limits lateral movement [1].

The Lifting Wing applies this fundamental engineering concept to the actual build process of a tall building and its life blood, the tower crane.

Previous research undertaken for "Britain's Tall Building Boom: Now Bust?" [2] and "The State of the Art of Building Tall" [3] provided a unique snapshot of the Britain's unprecedented demand for tall buildings in first quarter of 2007 to end of 2008 and the global state-of-the-art of the tall building industry over the first to third quarters of 2008. This research captured the industry's buoyant mood and strong belief in continual growth in demand for tall buildings, especially for those of "iconic" design. It also captured the industry's unexpected thirst for innovation in the build process over tried-and-tested approaches. The four key results were:

- The international construction industry is not keeping pace with the latest, cutting-edge design developments in tall buildings, and that the UK construction industry is not keeping pace with overseas construction industry developments;

- "Inclement weather (winding-off tower cranes)," consistently ranked one of the two highest construction risks, followed by "logistical problems (man and material access via hoist and crane)," "superstructure cycle times/speed of erection" and "façade installation," all directly related to wind and its effect on the tower crane;

- Tall building experts believe "construction programme surety" and "cost certainty" were the two most significant risks of a tall build. The most important attribute of a principal contractor was determined as "innovative build approach and the provision of an experienced tall building team," followed by "history of programme certainty," "logistics management efficiency," reinforcing the industry's thirst for innovation, as well as desire for logistical, programme and therefore cost certainty;

- Eighty percent of tall building experts interviewed would strongly embrace and promote the use of the innovative construction technique that reduces the effect of wind on tower crane material lifts on their tall building project.

The conclusion of that paper's research was that there was strong international desire for an innovative solution to critical construction problems, the most highly ranked of which was wind negatively affecting the build. Paired with the key desire of programme certainty and hence cost certainty, this clearly signposted that an innovative concept was needed to mitigate delays to the tall building programme duration by reducing the effect of wind on the critical path activities of the tower crane. This focused the final stage of the research on the design and testing of an innovative concept named the "Lifting Wing," aimed at directly addressing this industry need.

This paper describes the scientific advancement in applying aerodynamic theory, refined via modelling and testing, to a specific aspect of the building process of a tall building with potentially significant time and commercial benefits. The specific research undertaken in design, modelling and methodologically testing an aerodynamic shroud, was aimed at reducing the wind-induced load on a tower crane and the construction material being lifted, thereby allowing lifting in higher wind conditions, reducing the UK average of 40% down time for a tall building tower crane. This would therefore potentially reduce very costly wind-induced critical path delay to a tall building construction period.

Wind and its Effect on Tower Cranes—the Life Blood of the Tall Building

Tower cranes have come to symbolize the construction industry and perform an indispensable service in moving material components horizontally and vertically to their required positions. They are central to mid- and high-rise building projects [4]. They have become internationally recognised as a highly visible gauge of a city's economic growth. In the UK, if the view of London's skyline from the city to St Paul's Cathedral is unblemished by Wolff, Liebherr and Pontain cranes, then a slump is on the horizon [5]. In the US, the popularity of tower cranes has been slower to develop; however, in 2006, Miami was named "Crane City," as over 300 tower cranes were estimated to be working [6].

It is universally recognised that the tower crane's main weakness is the debilitating affect that high or gusting wind conditions can have on their ability to perform their critical construction role, hence there is a risk of delay to the tall building programme through a drop—or even halt—in the construction productivity rate. This delay can have huge commercial and reputational consequences for the builder if a tall building project is not completed and handed over in accordance with the construction contract dates.

There have been many technical advancements in computerisation, communication and control of tower cranes, the latest of which are integral to new cranes and available as retro-fit kit for older cranes [7], all aimed at improving productivity and safety. However, there have been no advancements aimed at the crane's oldest adversary—wind. The Lifting Wing aims to address this imbalance.

Wind forces exerted on the lattice structure of a tower crane and the construction load suspended from the crane hook directly affect the ability to safely operate and control a crane and its construction material load. The higher the wind speed, the greater the force exerted on the crane and load, and the greater the likelihood of having to shut down crane operations and hence site productivity on programme critical activities drops. The force exerted is wind pressure, caused by air particles travelling at speed and hitting a stationary object—in this case, the crane structure and its bulky suspended load. Wind pressure varies as the square of wind speed. Therefore, if wind speed doubles, the wind pressure increases by a factor of four. A relatively small increase of wind speed can therefore have a significant effect on the safe lifting operations of a tower crane.

Tower cranes are designed to international standards that specify the "in-service" wind speed that a crane must be able to withstand and operate safely. These are typically 14 m/s (31 mph) for mobile cranes and 20 m/s (45 mph, Beaufort Scale Gale Force 8) for tower cranes [8]. However, the reality of the construction site is that the Tower Crane Operator will decide to take the crane out of service at a wind speed significantly lower that the manufacturer's prescribed "out of service" speed, due to their increased difficulty in safely controlling the crane. This is recommended practice in the UK crane industry [9]. The primary reason for the inability to control the crane is due to the effect of the wind pressure on the construction load being lifted, rather than the crane structure itself. Wind pressure acting on the load suspended at the end of the tower crane's lifting cable results in increasing difficulty for the operator controlling the crane's operations of lift, swing, travel, lowering and landing of loads on a congested construction site. This causes a significant safety risk, not only for the crane operator, but for any operatives in the vicinity of the crane and its load. This effect results in the crane ceasing operations at relatively low wind speed with a relatively frequent occurrence, hence critical path programme activities are commonly delayed.

The Effect of Wind on Suspended Loads

Strong winds tend to gust rather than blow consistently. This is amplified in tall building construction site locations which are generally in, or adjacent to, built-up clusters in city centres. The neighbouring buildings tend to break up the relatively smooth flow of wind over open land and cause turbulent or *separated flow*. This turbulent flow of air across a tower crane and its load can result in an induced rotating (yaw) and swinging motion (drag caused by a gusting wind) on the suspended construction load, pushing it out of balance, increasing the radius from the centre of gravity of the crane and therefore the overturning moment on the crane, potentially making the tower crane unstable. For a relatively light load with a large surface area, such as formwork shutters for concrete frame buildings, steel floor pans for steel frame buildings or cladding panels, this situation will occur significantly below the tower crane's design wind speed.

For example, a wind speed of 14 m/s (30 mph) generates a wind load on a 2.5 m × 1.3 m (8 ft × 4 ft) standard formwork shutter of 372 Newtons (N). If the wind speed increases by circa 50% to 20 m/s (45 mph), the wind load rises to 740 N an almost 100% increase of load. If this wind blows from behind the crane, the load radius will be significantly increased, potentially overloading the crane. For example, a formwork shutter weighing 750 kg with an area of 3.25 m^2 and suspended on a 27 m cable will move 1.4 m from the vertical when subjected to a 14 m/s (30 mph) wind. Moving the load radius by this distance on a 35 tonne capacity crane with a 34 m main boom working at 18 m radius would reduce the rated capacity from 950 kg to 640 kg. If this occurs close to the lifting and radius limit of a tower crane, the result could be a catastrophic crane collapse.

This has occurred many times across the world with disastrous effect, the most famous of which is "Big Blue," a giant Lampson Transi-Lift crane that collapsed due to the effect of wind on its load whilst building Miller Park, the Milwaukee Brewers Stadium, USA [10], which was recorded by the Occupational Safety & Health Administration safety inspector on site the day of the collapse [11]. It had a rated capacity of 1500 tonnes and was lifting a load of 450 tonnes, well inside its maximum capacity. Upon investigation by independent specialist bodies, the concluded primary factor of the collapse was the high wind load acting on the section of roof being lifted and lack of consideration of those loads on the crane's rated capacity [12].

HYPOTHESIS

Conclusions from the earlier published paper summarised above [3] which signposted a widespread demand for innovation in the area of wind and its negative effect on the construction process, along with research undertaken in aerodynamic theory and site observations of the effect of wind force on a suspended load of a tower crane on many of the authors' construction projects, led to the idea of reducing the effect of this force by sheathing construction materials in an aerodynamic profile during lifting operations. This would reduce the wind force effect on the load, create more stable flight characteristics, ultimately reducing the loads imposed on a tower crane and thereby increasing the ability to lift safely in challenging wind conditions.

Various profiles were investigated to achieve the best compromise of two diametrically opposed requirements: that of an aerodynamic shape and the ability to allow large and irregular shaped construction materials to be encapsulated within the aerodynamic profile. A section of an aerofoil (a two-dimensional wing) in a horizontal orientation was ultimately selected, as established aerodynamic research shows that at low approach angles the air flow is able to follow the curve of the upper and lower surfaces of the aerofoil closely, then join smoothly towards the trailing edge, minimising eddies [13]. There remains a relatively high pressure region at the front, but the low pressure at the rear is much closer to atmospheric pressure, resulting in a resistance (coefficient of drag, CDrag) that is around 20 times less than a flat sheet and 10 times less than a cylinder profile [14]. CDrag is a dimensionless quantity that is used to quantify the drag or resistance of the Wing in air. The lower the CDrag, the less aerodynamic drag on the surface of the shape.

Figure 1 is a view from above a section of aerofoil and shows the smooth flow of air from left to right over the streamlined shape, but that flow separation occurs progressively as the aerofoil is turned at an oblique angle to the air flow (yaw angle). The Lifting Wing aerofoil design aims to prevent this "stall" effect by being freely suspended from the tower crane lifting cable, ensuring it is free to rotate and remain "nose to wind," presenting the minimal surface area to the prevailing wind direction, thereby minimising the effect of wind on the tower crane suspended load.

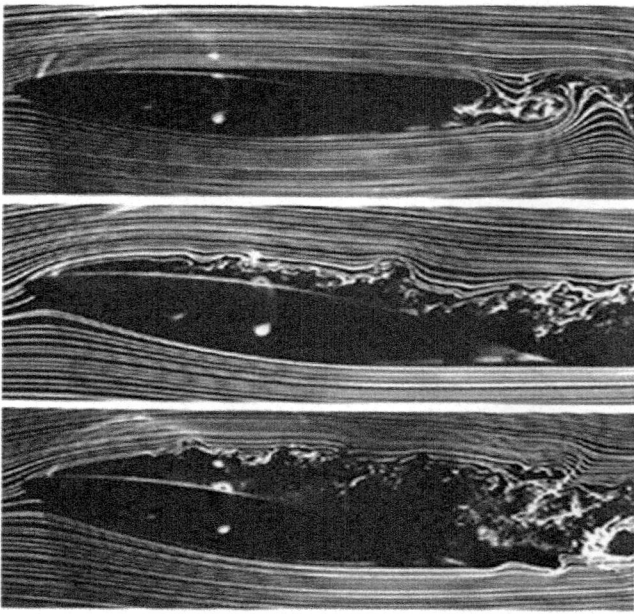

Figure 1: Increasing flow separation as yaw angle increases [14]. (Reprinted with permission from [14]. Copyright 2012 Prentice Hall).

As an aerodynamic ideal, the Lifting Wing design would follow a slim, streamlined aerofoil profile with a sharp trailing edge [15]. However, the practical consideration of ensuring typically large construction loads can be accommodated inside the profile outweighs the desire to reduce the drag (Cd) to an absolute minimum level. This results in an aerofoil profile that is wider than the ideal, but still aerodynamically efficient.

NACA Foil Design

The National Advisory Committee for Aeronautics (NACA) conducted extensive research into aerofoils from the 1930s, some of which are still utilised in aircraft manufacturing [16]. They are defined by four-digit wing sections:

- The first digit describes the maximum camber as percentage of the chord (the line between the leading and trailing edges);
- The second digit describes the distance of maximum camber from the aerofoil leading edge in tens of percent of the chord;
- The third and fourth digits describe the maximum width of the aerofoil as percent of the chord.

The XFOIL programme [17] was utilised to review 2D aerofoils between NACA 0012-50 to determine the most suitable profile that when extrapolated into a 3D shape would achieve a balance between aerodynamic efficiency and sufficient width to accommodate an array of typical construction load dimensions.

NACA 0035 (00 indicating that it has no camber, 35 indicates that the aerofoil has a 35% width to chord length ratio) was ultimately selected as the profile most suitable for the Lifting Wing design, balancing length and width to accommodate the largest, most commonly lifted tall building construction loads. An analysis was undertaken of materials most commonly lifted in the construction of typical concrete and steel-framed tall buildings. This analysis showed that metal floor pans or decking used as permanent formwork for concrete floors in the majority of steel-framed tall buildings, plus timber formwork, bundles of structural steel or concrete planks for concrete framed tall buildings (both commonly 1.2 m wide and up to 5 m long) can be inserted within the profile which would have a chord length of 6 m at full scale. The selected profile would also comfortably accommodate typical individual or loose loads such as mechanical and electrical services components, concrete kibbles and skips, edge protection screens, and palletised or bagged loads such as blocks, sand and cement. At full scale, the selected profile would accommodate these most commonly lifted items, whilst offering a relatively narrow frontal area, smooth flow path around the flanks to minimised flow separation and a sharp trailing edge to minimise drag and side forces otherwise exerted on the load and transferred to the crane.

The Lifting Wing

The full-scale Lifting Wing described by the NACA 0035 aerodynamic profile would be 6 m long × 2.10 m wide by 2.0 m high, built of a lightweight, high impact resistant clear plastic skin over a stiff, skeletal frame. It would be open at the top and bottom to allow it to be lowered over the load and for access to the lifting chains. It will be hung with three-point lifting chains attached to the crane hook and lowered by crane over the construction materials to be lifted. The load is then propped/strapped inside the Wing, restraining the load's position relative to the Wing. The Wing fully encapsulates the load, which is directly suspended from the hook of the tower crane. The Wing profile then gives the load an aerodynamically efficient, predictable and more controllable profile in high wind speeds. A smaller version would be made to accommodate smaller loads such as palletised and bagged loads, 3 m long and 1.5 m high.

Following established aerodynamic theory, the Wing would reduce the key drag load and pitching moment (which would cause the suspended material to swing fore and aft on a crane rope), along with side force and yaw (which would cause lateral oscillation of the lifted material) induced by the wind forces acting on the load being lifted. This is diagrammatically shown in Figure 6. The reduction of the effect of these wind-force-induced loads and a more stable "flight" of the lift should result in safer lifting of construction materials in higher and gustier wind-speed conditions than the current industry standard. The ultimate objective is to reduce the industry-accepted norm of 40% "down time" for the tower crane over the construction phase of a tall building due to "winding off." This would thereby save time on the critical path of the tall building construction programme and, hence, substantial costs. This theory was then tested by building a scale model of the Lifting Wing for wind tunnel testing.

AERODYNAMIC TESTING

Aim of the Testing Programme

Tests were conducted at Loughborough University's open circuit wind tunnel, the layout of which is shown in Figure 2 and the scale of which can be determined from Figure 4a. The aim was to compare the aerodynamic performance of a scale model of a typical rectangular, "brick"-shaped construction load (replicating a load profile most commonly lifted via a tower crane) against the aerodynamic performance of the scale model of the Lifting Wing in a range of wind speeds and yaw angles.

To ensure the test results are predicative of full-scale results, the tests were planned to be undertaken with a Reynolds number (Re) as close to the calculated full-scale Wing, Re of 8.2×10^6, calculated for a wind speed of 20 m/s, where Re = Inertia Force/Viscous Force = (Density × Velocity × Length)/ absolute coefficient of Viscosity. If the model has the same Re as the full-scale application, then they are dynamically similar [18]. The non-dimensional function of Fluid Viscosity, Density, Pressure, and Temperature will be the same for the model and full scale. However, Re sweep tests of both models showed the Re became invariant above 1.5×10^6, allowing the results obtained to replicate the full-scale Lifting Wing in wind speeds of up to 90 mph (current international standards for tower crane "in-service" wind speeds with no aerodynamic aid are up to 20 m/s or 45 mph).

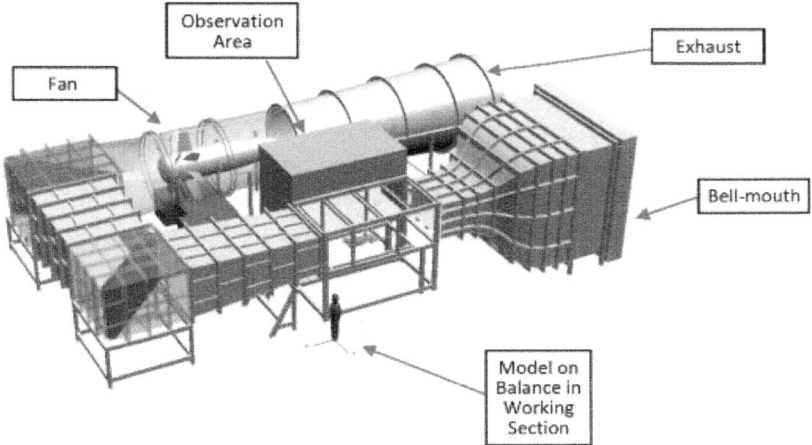

Figure 2: Loughborough university aeronautical and automotive engineering wind tunnel isometric.

Testing Method

The scale model of the Lifting Wing was built to an accuracy of ±1 mm, with the design based on the NACA 0035 aerofoil. The chord length was 600 mm, maximum width of 216 mm and height 200 mm, with a cross-sectional area of 0.0432 m². This equates to a 1:10 scale model of the full size Lifting Wing. The model construction was formed using a 2 mm-thin plywood sheet laid over and fixed to a slim CNC cut plywood spar frame at the top and bottom of the wing, as shown inFigure 3a,b.

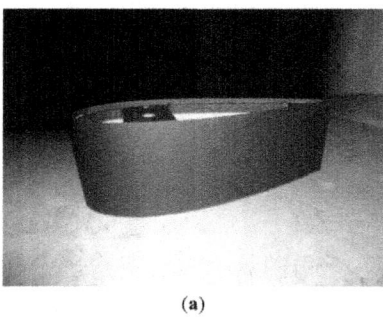 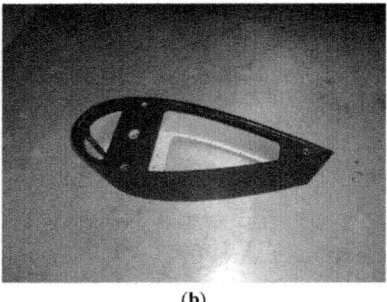

(a)   (b)

Figure 3: (a) Lifting Wing model and top mounting bracket; (b) Wing internal void, spar frame and brackets.

Similarly, the 1:10 scale model of the typical construction load, the "Brick", was built with the same technique, having a chord length of 600 mm, maximum width of 210 mm and height of 200 mm, giving a reference area of 0.043 m².

It was initially anticipated that multiple sets of results would need to be taken, depending on the accuracy and repeatability of the obtained results. However the first series of test results showed good accuracy and repeatability (within 5%) and a minor, but consistent level of asymmetry. This test series was run twice allowing the arithmetic mean to record the central tendency. The asymmetric tendency was subsequently determined as a feature of the tunnel and had been repeated in numerous wind tunnel test experiments undertaken by Aeronautical Researchers at Loughborough University and was quantified and accounted for, therefore deemed to be insignificant to the results.

The wind tunnel test allowed quantitative data for drag, pitch, side force and other relevant forces acting on the Brick model and the Wing model at a range of wind speeds and yaw angles to be compared. These forces and their directional impact on the Wing are shown in Figure 6. These tests were conducted in parallel with flow visualisation observations at key stages of the testing to cross check the quantitative results and the logic behind conclusions drawn. Additionally, a preliminary dynamic test was also undertaken as a third method of cross checking results obtained from the first two methods, giving qualitative information in the form of a visual display of the Wing under freely suspended conditions reflecting, as closely as possible, the conditions of the full-scale Wing suspended by a tower crane. However, results obtained were indicative only, due to issues with model symmetry and difficulty finitely levelling the model. The dynamic test will be refined and re-run in the next stage of research.

Test Summary

The objective of this test was to generate quantitative data for the model's drag (Cd), lift (Cl) and pitching (Cp) moments at varying degrees of yaw and wind speed. The wind tunnel test was designed to minimize systematic errors by considering and compensating for the most likely causes of error including model or tunnel asymmetry, error caused by the wind forces acting on the connection shaft between the model and the tunnel balance, plus random errors. The method of testing involved both the reference "Brick" model and the Lifting Wing being rigidly fixed by a steel connection shaft to the balance (Figure 4b), which is fitted into the floor of the working section of the tunnel. Once true zero (head-to-wind) position was established by undertaking a yaw sweep for each model, the tests were undertaken for each model in turn.

(a)   (b)

Figure 4: (a) LU AAE Wind Tunnel Bell-mouth and Exhaust; (b) Balance below Tunnel Working Section.

The reference areas of the models, the wind speed, barometric pressure, air temperature, drag, lift, side-force, pitching moment, yawing moment and rolling moments, plus their coefficients, were recorded by the tunnel computer data logger at a range of wind speeds from zero to 40 m/s. The model was then rotated (yawed) on the balance through two degrees away from true zero and all measurements recorded. This was repeated by further 2° increments up to ±20°, then 1 degree increments up to a maximum of ±25° yaw. Tests for each model were re-run after powering down the wind tunnel (effectively re-setting, or zeroing the tunnel and its data logger) to determine the repeatability of results. All results taken were within 5% of the initial result with no outliers, allowing the arithmetic mean to be utilised for the final result. Measurements for the Wing were compared to the reference Brick model, ultimately demonstrating the aerodynamic improvement of the Wing.

Test Method

- The steel connection shaft was mounted to the tunnel balance and the wind tunnel was run at 5 m/s increments from zero up to 45 m/s to determine forces due to shaft alone and allow balance results for each model to be adjusted for shaft effects. To refine these results, a replica support shaft of the same diameter as the one used to support each model was raised into the tunnel to a height of 450 mm. Each model was then attached to the tunnel roof via the original support shaft and lowered until it was just clear of the replica shaft fixed to the balance. This gave a more accurate balance reading of the shaft value to be subtracted from each model measurements;

- The Brick model was mounted on the steel shaft fixed to balance. The maximum velocity, V_{max} was established by running wind tunnel from 0 m/s at 5 m/s incremental speeds, whilst ensuring drag, lift, side-force, pitch, yaw and roll loads did not exceed 90% of the limit of the wind tunnel balance. This was repeated for the Wing model, resulting in a V_{max} of 40 m/s, with generated forces at just over 85% of the balance limit for the Brick model;

- A Reynolds Number (Re) sweep for the Brick at zero degrees yaw, over incremental wind speeds from 0 to 40 m/s was run allowing the calculation of the Re for the range of wind speeds, plotted to determine the minimum wind speed at which the Re becomes a constant (thus replicating full-scale results). This was repeated for the Lifting Wing. Resulting Re values shown in Figure 5, demonstrated that above Re of 600,000 there is relatively little Re effect and results are as close to full scale as possible;

- A series of tests for both the Brick and Wing were run, recording forces graphically shown in Figure 6, the results of which produced following graphs: Brick Yaw Angle *versus* CDrag for 30 m/s and 40 m/s; Brick Yaw Angle *versus* CLift for 30 m/s and 40 m/s; Brick Yaw Angle *versus* CSideforce for 30 m/s and 40 m/s; Brick Yaw Angle *versus* CPitch for 30 m/s and 40 m/s; Wing Yaw Angle *versus* CDrag for 30 m/s and 40 m/s; Wing Yaw Angle *versus* CLift for 30 m/s and 40 m/s; Wing Yaw Angle *versus* CSideforce for 30 m/s and 40 m/s; Wing Yaw Angle *versus* CPitch for 30 m/s and 40 m/s.

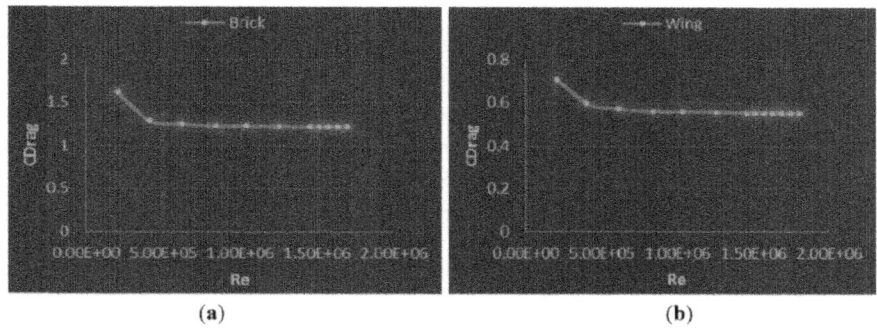

Figure 5: (a) Brick Re *versus* Cd; (b) Wing Re *versus* Cd.

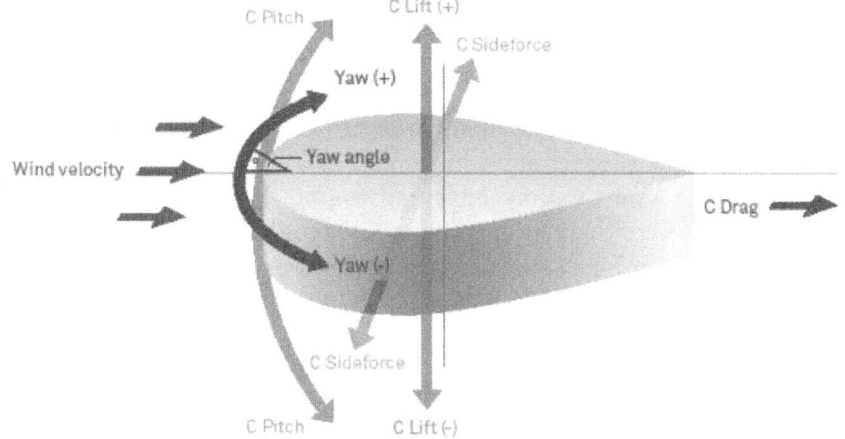

Figure 6: Wind force coefficients on the Lifting Wing.

Test Results

An overlay of the two crucial sets of results for the Wing and Brick Yaw Angles *versus* CDrag at 30 m/s and 40 m/s, and the Wing and Brick Yaw Angles verses CLift at 30 m/s and 40 m/s were graphically plotted, shown in Figure 7a,b.

The primary conclusions drawn from the CDrag overlay of the Wing and Brick are:

- The Wing profile had a dramatically lower overall drag profile (in excess of 50% less CDrag than that of the Brick), hence significantly less drag load would be induced on the cable and the crane, in all wind conditions;

- The Brick results plotted graphically exhibit a deep V, which shows a relatively large sensitivity to wind direction changes, which dramatically increase drag- and swing-induced loading, hence load on the crane. This feature is shown by comparing flow visualisation Figure 8a at zero degrees showing a wide flow attachment line one third back from the nose and Figure 8b at 10 degree offset, showing a more defined flow attachment line further forward, directly behind the front corner. This increases the size of the wake area and reverse flow behind the Brick, thereby increasing drag;

- By comparison, the Wing plotted results exhibit a smooth, shallow curve, showing relative insensitivity to changes in wind direction, with less drag- and swing-induced forces, hence a more stable flight. This is demonstrated by comparing the Brick Figure 8b and the Wing Figure 11a at 10 degree offset. This shows smooth attachment lines running to the sharp trailing edge of the Wing, limiting the separated flow, or wake area behind the Wing, hence low drag;

- The tendency for drag to increase as yaw angle increases tails off earlier with the Wing, reaching a maximum at around $\pm 12°$ (See Figure 7a) due to the sharp trailing edge and smooth flanks, whereas the Brick drag forces continue to increase as yaw angle increases to a maximum at around $\pm 18°$ as the wake area behind the Brick and reverse flow continues to grow. This demonstrates the improved stability generated by the Wing, reducing drag-imposed loads on the crane in higher wind speed and with changeable wind directions.

The primary conclusions drawn from the CLift overlay of the Wing (with Brick load inside the Wing) and Brick are:

- Lift forces generated on both models are less than a 10th of magnitude of drag forces and therefore its influence is likely to be less significant;

- The Wing profile has a lower overall lift profile (less than 1/4th of the lift of the Brick at higher yaw angles) hence significantly less rise-and-fall load would be induced on the cable and crane in higher wind conditions. This feature is most clearly demonstrated by comparing flow visualisation Figure 9a, the leeward side of the Brick at $-25°$. It shows a more pronounced flow along the top and bottom edges, which become more dominant at the higher yaw angle. In contrast, the Wing (Figure 12a) shows the leeward side of the Wing at $-10°$ (which was almost identical to the Wing at $-25°$). The Wing shows more fractured, multiple flow separation lines running from the nose toward the tail that drop away much earlier. These markedly differing flow features would explain the differing lift forces generated on each model;

- The Brick exhibits a sharp and deep W profile, which signifies sensitivity of this shape to increasing wind yaw angle, dramatically increasing lift- and fall-induced loading, hence load on the crane. This would result in a rotation of the load when freely suspended from a crane, causing safety issues when trying to fly and land the load safely;

- The Brick also shows increasing sensitivity to higher wind speed as the results for 30 m/s and 40 m/s diverge at higher yaw angles producing unstable flight characteristics as these factors increase;

- By contrast, the Wing exhibits a smooth, shallow curve, showing relative insensitivity to changes in wind yaw angle or wind speed, hence less rise- and fall-induced forces and more stable flight characteristics.

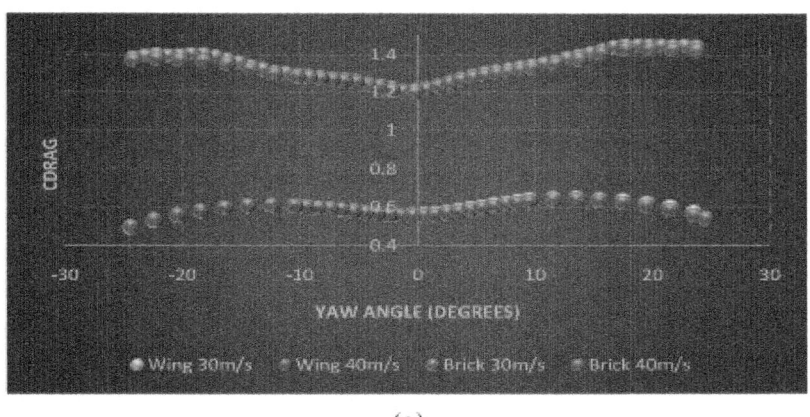

(a)

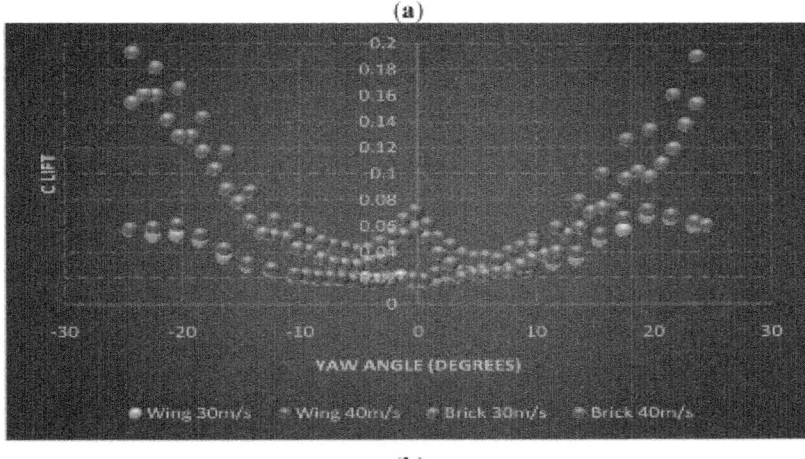

(b)

Figure 7: (a) Wing and Brick Yaw Angles *versus* CDrag for 30 m/s and 40 m/s; (b) Wing and Brick Yaw Angles *versus* CLift for 30 m/s and 40 m/s.

Wind Tunnel Test Conclusion

Each of these tests were run twice and results showed good repeatability of generated quantitative data for both model's drag and lift forces at varying degrees of yaw and wind speeds. The arithmetic mean was taken to give the central tendency as there were no outlier results taken (all values were within 5% of the initial the result). Side force and pitching moment were also measured in this method, but ultimately deemed less critical, being relatively similar for both models, with slightly less pitching moment generated by the Wing under extreme yaw angles and slightly higher side forces generated by the Wing at higher yaw angles, creating a restoring yawing moment (self-correcting characteristic), ultimately producing a stable flight in changing wind direction. These mean results demonstrated significantly improved aerodynamic characteristics of the Wing, resulting in significant reductions in critical forces generated by wind acting on the Wing, hence forces imposed on the tower crane at full scale. These results point toward the Wing assisting the tower crane operator in their control of the tower crane in higher wind-speed conditions experienced on a construction site, thereby delaying his decision to take the crane out of service at a wind speed significantly lower that the manufacturers prescribed "out of service" speed.

These conclusions were further tested by conducting flow visualisation analysis of the Wing and Brick at varying wind speeds and yaw angles in the wind tunnel, discussed below.

Flow Visualisation

A series of flow visualisation photographs of the Brick and Wing models were taken at key stages in the wind tunnel testing for each model to allow comparison of aerodynamic flow around the models. These were achieved by coating the Brick and Wing models with a mixture of titanium dioxide, paraffin and linseed oil, and capturing the resultant flows at true zero degrees, plus and minus 10° and plus and minus 25° yaw at varying wind speeds. A demonstrative selection of flow images are given in Figure 8, Figure 9, Figure 10, Figure 11 and Figure 12. Wind is flowing from left to right in all figures with exception of 11a,b, where it is right to left. Windward is a (+) yaw angle from true zero (head-to-wind flow), showing the side facing into the wind and leeward a (−) yaw angle showing the side in the wind "shadow".

Figure 8a is a flow visualisation photograph of the Brick at true zero to the wind flow and is viewed from the leading edge corner. The wind flow impacts on the flat face and spreads out towards all four sides of the Brick. The flow separates at the four edges and a large wake is formed behind the Brick.

This wake is responsible for the large coefficient of drag seen in the wind test results. The wide flow separation line running from the top to bottom of the Brick at the point where the vortices at each side of the Brick, created by the blunt nose, reverse the flow back toward the front of the Brick where it meets the wind flow spilling around the nose corner and become entrained in the wake. This causes the flow to stall and gravity then drags the mixture down. These flow patterns should occur on all four sides (excepting gravitational effect).

Figure 8b, taken from the same position, but with the Brick at +10° yaw (windward), shows the reverse flow separation line being pushed much nearer the front corner of the nose. This is caused by a more dramatic meeting of the vortex flow (which has increased force due to the +10° yaw) and the frontal flow spilling around the nose corner. It also shows flow detachment approximately mid-way along the Brick, with some flow being pushed toward the nose and some being pushed toward the rear of the Brick.

Figure 9a shows the leeward side of the Brick at −25° yaw. There is more pronounced flow along the top and bottom edges, which becomes more dominant at the higher offset angle. The flow has separated at the edge on the front face, but the reverse flow is now three dimensional, flowing towards both the front and side edges. The flow towards the sides meets flow spilling around from the top and separates at the white line and is entrained into the wake. The separation line at the lower edge is smaller due to gravity. This will contribute to the lift force seen in wind test results. The resulting wake will be more pronounced on this side.

Figure 9b shows the windward side of the Brick at +25° yaw. The flow is pushing from the front centre in three dimensions toward the top, bottom and rear trailing edges remaining attached along the flank. The resulting wake will be less pronounced on this side.

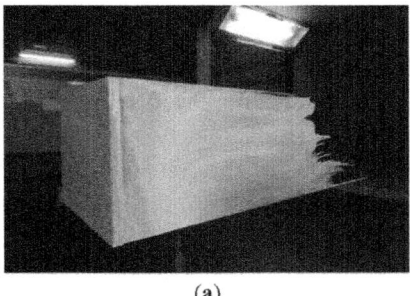 

(a)   (b)

Figure 8: (a) Brick at 0°, 40 m/s; (b) Brick at +10°, 40 m/s.

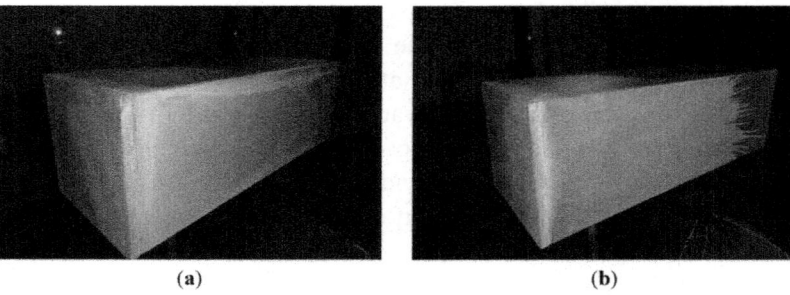

(a)   (b)

Figure 9: (a) Brick nose at −25°, 40 m/s; (b) Brick tail at +25°, 40 m/s.

Figure 10a shows the Lifting Wing model at 0°. The Wing has a very low aspect ratio and the flow around the trailing edge is highly significant. The extent of the separated flow producing the wake is much smaller than for the Brick, resulting in a significantly lower drag. The flow is attached from the nose and forms two strong flow separation lines running from the nose toward the tail. These drop away toward the bottom rail at the tail of the Wing as flow rate reduces and gravitational forces take over. The collection of mixture at approximately one third along from the nose of the Wing is at its widest point. This may indicate a nearing of flow separation at this potential transition point, but the flow successfully negotiates the curve of the Wing and continues in laminar flow along the Wing's surface until it nears the tail's trailing edge.

Figure 10b shows this effect from the tail view. It shows the flow reducing as it runs along the Wing and gradually falling under gravitational force as it nears the narrowest point, the trailing edge. It then separates, creating a relatively small wake.

Figure 11a shows the Wing at +10° yaw, where it exhibits a more singular flow separation line running from the nose toward the tail. This drops away more gradually, only hitting the bottom rail at the tail intersection point. The collection of mixture has moved further back from the nose of the Wing and is now behind the point of maximum Wing width. This indicates the potential transition point has moved further back due to the increased windward yaw angle, hence increased flow across this face of the Wing. Again, it does not actually separate at this point and continues toward the tail in laminar flow until it nears the tail trailing edge, but at a point further from the tail, indicating that the turbulent boundary layer is occurring earlier. The flow patterns are very similar to the 0° case which explains why the drag appears to be relatively invariant for the Wing.

Figure 11b shows this effect from the tail view and shows the flow reducing and falling under gravitational force as it travels to the trailing edge, but that it separates earlier to become turbulent flow.

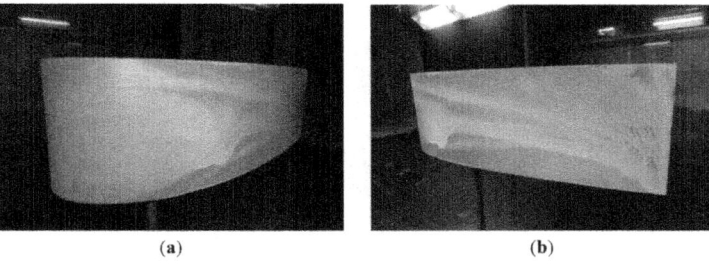

Figure 10: (a) Wing nose at 0°, 40 m/s; (b) Wing tail at 0°, 40 m/s.

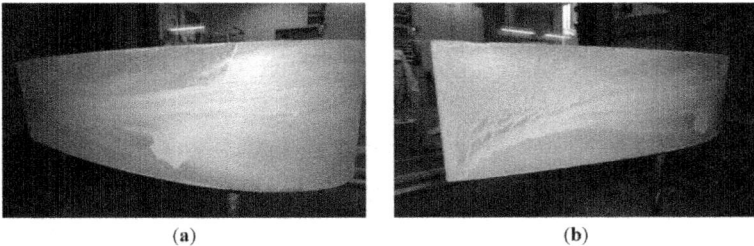

Figure 11: (a) Wing nose at +10°, 40 m/s; (b) Wing tail at +10°, 40 m/s.

Figure 12a shows the Wing at −10° yaw, leeward side viewed from the nose. The Wing now exhibits more fractured, multiple flow separation lines running from the nose toward the tail that drop away much earlier, demonstrating that the flow rate is much reduced across this face and gravitational forces take over earlier. The collection of mixture has moved further toward the nose and is now in front of the maximum Wing width position. This indicates the transition point has moved forward due to the more turbulent, reduced flow formed in the wind shadow. It now actually begins to partially separate at this point, whilst some flow does continue toward the tail in laminar flow but then separates at a point much closer to the midpoint of the Wing, indicating that the turbulent boundary layer is occurring much earlier. This effect would create the restoring turning moment in the Wing, ensuring it returns to a zero yaw position (nose-to-wind).

Figure 12b is the tail view and shows the flow reducing and falling under gravitational force much earlier as it travels along the Wing and separates earlier to become turbulent flow across the rear third of the Wing.

Flow visualisation pictures of the Wing at ±25° yaw showed no significantly differing patterns to the ±10° discussed above. This fact demonstrates that the drag is relatively invariant for the Wing, whilst exhibiting significantly less drag variation than the Brick.

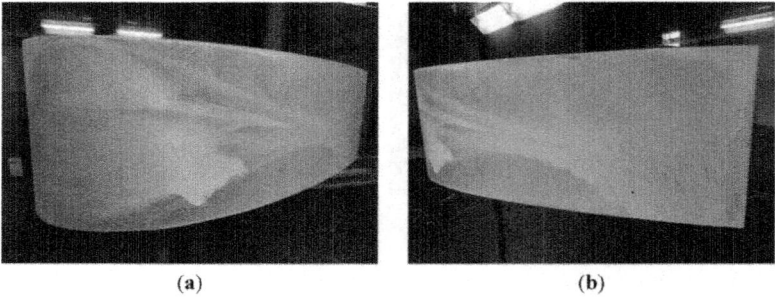

(a)   (b)

Figure 12: (a) Wing nose at $-10°$, 40 m/s; (b) Wing tail at $-10°$, 40 m/s.

Flow Visualisation Conclusion

These flow visualisations show a relatively clean, stable flow over the Wing at varying degrees of yaw, demonstrating stable and predictable aerodynamic behaviour. The significantly reduced drag of the Wing compared to the Brick, along with the Wing's invariance of drag at higher yaw angle, are the key factors in proving the ability of the Wing to operate safely in higher and gustier wind conditions than a standard construction load. These observations correlate with the quantitative data taken during wind tunnel testing and reinforce the characteristics of stable and improved aerodynamic behaviour of the Wing over the Brick. These results were further tested by conducting a dynamic test of the Wing suspended in the wind tunnel.

Preliminary Dynamic Test

The objective of this dynamic test was to conduct visual analysis of the Wing's aerodynamic characteristics under conditions reflecting, as closely as possible, suspension of the Wing from a tower crane cable in wind conditions likely to be experienced on a tall building site. No quantitative measurements could be taken during this test, as it was purely a visual analysis of the Wing's aerodynamic performance.

It was noted during this test that error in model symmetry and the inability to finitely level the model affected the results and would need further refinement to achieve an accurate replication of full-scale results.

The Wing model was freely suspended by three, 2 mm in diameter multi-strand steel cables, each with a 10 kg breaking strain. These were mechanically fixed to the top edge of the model, one directly above the centre of the nose and two equally positioned on the top edge either side of the Wing, behind the widest section of the Wing. The centre line of the three wires were over the centre of gravity of the model. These wires were sufficiently long to allow the Wing to be suspended in the centre of the tunnel working section, with the wires running through a hole in the roof of the tunnel and mechanically fixed externally to support the dead and live loads of the model during testing (Figure 13a). This suspension method replicates the envisaged method of suspension of the full-scale Wing from a tower crane.

A series of videos were taken to record the behaviour of the Wing under increasing wind speeds from 0 to 12 m/s. These tests were then repeated with loads added inside the Wing (1 kg metal plates fixed inside the wing profile) to replicate 1, 2 and 3 tonne loads on a full-scale Wing (Figure 13b). Observations were made on Wing stability and flight behaviour from the side and roof windows of the tunnel working section.

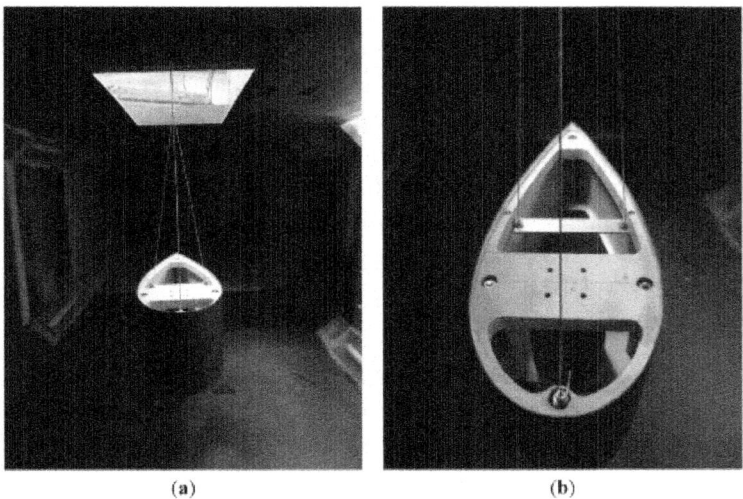

(a)   (b)

Figure 13: (a) Wing Suspended for Dynamic Test; (b) Wing with Internal Load.

Dynamic Test Observations

The test was initially run with no internal load and videoed from the side window of the tunnel. The Wing remained relatively static as the wind speed was increased from 0 m/s to 9 m/s, swinging slowly back by approximately 5° from the vertical as wind speed increased to 9 m/s. At 10 m/s the nose of the Wing was observed to begin to move horizontally from left to right, stop and then return from right to left through the head-to-wind at 0° yaw. This repeating oscillation increased in yaw angle as the wind speed was increase to a maximum of 12 m/s, whereupon the nose of the model, viewed from above, moved left to right whilst swinging forward and back, describing a repeating infinity (∞) symbol movement over a distance approximately equal to half the length of the model (300 mm). This oscillation reduced as the wind speed was reduced to 9 m/s, whereupon the model became relatively static again, holding the 5° inclined position.

This test was repeated with a load of 2 kg fixed inside the Wing. It repeated the pattern of the first test, with the exception that the oscillation began at the increased wind speed of 11 m/s and diminished as the wind speed was reduced below 11 m/s.

Finally, a load of 3 kg was fixed inside the Wing, again repeating the pattern of the first and second tests, with the exception that the oscillation began at an increased wind speed of 12 m/s and diminished when wind speed was reduced below 12 m/s.

The initial movement of the nose from left to right was deemed to be caused by a lack of absolute symmetry of the model and it being slightly out of level horizontally due to unequal lengths of its three suspension cables. These small errors create a gradually increasing turning moment on the model as the wind speed increases. However, this also demonstrates the Wing's self-correcting characteristic, producing stability of flight at full scale, as this would ensure a slowly correcting nose-to-wind position of the Wing, desirable in changeable, gusty wind conditions typified on congested city-centre tall building sites.

Implications of Results on Wing Design

This preliminary dynamic test demonstrated that the full-scale Lifting Wing would need to be made symmetrically, ideally utilising vacuumed formed thermoplastic technology or moulded carbon-fibre-reinforced polymer, plastic or thermoplastic giving the added benefits of a higher strength-to-weight ratio and greater ability to withstand impact deformation. Residual error could be corrected by adding a top mounted vertical stabilising fin, fixed above the trailing edge of the Wing. The dynamic test also demonstrated the need for finite adjustment of suspension cables to ensure truly level flight. Following this paper's publication, this dynamic test will be further refined by the introduction of turnbuckles on each of the three suspension wires above the tunnel, allowing finite adjustment of each cable length, and hence achieving true horizontal suspension of the model in the tunnel.

This test also demonstrated the proportional relationship of increasing load to more stable flight—the greater the load carried inside the Wing, the less effect the non-symmetrical features of the model had on the stability of the flight in increased wind speed. It also proved that the ultimate wind speed in which stable flight could be achieved would be directly related to the size of the load carried inside the Wing.

OVERALL CONCLUSIONS

The wind tunnel test quantitative data correlates with the flow visualisation and preliminary dynamic test observations. These reinforce the primary Wing characteristics of reduced drag in excess of 50% lower than the Brick and of side forces on the Wing creating a restoring moment when flying in changeable wind direction conditions, giving a desirable nose-to-wind behaviour. These key characteristics combine to reduce induced loads on the tower crane and produce stable improved aerodynamic behaviour of the Wing when compared to typical construction loads.

This demonstrates that the Wing achieves its primary purpose of increasing the ability to lift construction materials safely in higher and more gusty wind-speed conditions than is currently achievable. Therefore, the Lifting Wing design, if used on a tower crane of a tall building, should create a valuable contribution in mitigating the effect of wind causing critical path delay during the construction of a tall building, potentially reaping substantial time and cost savings. This knowledge and benefit could be transferable internationally as, without exception, tall buildings across the world are built using tower cranes which are negatively affected by wind during the build period, delaying completion, frustrating builders from completing on time and budget and ultimately, owners from occupying their new tall buildings. These positive results will be further demonstrated by future studies utilising a full-scale Lifting Wing on a tower crane, discussed in the following section.

FURTHER WORK

Following running the refined dynamic test discussed above, the final stage of the Wing development will be undertaken with assistance from the authors' sponsoring company involving the construction of a full-scale Wing and its dynamic testing utilising a Saddle Jib or Luffing Jib Tower Crane. In this test, an experienced tower crane operator will lift a rectangular "Brick"-shaped reference load in wind conditions approaching industry-recognised winding-off speeds. The load will then be placed inside the full-scale Wing and lifted in the same wind conditions. The operator will note flight characteristics of each lift and determine the increased wind speed in which the Wing can still be lifted safely. This qualitative analysis will rely on the feedback from the operator, rather than on any measured force data. However it is exactly this operator analysis that is used across the industry to determine the safe limit of lifting by cranes on every site the world over. If tower crane operators feel the Wing allows extended lifting in higher wind conditions, then it will have succeeded.

An international patent has been applied for covering the Lifting Wing and the research that has been undertaken to date.

AUTHOR CONTRIBUTIONS

This paper describes an element of the doctoral research conducted at Loughborough University in partnership with Lend Lease for the award of Engineering Doctorate. Ian Skelton was the Research Engineer and undertook the majority of the primary work. The co-authors of this paper are the EngD supervisory team, augmented as the project progressed due to staff movements. All supervisors contributed to manuscripts as the paper was being developed and reviewed. Peter Demian was the principal supervisor from the beginning of the project. Jacqui Glass was the second supervisor at Loughborough University at the time the experiments for this paper were conducted and the paper was written. Dino Bouchlaghem and Chimay Anumba were supervisors during their time at Loughborough University, and remained active contributors after moving to other organisations.

REFERENCES

1. Baker, W.F. S.O.M. Structural Engineering Partner. In Proceedings of the World's Tallest Building, Burj Dubai, UAE, Seoul, Korea, 10–14 October 2004.

2. Skelton, I.; Demian, P.; Bouchlaghem, D. Britain's tall building boom: Now bust? *Proc. Inst. Civ. Eng. Struct. Build* 2009,*162*, 161–168.

3. Skelton, I.; Demian, P.; Bouchlaghem, D.; Anumba, C. The State-of-the-art of Building Tall. In *Challenges, Opportunities and Solutions in Structural Engineering and Construction*; Taylor & Francis Group: London, UK, 2010.

4. Shapria, M.; Lucko, G.; Schexnayder, C. Cranes for Building Construction Projects. *J. Constr. Eng. Manag.* 2007, *133*, 690–700.

5. Morby, A. Dreaming Spires for the Towering Giants. Available online: http://www.cnplus.co.uk/news/dreaming-spires-for-the-towering-giants/893729.article (accessed on 23 May 2014).

6. Shiffler, D.A. Crane City. *Am. Cranes Transp.* 2012, *2*, 21–25.

7. Rosenfeld, Y.; Shapira, A. Automation of existing tower cranes: Economic and technological feasibility. *Autom. Constr.* 1998, *7*, 285–298.

8. International Standard ISO 4302-1981. Cranes—Wind Load Assessment. Available online: http://www.iso.org/iso/home/store/catalogue_ics/ catalogue_detail_ics.htm?csnumber=10156 (accessed on 16 May 2014).

9. Construction Plant Hire Association (CPA). *Tower Crane Interests Group and Health & Safety Executive the Tower Crane Operator's Handbook*; Health and Safety Executive (HSE): London, UK, 2008; pp. 27–28.

10. Ross, B.; McDonald, B.; Vijay Saraf, S.E. Big Blue Goes Down. The Miller Park Crane Accident. *Eng. Fail. Anal.* 2006, *14*, 942–961.

11. OSHA. Big Blue Crane. Video by Occupational Safety & Health Administration safety inspector on site the day of the collapse. 1999. Available online: http://www.osha.gov/video/bigblue.wmv (accessed on 25 November 2013).

12. Riewestahl, B. Lampson Transi-Lift "Big Blue" Crane Collapse E.I.T., M.S., Penn State, 2010. Available online: https://www.youtube.com/ watch?v=zRvQlEbJnqc (accessed on 23 May 2014).

13. Anderson, J. *Fundamentals of Aerodynamics*; McGraw-Hill: New York, NY, USA, 2010; pp. 23–24.

14. Kermode, A.; Philpot, D.R.; Barnard, R.H. *Mechanics of Flight*; Prentice Hall: London, UK, 2012.

15. Dole, C.E.; Lewis, J.E. *Flight Theory and Aerodynamics: A Practical Guide for Operational Safety*; John Wiley & Sons: New York, NY, USA, 2000; p. 351.

16. Allen, J.E.E. Aerodynamics. In *The Science of Air in Motion*; University of Michigan: Michigan City, MI, USA, 1999; pp. 83–85.

17. Drela, M. *XFOIL: An Analysis and Design System for Low Reynolds Number Airfoils*; Springer: Berlin, Heidelberg, Germany, 1989.

18. Barlow, J.B.; Rae, W.H.; Pope, A. *Low Speed Wind Tunnel Testing*; Wiley-Interscience: Ann Arbor, MI, USA, 1999; pp. 143–149.

Chapter 5

A MULTI-CRITERIA DECISION SUPPORT SYSTEM FOR THE SELECTION OF LOW-COST GREEN BUILDING MATERIALS AND COMPONENTS

Junli Yang, Ibuchim Cyril B. Ogunkah

Department of Construction, School of Architecture and the Built Environment, University of Westminster, London, UK.

ABSTRACT

The necessity of having an effective computer-aided decision support system in the housing construction industry is rapidly growing alongside the demand for green buildings and green building products. Identifying and defining financially viable low-cost green building materials and components, just like selecting them, is a crucial exercise in subjectivity. With so many variables to consider, the task of evaluating such products can be complex and discouraging. Moreover, the existing mode for selecting and managing, often very large information associated with their impacts constrains decision-makers to perform a trade-off analysis that does not necessarily guarantee the most environmentally preferable material. This paper introduces the development of a multi-criteria decision support system (DSS) aimed at improving the understanding of the principles of best practices associated with the impacts of low-cost green building materials and components. The DSS presented in this paper is to provide designers with useful and explicit information that will aid informed decision-making in their choice of materials for low-cost green residential housing projects. The prototype MSDSS is developed using macro-in-excel, which is a fairly recent database management technique used for integrating data from multiple, often very large databases and other information sources. This model consists of a database to store different types of low-cost green materials with their corresponding attributes and performance characteristics. The DSS design is illustrated with particular emphasis on the development of the material selection data schema, and application of the Analytical Hierarchy Process (AHP) concept to a material selection problem.

Details of the MSDSS model are also discussed including workflow of the data evaluation process. The prototype model has been developed with inputs elicited from domain experts and extensive literature review, and refined with feedback obtained from selected expert builder and developer companies. This paper further demonstrates the application of the prototype MSDSS for selecting the most appropriate low-cost green building material from among a list of several available options, and finally concludes the study with the associated potential benefits of the model to research and practice.

INTRODUCTION

As the green building movement begins to sweep through the housing construction industry, the application of cost effective and energy efficient building materials has become necessary in today's demanding economic market [1,2]. Recent discussions on the need to lower the growing demand for conventional sources of energy have highlighted the value of using low-cost green building materials and components, given their lower cost and energy requirements [3,4]. Evidence from previous studies has proven that implementing such products in construction has the potential to not only reduce health and environmental effects, but to also bring savings from energy, maintenance, and operational costs [5-9]. Yet, research has consistently shown that the patronage for such materials in housing construction is still at a very low level in comparison to many other conventional building materials [8,9].

Recent studies [10,11] argue that several attempts to adopt low-cost green building materials for housing design projects have generally been viewed as challenging, given that most designers are vaguely informed about the full life-cycle impacts of such products. They note that information relating to the impacts of such building materials in the housing construction sector appears to be less available, as evidence [11,12] indicates that only a small proportion of design and building professionals seem to have sufficient knowledge that could allow effective decisionmaking. Ashraf [11] and Zhou et al. [12,13] suggest that maximizing their potential use in the housing industry requires seamless access to appropriate informed information and full understanding of the various options available, so as to inform decision trade-offs at the design stage. Despite the availability of accurate and reliable data, Seyfang [14] and Malanca [15] however, noted that most designers are found to make decisions regarding the selection of such materials on the basis of their past experience.

They observed that inexperienced designers generally engage the traditional mode of selection, by relying on subjective individual perceptions of values and priorities in the material selection process, which rather than facilitate or drive their design ideas, appear to do the opposite thereby limiting creativity and sometimes resulting in considerable frustration [16,17]. Trusty [18-21] & Woolley [22] further disclosed that existing databases on such materials and their formats are not designed to efficiently and directly provide such information to decision makers. They note that the available data on such materials are normally in the read-only format, and are stored in various operational databases that are not easily accessible to decision makers in usable forms and formats. As a result, decision-making failures during the planning and design stage(s) of low-cost green housing projects hinder their use in terms of their industrial capacity utilisation in the housing industry. While several studies [14,15,20] have emphasized the relative importance of information access in aiding well-considered and justifiable material choices during the early stages of the design process, Wastiels et al. [16] argue that the existing material selection method focuses mainly on limited aspects of such materials, in terms of their properties and factors that influence the decisionmaking process. Quinones [17] asserted that some lowcost green building materials, for example, contain high embodied energy that leads to ecological toxicity and fossil fuel depletion impacts during their manufacturing phase. She argued that ignoring the relevant factors or properties of any of such materials during the crucial material selection phase could reduce the effective life of that product to less than half of its normal effective life span. Moreover, Seyfang [14] and Trusty [18-22] argue that choosing the right materials for a particular project can be a very complex decision-making task, given that the selection process is influenced and determined by numerous preconditions, decisions and considerations. They suggested the idea of a decision support system (DSS) as a useful aid in making quick and critical decisions during crucial material selection process. They stressed that the considered approach to encourage the wider scale use of low-cost green building materials in mainstream housing should enable design professionals to have easy access to adequate information on the available options, hence, making the selection results more reasonable and bringing more standardization to the material selection decision-making process at the design stage.

They went on advising that whatever method is employed must be such that it allows comparison of not only the cost or technical performance of such materials, but also able to take into account several decision-making criteria, so as to derive conclusive and valid evidence of the differing impacts of various material alternatives. While there seem to be no compelling evidence of technical research on a holistic approach used by design professionals for the evaluation and selection of building materials, previous material assessment models such as the Leadership in Energy and Environmental Design (LEED) and Building Research Establishment Environmental Assessment Methods (BREEAM), have shown great promise for guiding evaluations of material predictor performance [23]. The findings of the main research study yet, criticised and noted the flawed existing support systems for being partially objective and fraught with problems of fairness [24]. The study revealed that existing methods are found wanting in that they are culturally implicit, and that such methods or tools treat the sustainability [of the] wider built environment as simply a matter of energy and mass flows with little or no regard to the socio-economic, technical, emotive and political dimensions of sustainability [24]. It further revealed that individual country teams establish scoring weights subjectively when evaluating building products, which often pose problems when applied to other regions [25,26]. The analysis of the study however, showed little evidence to justify the assumption that there are tools of demonstrable reliability for designers to assess the sustainability and suitability of such materials or products or their applicability and utility for their potential use in the design of low-cost green housing projects. Hence, a more reliable method is needed to aid design and building professionals in the selection of such building materials and components for low-cost green residential housing projects. Consequently, to promote more informed decisionmaking in the selection of low-cost green materials bothindividually and as assembled building components, a Decision Support System (DSS) is presented in this paper as an aid to design and building professionals. The objective of this study is to support decision-makers in selecting low-cost green building products that are environmentally, socio-culturally, technically and economically balanced through a proposed conceptual system. The model is to facilitate the integration of more sustainable materials into future designs by helping designers quantify how they compare to materials already permitted under existing codes, using the concept of the Analytical Hierarchy Process (AHP). The AHP approach is designed to be practical, as it combines environmental, technical, socio-cultural and economic performance into a single performance value that is easily interpreted. In the following sections, the reviews of existing technological approaches are summarised and the main findings and themes to emerge from the literature review and the fieldwork seminars and interviews are reported. Then a step-by-step methodology is presented to illustrate the different stages of the DSS model development. Finally, the application of the prototype DSS for selecting appropriate floor material for a residential project in the London Borough of Sutton is demonstrated. The final section concludes the study and suggests areas for further research.

TECHNOLOGY IN MATERIAL SELECTION: REVIEW

For the past ten years, proven and commercialized technologies have been developed to promote environmental awareness amongst built environment professionals [18,22,23]. Empirical research validates that various studies on building material selection support systems have developed in size and specification within the last ten years [24-26]. Castro-Lacouture et al. [26] note that the application of green building support/assessment tools has been widely accepted as an effective and useful way of promoting green housing construction in the housing construction industry. Keysar and Pearce [27] and Bayer et al. [28] however, argue that the contexts in which building environmental assessment methods now operate, and the roles that they are increasingly playing, are qualitatively different than earlier expectations. They note that material assessment tools are now classified based on the type of analysis they perform, such as product, assembly, or whole building analysis, or classified as region-specific tools, either considered based on the life-cycle phases they cover, or on the required skills necessary to operate the tool. While there is clearly an urgent need for new technologies to optimise the use of low-cost green building materials, it is also true that there are many technologies or systems already in use [25-28]. The first real attempt to establish a comprehensive means of simultaneously assessing a broad range of environmental considerations in building materials was the Building Research Establishment Environmental Assessment Method (BREEAM) [28].

The BREEAM tool assesses the environmental impacts of over 150 various materials and components most commonly used in home construction. The tool takes environmental issues into account, then adds measurements and user-defined weighting to arrive at environmental impacts, measured as "Eco-points" for each building material being assessed. Twelve different environmental impacts are individually scored, together with an overall summary rating, which enables users to select materials and components according to overall environmental performance over the life of the building. This scientifically accepted program however, focuses only on the environmental performance of products rather than environmental, social and financial considerations going hand in hand as parts of the material evaluation and selection process. With emphasis on the Leadership in Energy and Environmental Design tool (LEED), Keysar and Pearce [27] conducted a detailed evaluative study comparing the effectiveness of five different relative importance indices for selecting appropriate material selection tools such as: relative advantage; compatibility; complexity; trialability; and observability, with the goal of improving the sustainability of materials for capital projects.

Here, materials such as; regionally manufactured materials, materials with recycled content, rapidly renewable materials, salvaged materials, and sustainably forested wood products are selected based on credit scores. Analyses of their study however, revealed that the LEED model for example specifically requires an energy model, a task often handled by a specialist within a design firm or outsourced to a third party specializing in energy modeling. Due to the inflexibility inherent in the application of first generation tools, and since they tend to require greater technical expertise to implement, many different tools of the second generation group have also been launched to address these limitations. Among this category is the ATHENA estimator. This has been one of the most popularly used material data-analytic models that analyses over 1200 building material and assembly combinations [28]. It allows the users to look at the life cycle environmental effects of a complete structure or of individual assemblies and to experiment with alternative designs and different material mixes to arrive at the best scenario. Bayer et al. [28] noted that the major drawbacks to this tool are the fixed assembly dimensions, software cost, the cost and required skills to use it, the limited options of designing high-performance assemblies, and the overall incomplete assessment of whole buildings environmental impacts [28-30].

With the identified setback associated with ATHENAestimator, The National Institute Standards and Technology (NIST) developed the Building for Environmental and Economic Sustainability (BEES®) 4.0. This model provides a cradle-to-grave product-to-product comparison of over 230 building products based on manufacturer and supply company information [28-30]. The impact categories are weighed, normalized, and merged into a final environmental performance score, to generate a single measure of desirability for product alternatives by combining qualitative and quantitative data. The BEES 4.0 model is however, not capable of providing data for a full LCA of a complete building product, as it only produces data for a limited amount of building materials and evaluative factors [28-31]. These singleattribute claims ignore the possibility that other life-cycle stages or environmental impacts can yield offsetting impacts. Other limitations include; limited product options, limited use for local/regional impact materials and devaluating weighing process [17]. Trusty [18-21] argued that these sets of first and second-generation tools less often consider any of the MultiCriteria Decision Methods available to solve MCDM problems, adding that some systems do not even consider Life Cycle Cost (LCC) and other performance criteria simultaneously or completely.

Moreover, he claimed that the existing performance requirements/criteria approach used in such tools tend to rely on immeasurable characteristics in demonstrating the extent of sustainability in a product, which makes them over-burdensome to implement and communicate. Since the highlighted material assessment tools were developed primarily to be used in different countries, and the data sources used by each tool differed, further efforts have been undertaken to develop knowledge-based or expert DSS for assistance in material selection. For instance, Rahman et al. [32,33] developed an integrated knowledge-based cost model for optimizing the selection of materials and technology for residential housing design using Technique of ranking Preferences by Similarity to the Ideal Solution (TOPSIS). The system is developed to assist architects, design teams, quantity surveyors and self house builders to make decisions for the design from early stage to detailed design stage by ranking the performance and cost criteria of technologies and materials. Loh et al. [34] however, criticised the tool for providing partial assistance in the material selection process of the whole building design as it only considers the cost of roofing materials. They argue that material selection process depends on a number of other factors such as the location, zoning and environmental regulations, demographic characteristics, etc. that are not considered in their system. They note that the TOPSIS approach adopted does not only lack the ability to eliminate bias in the selection process but also unable to allow fairer trade-off process. Loh et al. [34] emphasise that strategic selection of sustainable materials and building design prior to the building construction is crucial to increasing building life cycle energy performance. They argue that stakeholders involved in the early design process often have conflicting priorities for both building design and construction materials. They developed an environmentally focused decision support system in the form of an Environmental Assessment Trade-off Tool (EATT), which supports the development of the ideal building design and materials combination that meets stakeholders' requirements. It is designed to assist users select the most appropriate material among a set of candidate materials based on the analytical hierarchy process (AHP) concept of decisionmaking, since AHP technique has the robust ability to handle the complexities of real world problems, and to deal formally with judgment error, which is distinctive of the AHP method.

The system rank orders a set of preselected, technically feasible materials using different decision factors with and without tangible values, such as a clients favour over a particular building design, publicity potential of the building design, life cycle cost, capital cost and energy performance of different materials and building layouts. Zhou et al. [12] argued that the approach adopted by Loh et al. [34] lacked in robustness as it does not take into account the full-life cycle impacts of newly-accepted building products, and did not specify the sort of materials under studied. Zhou et al. [12] developed a decision support multiobjective optimization model for sustainable material selection. The material selection tools and material data sheets provide extensive information that includes factors such as cost, mechanical properties, process performance and environmental impact throughout the life cycle based on expert knowledge. Wastiels et al. [16], confirmed that the tool, however, lack the considerations or descriptions to evaluate the intangible aspects of building materials, which are important to architects. They also criticised the selection methodology for being highly restrictive to a limited range of factors and incompatible with other stakeholders. Ashby and Johnson (2002) introduce "aesthetic attributes" in the material properties list for product designers when describing material aspects such as the transparency, warmth, or softness. Within the discipline of architecture, however, the intangible qualities of materials are not described and mapped within the current design models.

No selection framework was provided to support the implementation of a system. Wastiels et al. [16], proposes a qualitative and quantitative framework to support informed decisions based on physical aspects' and "sensorial aspects" of building materials, but without the tools integration and computerisa-tion as done by Zhou et al. [12]. In the presented framework, no pronouncement is made upon how sustainable considerations from these different categories could influence each other, and what MCDM approach could possibly be used if developed. A similar study by Ding [35], developed a comprehensive assessment decision support system that measures the environmental characteristics of a building product using a common and verifiable set of criteria and targets for building owners and designers to achieve higher environmental standards. Upon analysis it was found that the assessment for her study focused heavily on environmental issues rather than the broader social, cultural, technical and economic aspects of sustainable green construction. Keysar & Pearce [27] cited extensive research literature describing how material selection tools facilitate the innovation diffusion process and radical decision-making transformation. They however, note that most of the examined models make choices that result in "fabricated assemblies of standardized performance attributes", implying that they do not choose for materials but rather for 'material systems'. Hopfe et al. [36] conducted a study that assessed the features and capabilities of six software tools to screen the limits and opportunities for using BPS tools during early design phases. The tools classification was based on six criteria namely the capabilities, geometric modeling, defaulting, calculation process, limitation and optimization. However, the authors did not report what methodology was used to compile these criteria. A cost modeling system for roofing material selection was further proposed in Perera and Fernando [37].

Several factors were identified and considered in the selection process. Results demonstrated large inconsistency in the evaluation process. No particular reference was made to the selection methodology. Other influencing reviews within the scope of this study include Mohamed and Celik [38] who proposed a computerised framework that is responsible for materials selection and cost estimating for residential buildings where users are able to choose their preferred one from list of materials without evaluation and synthesis of multiple design criteria and client requirements. No mention was made about the MCDM technique used for evaluating the list of materials selected and their respective quantities. Mahmoud et al. [39] suggested a method for the selection of finishing materials that covered floors, walls and ceilings and integrates cost analysis at the appropriate decision points, but without the selection information requirements or methodology as proposed in this study. Lam et al. [40] carried out a survey on the usage of performance-based building simulation tools. His study examined the relative impacts and limitations of knowledge-base tools in decision-making. Murray argues that while there is a natural tendency for design and building professionals to focus on the scientific and technological aspects of green and sustainable construction, their approach does not necessarily maximise the positive contributions professionals have to offer if tools are designed to replace professional judgment in the choice of materials. Murray suggests that this is because tools cannot address the intrinsic motivations people need if they are to embrace the positive changes sustainability requires. He continues that limiting the assembly of buildings to the specification of systems would impede the discovery of design opportunities inherent in materials themselves. Similar patterns of consistency, and lack thereof, have also been obtained [for detailed reviews see 17,24-27,31]. By highlighting the different green building material assessment tools, it can be deduced that existing tools are dispersed and based on individual initiatives without a unified consensus based framework [41,42]. It is apparent that each tool has its own unique application. While each tool could be called an LCA tool, there was little consistency in the methodologies used from one tool to another. In addition, while one tool considered the building as a system, other tools considered primarily the product's individual attributes rather than how that specific product performed within the building system [42]. A key question therefore, is whether current assessment methods that were conceived and created to specifically evaluate the environmental merits of conventional building materials can be easily transformed to account for a qualitatively different set of materials. Giorgetti & Lovell [43] for instance have reported the sub-optimal performance of existing tools.

They argued that the subjective values and priorities of the authors of the assessment scheme largely dictate the technical characteristics of the systems, and currently represent the major focus of discussion. They suggest that it is necessary for potential users to analyse the local situation and identify the adaptability of using any tool before applying a universal green building assessment tool to a specific country and region. They warned that some existing tools such as BREEAM, LEED, and even current expert tools might potentially institutionalize a limited definition of environmentally responsible building practice at a time when exploration and innovation should be encouraged in another region. However, in all the reviewed studies, no efforts to develop a DSS that associates with the corresponding attributes and performance characteristics of low-cost green building materials and components, starting from the broad list of available options in the database to the final selection of the most appropriate material, were found in the existing literature [43,44].

The findings of the review have shown that each of the indices applied in developed regions to deal with issues associated with the impacts and performance of low-cost green building materials in other regions have proven unsatisfactory [44,45]. This finding is premised on the fact that most existing material selection systems have been designed by countries with more developed economies such as the UK, where the scale of social issues and lack of access to resources is simply not as critical as observed in the developing nations [45,46]. The setbacks that associates with the tools reviewed in this research thus, highlights the opportunity for developing a Material Selection Decision Support System (MSDSS), to better address the specific needs and attributes specific to the use of low-cost green materials for tool adopters new to green housing. The following section briefly highlights the aim and objectives of the study. It extensively describes specific methods adopted for each task in Section 3.1.

RESEARCH METHODOLOGY

In order to identify the key selection factors or variables that formed the basis for the development of the prototype multi-criteria decision support system (DSS), suitable clusters of research approaches were considered in the research exercise, some of which include: exploratory literature reviews, networking with domain experts and practitioners, series of questionnaire surveys and knowledge-mining interviews [47]. Table 1 provides an overview of the research aim, objectives and the methodology undertaken in four major stages.

Research Design

To provide a clear theoretical framework for the relatively new area of study, and develop preliminary ideas on issues specific to the research theme within the context of decision-making associated with the impacts of low-cost green building materials and components in housing construction, this study reviewed relevant literature through synthesis and analysis of recently published data, using a range of information collection tools such as; books, and peer-reviewed journals from libraries and internet-based sources. Recognising the limitations of the literature review in terms of examining current research thinking in respect of decision support systems for the selection of low cost green building materials and components, a preliminary research study was undertaken to check and validate prior assumptions in the background and review sections. In order to build upon knowledge gained from the literature review, and recognising the limitations of the preliminary research survey in terms of examining current research thinking in respect of decision support systems for low cost green building materials and components, a mixed method was adopted for this study.

This was followed by in-person interviews to further clarify and elaborate on less detailed and pertinent issues associated with the use low-cost green building materials. The in-depth interviews consisted of 10 participants, who involved a sample of practicing architects, engineers, material specifiers, and a host of building professionals-who influence material choice decisions in the UK housing construction industry. This approach was used to examine the potentials of the proposed MSDSS, (being a tool for the assessment and evaluation of low-cost green materials). It further investigated the effectiveness of design and decision support tools, as well as identified requirements of Life Cycle Assessment (LCA) tools for design decisions at the various stages of the design process. Consequently, a quantitative questionnaire was developed as the result of the analysis of the results from the interviews.

In order to elicit the "most important" factors, a questionnaire survey was conducted among the executives of some selected builder/developer firms. They were asked to rank order from a list of factors (compiled from existing literature on the topic and after initial consultation with some of the executives) based on their judgment and experience. The executives were also asked to indicate desired features they would like to have in a DSS for low-cost green material selection. Since the respondents were widely dispersed, and because it was anticipated that building professionals would be more likely to reply and cooperate with a less time-consuming research method, giving the constraints of time, wider coverage, and budget, it was therefore, decided that a questionnaire sent and returned by email would be the most convenient way of collecting the required data.

The inclusion of qualitative open-ended questions provided respondents a chance to express their views more freely. The target groups of respondents were also taken from a database or directory of building professionals provided by the UK, China, Canada, South Africa, Brazil and US Green Building Councils (GBCs). The selection approach followed the random sampling technique to avoid bias and uneven sample sizes amongst different professional groups, and ensure uniformity, consistency and quality of data. To facilitate the response rate, snowball sampling was also adopted, where the approached respondents were asked to distribute the questionnaire to their colleagues and partners within the field [47]. The selection of South Africa and Brazil for the analysis was due largely to their great similarities in social, economic, and geopolitical terms, and likewise their developed counterparts. In a similar vein, the choice of building experts within the selected countries was as a result of their expertise and advancement in the use anddevelopment of green building tools (as they have had the most uptakes in both geographical regions and being part of an emerging market).

Table 1: Basic summary of the research methods

AIM	To develop a decision support system (DSS) that will provide designers with useful and explicit information associated with low-cost green building materials and components, to aid informed decision-making in their choice of materials for low-cost green residential housing projects.		
Stage	Objectives	Tasks	Method
1: REVIEW	1.Examine current views on themes related to decision-making associated with the use of low cost green materials in the housing industry, to identify new ideas & issues arising from the study	**Step 1.** Reviewed relevant literature through synthesis and analysis of recently published data, using a range of information collection tools such as; books, peer-reviewed journals, and articles from libraries and internet base sources	AA,
	2. Review various DSSs currently used at national and international levels for the selection of materials to identify knowledge deficits and the potential benefits associated with their use	**Step 2.** Carried out a preliminary research study with leading researchers who influence the selection of building materials in the field of housing construction	AA, QS, INT
		Step 3. Conducted a pilot study, by deploying a test-questionnaire to a small sample of researchers who possess relevant knowledge on issues specific to the use of low cost green materials using the email addresses taken from the databases of recognised building construction companies and research institutions	
2: DATA COLLECTION &SYNTHESIS	3. Conduct surveys and interviews with building professionals, to identify the potential factors or variables that influence the informed selection of low cost green building materials and components	**Step 4.** Conducted the main survey, by administering the revised questionnaire through email contacts taken from databases of interested registered building professional groups, who influence the selection of construction materials from throughout the construction value chain	AA, QS, INT
		Step 5. Conducted in-person interviews with interested building professionals who influence material choice decision in housing construction using audio recording system to avoid re-contacting the respondents or falsification of information	
		Step 6. Carried out inspection on available expert systems most commonly used in building firms in the UK, USA, China etc. by interviewing experts, with years of experience in the industry, who have implemented or used such systems and directly observing how they function when in operation	
3: DATA ANALYSIS	4. Evaluate and establish the weighted importance of the key factors or variables that will help to determine the relative impacts of the different choices of building materials and components	**Step 7.** Analysed the information and report gathered from the survey exercise(s) using a suite of statistical analytical programs, and various quantitative data analytical techniques	AA, QS M
	5. Develop a system to integrate the necessary information appropriate to the informed selection of low-cost green building materials & components	**Step 8.** Assembled the key components by synthesising the relevant databases to be incorporated in developing the proposed DSS model.	AA, QS, M
		Step 9. Developed the main structure workflow of the proposed system by creating links among the various databases,	
		Step 10. Inputted relevant data to test the internal links to know what needed to be measured within the system, and checking the output of the results against easily calculated values	M
4: DEVELOPMENT	6. Test the functionality of the proposed approach; and validate the effectiveness by applying it to a building material selection problems using a series of case study residential building projects in the UK	**Step 11.** Conducted experts survey by deploying a sample of the prototype system via email of those who participated in the main survey, using feedback questionnaires as a quicker and cost effective means of assessing respondents' judgments about the system	QS
		Step 12. Made necessary changes based on the feedback from the survey	M
		Step 13. Validate the modified prototype system using a series of completed building projects in the UK, by comparing the outputs from the algorithms to monitored data from the completed building	M, CS

KEYS: AA (Archival analysis); INT (Interview); CS (Case study); QS (Questionnaire Survey); M (Modeling).

To receive a reasonably sized sample, 500 surveys were sent out by email, over a two-month period of March and April 2012. Using a progressive approach of data collection, a total of 250 respondents returned the completed survey, representing a response rate of 50%. The response rate was accepted as the normal ranges between 20% - 30% were found in most of the construction industry related research [33,34]. Prior to distribution, the questionnaire was pre-tested for comprehensibility by consulting five academics at two universities [47]. A number of changes were suggested and implemented.

Respondents were also invited to post their ideas about current limitations or improvements that should be avoided or integrated in the development of the proposed MSDSS model at the later part of the questionnaire. The questionnaire also examined the adequacy/inadequacy between traditional manual approach of material selection and computer-aided decision support tools. One of the group's participants commented that one of the hallmarks of good science is that a result can be tested independently and proven to be right or wrong in the latter method. The analysis of the questionnaire survey and interviews provided a list of "key" decision-related factors having significant impacts on the process of material selection for residential development as shown in Section 4.1.1.

Research Findings

The results of the study however, revealed the following.

- Many existing decision support systems in the developed countries do not have the appropriate performance threshold for addressing the most relevant issues specific to less developed countries;

- Current DSS models are unable to relate to matters associated with the informed selection of materials that are commonly used for housing projects in countries with rather less-mature markets;

- The lack of informed knowledge by building professionals in terms of the principles, characteristics, and best practices relevant to the use of low-cost green materials at the design stage, has been identified as a common constraint peculiar to their wider-scale use in the housing industry;

- The majority of building professionals still regard cost and environmental factors as conventional project priorities when selecting building materials or components, but rarely consider the implications of social, political, technical, sensorial, legal and cultural factors in their choice of materials; and finally,

- The majority of low-cost green building materials are yet to be certified under the building regulations, standard specifications and codes of practice; and most importantly,
- There are no demonstrable and compelling evidence of technical research on a holistic approach used by design professionals for the evaluation and selection of low cost green building materials and components at the design stage.

The results of the study thus, provided the platform that suggested the need for a system that could aid informed decision-making to improve understanding, and enhance the effectiveness of actions to implement and promote the wider-scale use of low-cost green building materials and components at the core of the construction business process. In light of their feedback and useful suggestions from building experts who partook in the study, the following portions of the DSS model were either readjusted or improved.

- Easy searchable material selection inputs database;
- Ability to add/remove material selection features with ease;
- Ability to make custom reports;
- Ability to easily navigate all components with ease;
- Comprehensive "HELP or USER INSTRUCTIONS" menu explaining what the tool is doing;
- Being able to understand the material selection process through the lens of non experts;
- Ability to perform trade-off analysis to compare different material options;
- Clarity on the algorithms used to perform the simulations; and Real-time results;
- Data input forms to ensure easy and consistent data input; and,
- Having a huge amount of customizability in terms of output.

After the improvement, the system was shown to the same participants, and minor adjustments were made on the basis of second feedback. In the following sections the proposed MSDSS selection methodology is discussed, and a conceptual framework for the decision support system based on the methodology is presented. Subsequently, the MSDSS model is applied to a hypothetical but realistic material selection problem to rank order the candidate materials for selecting the most appropriate one.

SYSTEM DEVELOPMENT

For this research, AHP was selected for its simplicity and due to the fact that it can be easily implemented using any spreadsheet software application such as the MS Excel, as it possesses a powerful macro language that is essential since a menu driven interface had to be developed. Since the intention of the research was not to develop a commercial software product, Macro-in-Excel VBA (MEVBA) was utilized for the following reasons:

- Macro-in-Excel VBA (MEVBA) has the capabilities to perform all necessary calculations and is common enough that most people are familiar with it;

- It has the ability to write scripts that could automatically convert material data from any graphic table format to an appropriate condensed data table (hidden from the user's view) to allow quick and reliable indexing of material data;

- The Macro-in-Excel VBA framework has the code that makes Windows forms work, so any language can use the built-in code in order to create and use standard Windows forms;

- Makes the application easier to maintain; With MEVBA, codes were easily built into the form or report's definition, since the DSS model contained a large number of macros that respond to events on forms and reports; which would have been difficult to maintain using any other application;

- With Macro-in-Excel VBA it was easy to step through a set of records one record at a time and perform an operation on each record;

- Macro-in-Excel VBA helped to supply a standard security mechanism, which was made available to all parts of the MSDSS data application model;

- Enables the developer to create his own functions: The MSDSS contains a series of mathematical model and computational algorithmic procedures that provided a basis for computing the green development index of material alternatives within an integrated decision-support framework or tool(s).

- Ability to mask error messages during the tests run;

- Enables the system to quickly analyze existing data to discover trends so that predictions and forecasts can be made with reasonable accuracy;

- Allows for extensions and expansions: since the components of the framework are modular, meaning that each may be developed independently, and data may be added as it is acquired to supplement the knowledge and databases, macro-in-excel was used to achieve that goal

MSDSS Database/Data Warehouse Design

The data warehouse design constitutes the major portion of the MSDSS development and hence will be explained in detail in this section. The data warehouse design essentially consists of four steps as follows: Step 1: Identifying the key influential factors that will impact on the choice of materials;

Step 2: Designing the material selection methodology framework and identifying the objectives of each step;

Step 3: Designing the various components of the MSDSS model and defining their features and functions;

Step 4: Defining the workflow selection methodology and analytical procedure of the actual prototype MSDSS model.

Identifying the Key Influential Factors

In order to identify the relative importance of the subcategorical factors or variables based on the survey data, ranking analysis was performed. Five important levels were transformed from Relative Index values: Highly Significant Level (H) ($0.8 \leq RI \leq 1$), High-Medium Level (H–M) ($0.6 \leq RI < 0.8$), Medium Level (M) ($0.4 \leq RI < 0.6$), Medium-Low Level (M–L) ($0.2 \leq RI < 0.4$), and Low Level (L) ($0 \leq RI < 0.2$).

From the results of the analysis, 40 factors were identified under the "Highly significant" level for evaluating low-cost green building materials with an RI value ranging from 0.952 to 0.806 and a total of 15 factors, were recorded to have "High-Medium" importance levels with an RI value ranging from 0.795 to 0.652. The analysis of the main survey identified a total of 55 key influential factors out of 60 initial factors as important components of the material selection process. "Life Expectancy" was ranked as the first priority in the technical category with an RI value of 0.952, and it was also the highest among all factors and was highlighted at "High" importance level. "Resistance to fire" was also rated high in importance among the selection factors. "Maintenance Cost" was ranked third in importance. It was clear from this research that there is a perception of ambiguity surrounding the long-term maintenance of low-cost green building materials.

This is not entirely any surprise given that maintenance free buildings are increasingly sought after by clients, anxious to minimise the running costs associated with buildings.

"Life Expectancy" was ranked as the first priority in the technical category with an RI value of 0.952, and it was also the highest among all factors and was highlighted at "High" importance level. "Resistance to fire" was also rated high in importance among the selection factors. "Maintenance Cost" was ranked third in importance. It was clear from this research that there is a perception of ambiguity surrounding the long-term maintenance of low-cost green building materials. This is not entirely any surprise given that maintenance free buildings are increasingly sought after by clients, anxious to minimise the running costs associated with buildings. "Life-cycle cost" has been, and will continue to be, major concerns for building designers, as well as important traditional performance measure. Among the top 20 ranking factors, it was observed that only one factor from the environmental category out of the list was ranked high among the selection factors.

This again suggests that environmental issues within the context of the developing countries are not strongly considered despite the high environmental awareness exhibited by design and building professionals in developed regions. This finding also corroborates the initial observations of various studies [14,15] repeatedly highlighted in the background and literature studies. They suggest that the problems within the developing regions are characterised by mainly social and economic issues, unlike the developed regions where the scale of social issues and lack of access to basic resources are simply not much of a problem as it is in the developing world. From Figure 1, a total of 15 factors, consisting of 12 site factors, 1 socio-cultural factor, and 2 sensorial factors, were recorded to have "High-Medium" importance levels. Although these 15 variables were in the same importance level category, the "building orientation" factor within the "general/site category" (average RI = 0.652) was considered to be the least important variable compared to the factor "Glossiness" under the "sensorial category" (with an average RI = 0.774), and "material availability" still under the "general/site category" (with an average RI = 0.795). However, it should be noted that site factor accounted for 75% in the "High-Medium" importance level. The result is an example of evidence pointing to the trend that environmental and perhaps site issues are no longer considered as the most important factors for material selection in housing projects, especially within the context of the less developed regions. Some factors in the three categories were ranked relatively higher in the "High-Medium" level. For example, "material availability (GS1)" was rated as first in the general/site subcategory, and ranked as thirty-fifth in the overall ranking with an RI value of 0.795. An interesting observation from the results is that none of the criteria fell under the medium and other lower importance level. This clearly shows how important the factors are to building designers in evaluating low-cost green building materials. All factors were rated with "High" or "HighMedium" importance levels. However factors such as Compatibility with other materials, Skills availability, and UV resistance fell within the medium-low level. The findings of the analysis asserted that the criteria with medium or low RI does not mean they are not important for selecting materials, but rather created an opportunity to highlight the relative importance of the key criteria from their vantage points. The following shows a framework consisting of the key factors in their order of importance.

Designing the MSDSS Selection Methodology

The diagram shown in Figure 2 demonstrates the conceptual framework of the selection methodology for the decision support system. Table 2 describes a step-bystep procedure of the selection methodology for the material selection decision support system. Section 4.4 presents various components of the MSDSS schema or model.

Designing the Features of the MSDSS Model

The next stage of the model development was to design the various features of the databases containing the logic and showing relationships between the data organized in different modules. Each module contains the physical information and contents needed to aid in the material evaluation and selection process.

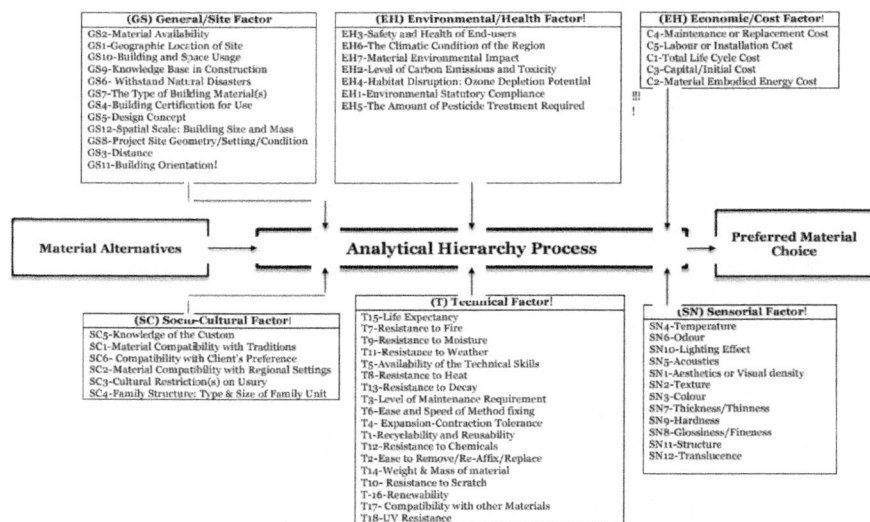

Figure 1: Ranked factors for measuring the impacts of low-cost green building materials.

Table 2: Description of the selection methodoly

OBJECTIVE	TASK
1. Define or state overall objective/goal	The first step of the methodology is to define the main goal of the intended task.
2. Identify Set of all Possible Material Alternatives to be Assessed	After defining the main goal of the task, the next step is to generate the set of all possible alternatives that are available for selection based on the decision-making parameters. In the material selection process, this comprehensive set of alternatives includes all construction materials and components currently in the database, and the market in context.
3. Prune all infeasible alternatives from set	The third step is to reduce the complete set of alternatives by eliminating/pruning those alternatives, which are clearly infeasible for the intended application from the database consisting of all materials, based on classifications of materials according to the Construction Standards Institute (CSI) Divisions, and material heuristics. For example, if the element under consideration is a structural beam, materials such as roofing sheet and glass are automatically pruned from the set of possible alternatives under consideration, since none of these materials fall under the CSI structural divisions. This should result in a subset of alternatives, all of which would be feasible choices for the intended application. The "pruning" approach is used rather than allowing the user to select feasible materials from the whole set because users tend to overlook alternatives which might be unfamiliar to them but are nonetheless feasible.
4. Evaluate Remaining Alternatives	The fourth step in the methodology is to evaluate the feasible alternatives using the AHP model such that a ranking can be developed according to the relative importance of the material for the intended application.
• Weight Attributes (Decision Factors)	• First, the decision maker weights each factor or variable according to the relative importance that the decision factor or variable holds for the decision maker. It involves the decision-maker replacing probabilities with user weightings for each factor or variable to supplement, not replace, his judgment.
	• Second, values for each of the factors or variables are determined for each material with regard to the manufacturer's information & details of the material or component contained in the material database, and then, a normalized value between zero and one is calculated for each factor value.
• Calculate Values for Attributes	• After weights have been established and values calculated for each attribute against a set of materials or components, the weights and normalized values are multiplied and summed to create an index of preference for that alternative(s).
• Amalgamate Weighted Attributes	
• Develop Ranking	• Then, a list of alternatives ranked according to the relative importance of the factors or variables is then presented.
5. Review Ranking of Alternatives	When the indices of factors or variables have been calculated for all feasible alternatives, a ranking is developed sorting the alternatives according to each utility value based on the AHP model of decision-making. The alternative with the highest utility value is recommended from the ranked list of potential materials for each design/building element.
6. Select Alternative Based on Ranking	The decision maker may then either elect/decide to select the highest ranked alternative, or choose another alternative from the set based on his professional judgment.
7. Proceed to Next Design Elements	The decision maker satisfied with the selection process, then proceeds to the next design/building element.

The conceptual model/framework of the prototype MSDSS tool consists of a number of interconnected modules/features. A logical model illustrating the developed DSS for material selection is shown in Figure 3. Table 3 describes the functions of each component of the MSDSS model.

How the System Works

The following steps explain how the prototype MSDSS model works during the material evaluation process. Step 1: The load manager provides the user with a list of design elements from the "Design Elements" module, and then prompts the user to select the design element of his/her choice in accordance with the terms and specifications of the Construction Standards Institute (CSI) Divisions;

Step 2: The User then selects the particular design element needed for the intended task from a list of design elements (as broken down by the Construction Standard Institute Division);

Step 3: User then enters values for the relevant parameters to answer prompts about areas and dimensions of the selected design element, and then sets the threshold values in the material knowledge base

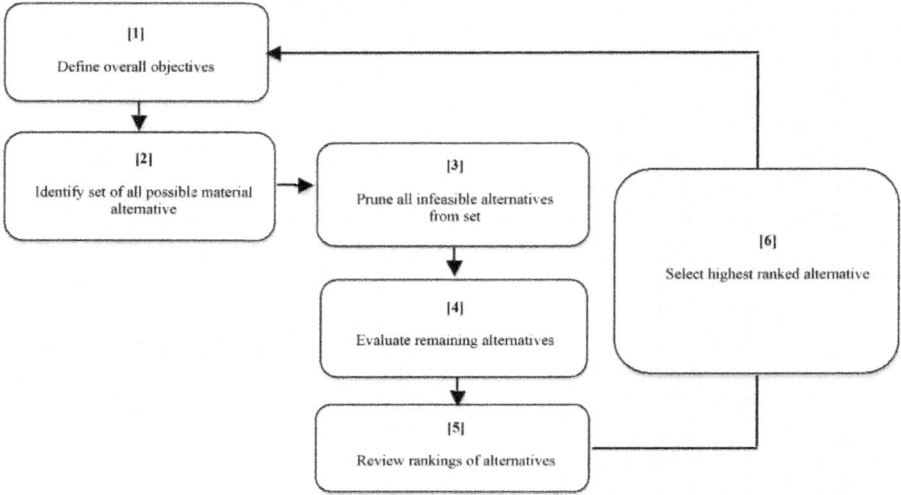

Figure 2: Selection methodology for the MSDSS model.

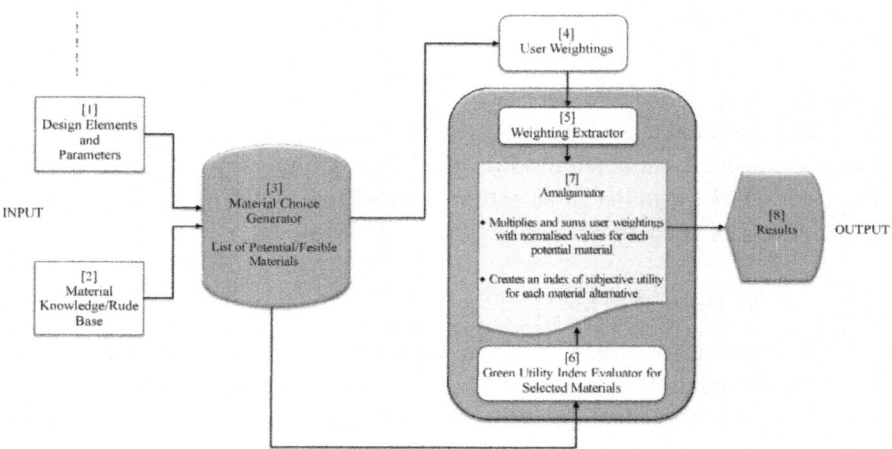

Figure 3: Conceptual framework of the MSDSS model.

Step 4: The system validates the design parameters and threshold details entered by the user, and then generates the set of all feasible material alternatives that are available for selection, (which includes all categories of construction materials contained in the materials database);

Step 5: After a set of feasible material alternatives has been generated for the "particular design element", the system through the "Weighting Score Extractor Module" prompts the user to obtain weightings for the desired parent and sub-factors according to the relative importance that each factor or variable holds over another based on the decision maker's preference of value;

Step 6: After weights have been established and values calculated for each factor for a particular material, the weights and normalized values are multiplied and summed to create an index of subjective utility for each alternative;

Step 7: The alternative with the highest utility value is recommended by the system;

Step 8: The user reviews the system's recommended choice for each element in the "Result" module, and then either selects the highest ranked alternative, or chooses another alternative from the set based on professional judgment and/or the system's recommendation.

Step 9: The user may choose to generate a printout report or graphical representation of the list of selected materials and green utility indices if desired.

Step 10: The selection process then proceeds to the next design element.

Figure 4 presents a graphical representation of the system workflow. An illustrative example of the AHP concept is displayed and explained in following section to demonstrate the selection process by applying the prototype MSDSS model to a hypothetical case study design project.

APPLICATION

The following example illustrates the selection process of floor covering products. It selects the best one among three alternatives. The prototype MSDSS, developed using the AHP technique, was used to select the most appropriate residential building floor material for housing development in the city of London, located in the Sutton County of London. The results demonstrate the capabilities of the MSDSS system in a real-life but hypothetical application scenario. In the following section this process of application is described and discussed.

A Hypothetical Study Case

The next stage of the model development was to design the various features of the databases containing the logic and showing relationships between the data organized in different modules. Each module contains the physical information and contents needed to aid in the material evaluation and selection process. Table 4 summarizes the details for the three options of flooring materials for the proposed residential low-cost green housing project. The description of the three options in Table 4 was based on the standard practices and construction details commonly used in the housing construction industry. These three (3) floor materials described above will be analysed amongst a host of other material alternatives for the selection of a more sustainable option. In other words, this section will analyse the problem using the MSDSS model, which relies on the use of the AHP mathematical multi-criteria decision-making technique, to identify and decide which material is the most sustainable and suitable flooring material in this case. To achieve this goal, the MSDSS model was sent to 10 expert evaluators who had the following qualities:

- Considerable amount of knowledge in material analysis based on the AHP concept;
- Used a wide range of green building assessment tools for material selection; and
- Taken part in the previous survey.

The aim of this exercise was to compare their view of the prototype MSDSS model with existing models in terms of their usability, flexibility, and interoperability attributes using the concept of the Analytical Hierarchy Process (AHP).

Table 3: Functions of the features of the MSDSS model

MSDSS Features	Functions
1. Design Elements and Parameters	This feature provides users with a range of building design elements and their respective parameters
2. Material Rule Base	This feature articulates the listing of individual materials in prescribed sequences, gradually eliminating candidate materials based on their inability to meet stated material selection heuristics/rules.
3. Material Choice Generator	This feature contains the material/component database, which generates the set of all possible material alternatives that are available for selection.
4. User's Weightings	Sets preferred weighting value for all attributes to compare with.
5. Weighting Extractor	This feature queries the user to obtain weightings for the factors, based on the user's preference of value on a scale of 1 - 9.
6. Material Index Evaluator	The material index evaluator calculates values of the selected factors or variables for each feasible material choice
7. Amalgamator	Here the user's weightings are amalgamated (*i.e.* multiplied and summed) with the factor values or weightings for each potential material, resulting in a relative ranking of the feasible materials for each element.
8. Results	- This component provides the ability to view the processed data, and to generate reports. It allows the MSDSS model User Interface to communicate with the user; and also connects all the reports and queries that are generated in the Monitoring databases to the corresponding project files.

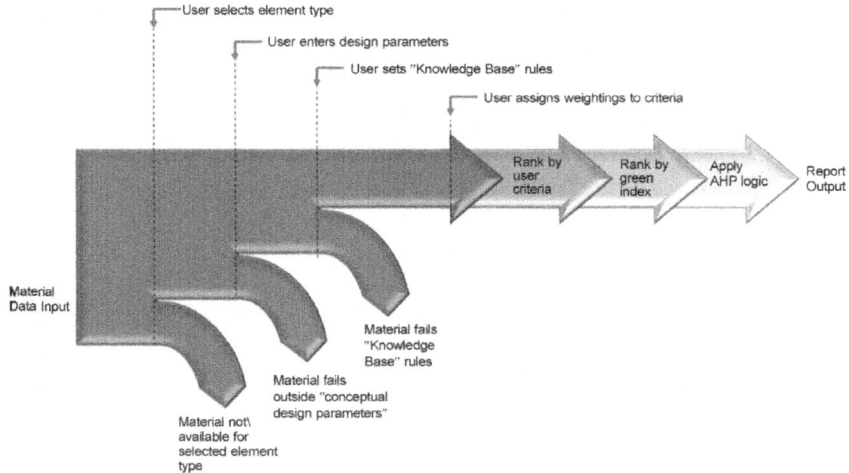

Figure 4: Workflow of the MSDSS model.

Table 4: Summary of the flooring options

Description	Material A	Material B	Material C
Design Element Type	Paneled Flooring	Laminated Flooring	Concrete Flooring
Building Type	Residential	Residential	Residential
Material Type	Bamboo XL laminated Split Paneled Flooring	Reclaimed/Recycled Laminated Wood Flooring and Paneling	Fly Ash Cement concrete Floor Slab
Size of Materials	230 mm × 150 mm	50 mm × 6000 mm	900 mm × 900 mm

Rationale for Adopting the AHP Concept

The study adopted the use of the AHP technique to investigate the interrelationships amongst various criteria and low-cost green material alternatives due to the following reasons:

- AHP is a method that is conceptually easy to use, and decisionally robust to handle the complexities of real world problems;
- It does not require the very strong assumption that the stakeholders make absolutely no errors in providing preference information;
- It has the ability to deal formally with judgment error, which is distinctive of the AHP method;
- The AHP method provides the objective mathematics to process the unavoidably subjective preference inherent in real- world evaluations;
- Possesses an inherent capability to handle qualitative and quantitative criteria important for sustainable material selection; and finally,

- Can enable all members of the evaluation team to visualize the problem systematically in terms of parent criteria and sub-criteria.

Figure 5 shows the flowchart of the material selection computational analysis technique based on the concept of the Analytical Hierarchy Process model. The following sections present details of the evaluation exercise.

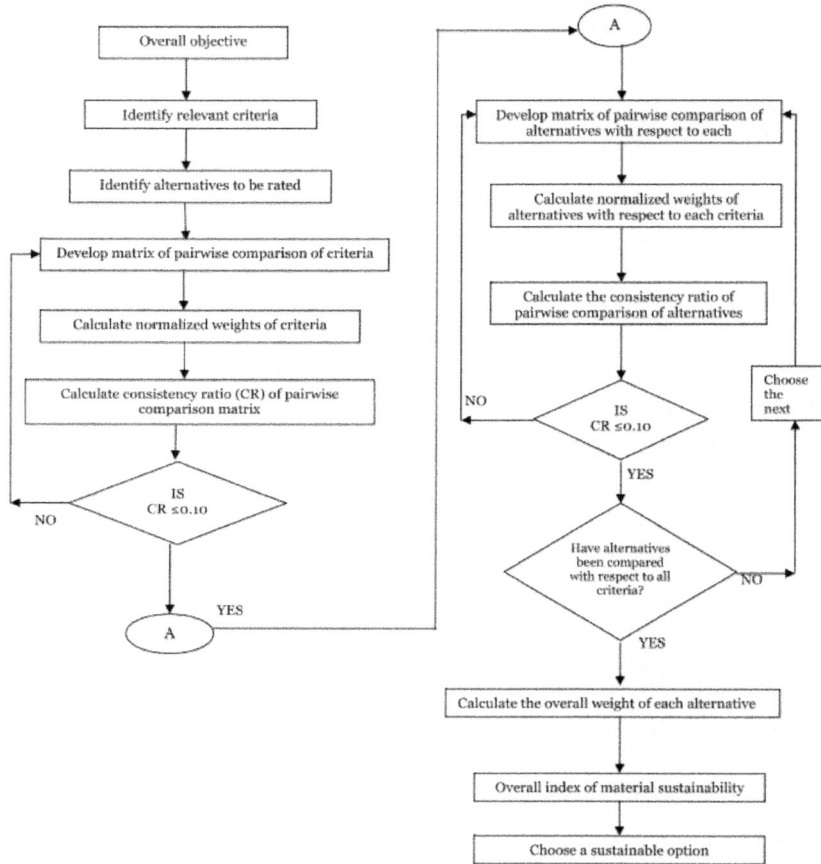

Figure 5: Flowchart of the AHP concept.

Applying the AHP Model to the Problem

According to Reza et al. 48., AHP is a subjective MCDM method that does not necessarily involve or rely on a large sample for its analysis. To better illustrate the procedure of the AHP technique of decision-making, with reference to the case presented in Section 5.1, a complete example of applying AHP to the problem of material selection is provided here based on evaluators' results.

Twenty (20) respondents representing various fields of the housing construction industry, and who had fore knowledge of the AHP procedure were selected to participate in the AHP survey.

By evaluating the consistency level of the collected questionnaires, 5 questionnaires out of the 10 received had acceptable consistency and were entered into the system. In order, to avoid arbitrary and inconsistent answers in the data, the mean values of five (5) out of the ten (10) respondents were used to fill out the pair-wise comparison matrices for the parent and sub-factors. The package included the model, evaluation questionnaire and a cover letter stating the purpose of the research, the validation process and what was expected of them. To conduct the exercise, the study adopted Chua's et al. [49]. approach based on a number of suggestions as follows:

- A document that reminded and explained the overall aim and objectives of the study to the respondents, followed by a step-by-step demonstration of its operation;

- A demo illustrating a practical exercise. This allowed the evaluators the experience of using the system ensued. During the practical assessment session of the demo, evaluators were able to see the controls and get a general overview of the MSDSS interface;

- An illustrative example of the objective and methodology of the AHP technique based on the instructions in the demo, to guide and illustrate to every respondent on how to browse and conduct analysis;

- After the introduction, a feedback questionnaire was forwarded to the evaluators;

- After each evaluation, each evaluator highlighted their experience(s) and provided feedback on the feel of the system, with special attention to the problems that they encountered during the evaluation process; \

- Finally, a reflective or post-user questionnaire was completed to obtain feedback;

- Evaluators were asked to answer each statement or question relating to the model in the questionnaire based on their personal view(s);

- They were also asked to assess the importance of the system based on their perception. Evaluators were also asked to add general comments on the system, and provide feedback on the applicability of the prototype system in assisting in specific material selection problems during their experience and other ways of improvement;

- Problems uncovered or areas that proved difficult to understand during the evaluation process were immediately modified so that it did not arise in subsequent sessions, as this procedure followed each evaluation;

- The respondents were instructed of the relevance of observing consistency in their answers whilst using the MSDSS model;
- The questions relating to different aspects were presented in different sections. This helped respondents to focus on one aspect at a time.

The following sections exemplify the process.

Decomposition of the Decision Problem

The evaluation exercise provided users with the opportunity to define the problem. Figure 6 shows the exemplary hierarchy of the problem. The goal is placed at the top of the hierarchy. The hierarchy descends from the more general or parent factors in the second level to subfactors in the third level to the alternatives at the bottom or fourth level as shown in Figure 6). To select a suitable choice among alternatives, the users were instructed to define the decision factors needed for the analysis. In other words, the users determined which alternative could be the best choice to meet the goal considering all the selected decision factors or criteria displayed in Figure 6. The first step of the methodology (as illustrated in figure 2) was to define the main goal of the intended task, by identifying the design element needed for the analysis, and inputting the relevant dimensional scale for the suggested design element (see Figure 7(a)). After defining the main goal of the task, the next step was to generate the set of all possible alternatives that were available for selection with reference to the decision-making parameters as shown in Figure 7(b). At this stage the users are prompted or alerted by the MSDSS model to identify a set of feasible floor material alternatives based on a range of material selection heuristics/ knowledge-based rules. The goal is to choose a suitable floor material among options for the project case described in Section 5.1.

Pair-Wise Comparison of Parent Factors

After selecting the design element, and identifying a set of feasible alternatives using the material selection heuristics/knowledge-based rules, the respondents were made to perform pair-wise comparisons following the demo instruction guide of the MSDSS model. This included the analysis of all the combinations of parent factors and sub-factors relationships. The sub-factors were compared according to their relative importance (based on the ratio scale proposed by Saaty [50-55], with respect to the parent element in the adjacent upper level. After performing all pair-wise comparisons by the decisionmakers, the individual judgments were aggregated, basing its analysis on the geometric mean technique as Saaty suggested [52-55].

Pair-Wise Analysis of the Parent Factors

To avoid arbitrary and inconsistent answers in the data obtained from the 10 participants who consented to partaking in the study, the mean values of five (5) out of the ten (10) respondents were used to fill out the pair-wise comparison matrices for both the parent and sub-factors. The pair-wise comparison matrices obtained from 5 respondents were combined using the geometric mean approach at each hierarchy level to obtain the corresponding consensus pair-wise comparison matrices [54-56]. Using the verbal/ratio scale shown in Figure 8, respondents obtained weightings for each parent factor, based on the preference of value(s) on a scale of 1 - 9. The MSDSS model then automatically translated each of the matrixes into the corresponding largest eigenvalue problem and was solved to find the normalised and unique priority weights for each factor (as shown in Figure 9). Going by Saaty's [55] rule, the judgment of a respondentis accepted if the Consistency Ratio (CR) ≤ 0.10. In cases were the results of the respondents were not consistent, the participants were alerted or prompted by the model to carefully re-evaluate the factors until consistency was achieved.

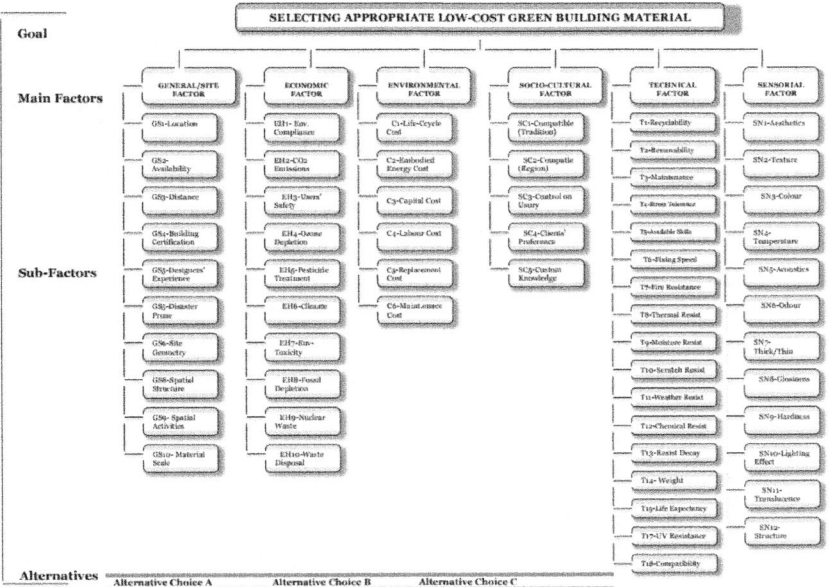

Figure 6: Hierararchy of the material selection phases.

Figures 9 and 10 represent the principal matrix of comparison, which contains the comparison between main/parent factors in relation to the overall objective of the problem (i.e., the selection of a sustainable low-cost green building floor material). From Figure 9, it is possible to observe that factor SC is 3 times more important than factor EH. As a logical consequence, factor EH is 3 times less important than factor SC. It is also possible to observe that the elements in the principal diagonal are always equal to 1. In other words, the weight of a criterion in relation to itself, obviously, is always 1. From Figure 9, it is also possible to observe that comparing Socio-cultural [SC] and Technical [T] factors, the participants slightly favoured Technical aspects of the products [T], thus arrived at an average value of two (2), derived from the mean calculation of the five respondents. Comparing Socio-cultural [SC] impacts with Sensorial [SN], participants somewhat considered Sociocultural [SC] as more relevant in their choice of materials than the emotive or sensorial [SN] aspects of the products, thus arriving at a mean score of 2. Comparing Technical [T] and Sensorial [SN], Technical [T] issues where proven to be more relevant or more slightly favoured than others making it the most dominant factor of the three. Based on their preference values, the system automatically creates a reciprocal matrix on the opposite end as the case may be. At this stage (as shown in Figure 11), ratio scales are defined for pair-wise comparison of the main or parent factors using the ratio scale of 1 - 9. As mentioned earlier, the decision makers obtained values for each parent factor based on their aprioristic knowledge and individual weighting preference. Here, the AHP main criteria matrix is then automatically developed by comparing the relative importance of one parent factor over the other as shown above in Figure 11. Next, the parent criteria matrices are normalised (by dividing a cell value by the sum of each column) and then checked for consistency using Eigen values asshown in Figure 12. A local priority vector score is then generated for the matrix of judgments by normalizing the vector in each column of the matrix (i.e. dividing each entry of the column by the column total) and then averaging over the rows of the resulting matrix [55]. The normalized eigenvector shown in Figure 12 represents the relative importance of each parent criteria.

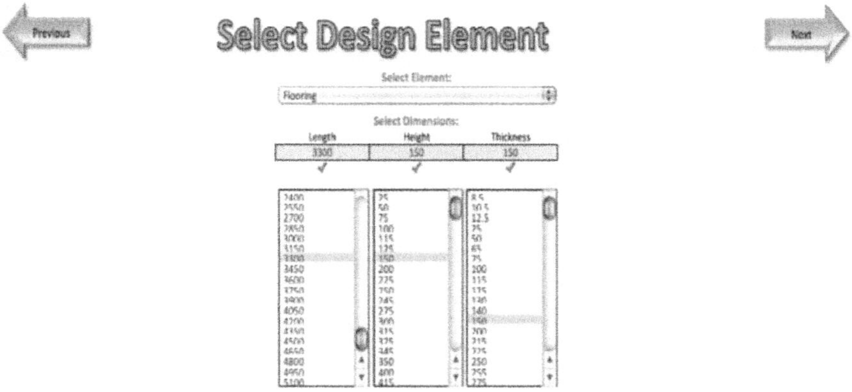

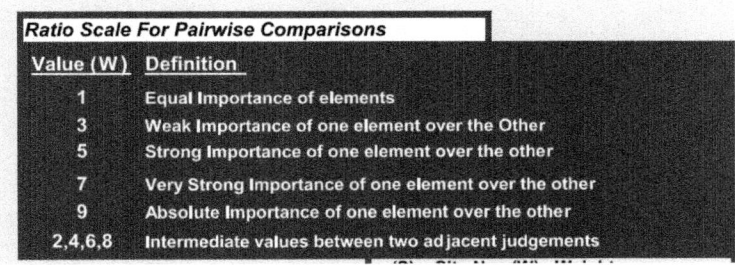

Figure 7: (a) Dimensional scale for the elected design element; (b) Selection rules for the elected design element.

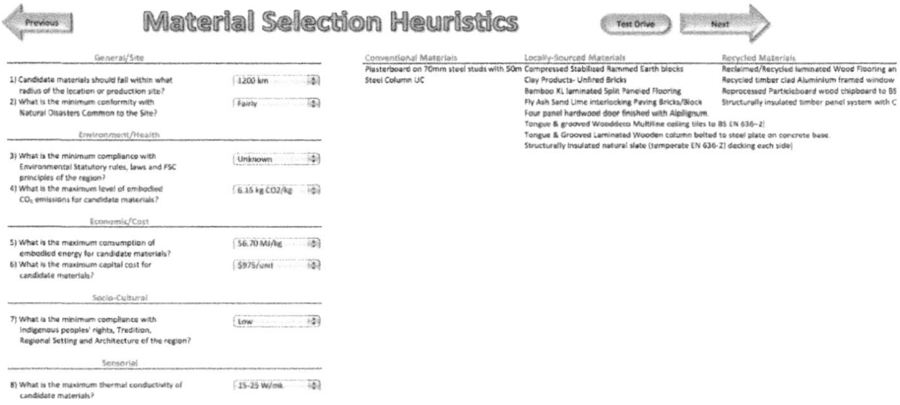

Ratio Scale For Pairwise Comparisons	
Value (W)	**Definition**
1	Equal Importance of elements
3	Weak Importance of one element over the Other
5	Strong Importance of one element over the other
7	Very Strong Importance of one element over the other
9	Absolute Importance of one element over the other
2,4,6,8	Intermediate values between two adjacent judgements

Figure 8: Ratio scale for pair-wise comparison of factors.

 Previous

Property Category Weightings

CR ✓ 0.08

User Compulsory:
◄ Adjust Sliders to Indicate Preference ►

Demo   Reset

User Compulsory:
◄ Adjust Sliders to Indicate Preference ►

General/Site	⌐ 0.03		

General/Site	⌐ 1/3	◄ ► ⌐3	Environment/Health
	⌐ 1/6	◄ ► ⌐6	Economic/Cost
	⌐ 1/9	◄ ► ⌐9	Socio-Cultural
	⌐ 1/8	◄ ► ⌐8	Technical
	⌐ 1/9	◄ ► ⌐9	Sensorial

Environment/Health	⌐ 0.07		

Environment/Health	⌐ 1/3	◄ ► ⌐3	Economic/Cost
	⌐ 1/3	◄ ► ⌐3	Socio-Cultural
	⌐ 1/3	◄ ► ⌐3	Technical
	⌐ 1/6	◄ ► ⌐6	Sensorial

Economic/Cost	⌐ 0.12		

Economic/Cost	⌐ 1/3	◄ ► ⌐3	Socio-Cultural
	⌐ 1/3	◄ ► ⌐3	Technical
	⌐ 1/2	◄ ► ⌐2	Sensorial

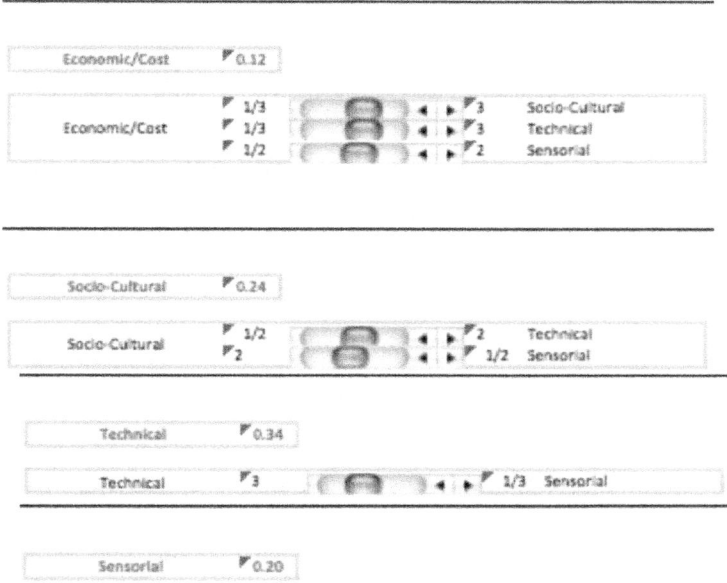

Figure 9: Consensus pair-wise comparison of main factors.

Figure 10: Consensus pair-wise comparison of main factors.

Weighted Criteria Matrix						
	General/Site	Environment	Economic/Cost	Socio-Cultural	Technical	Sensorial
General/Site	1.00	0.33	0.17	0.11	0.13	0.11
Environment/Health	3.00	1.00	0.33	0.33	0.33	0.17
Economic/Cost	6.00	3.00	1.00	0.33	0.33	0.50
Socio-Cultural	9.00	3.00	3.00	1.00	0.50	2.00
Technical	8.00	3.00	3.00	2.00	1.00	3.00
Sensorial	9.00	6.00	2.00	0.50	0.33	1.00
Total	36.00	16.33	9.50	4.28	2.63	6.78

Figure 11: Results of pair-wise analysis of parent factors.

Based on the calculation in Figure 11, the relative priorities of the parent factors in the final selection of a sustainable floor material were calculated as displayed in Figure 12. The resulting local priority vectors were given as: (GS = 0.030, EH = 0.070, C = 0.120, SC = 0.240, T = 0.340, and SN = 0.200) as shown in Table 5. In order to measure the level of consistency of the matrix for the parent factors, the consistency index (CI) was then calculated at 0.103 (see Figure 11). The random index (RI) was also taken into consideration and values calculated at this stage of the evaluation exercise. According to Saaty (2008), for matrix of order 6, the RI is 1.24 (see Table 6). Given the two values (consisting of both the consistency index (CI = 0.103) and the relative index (RI= 1.24), the CR was then calculated as: CR = CI/RI = 0.103/1.24 = 0.083 (see Figure 11).

Normalised Average Criteria Matrix								
	General/Site	Environment/Health	Economic/Cost	Socio-Cultural	Technical	Sensorial	Av.	λ_{MAX}
General/Site	0.03	0.02	0.02	0.03	0.05	0.02	0.03	0.934297901
Environment/Health	0.08	0.06	0.04	0.08	0.13	0.02	0.07	1.113775203
Economic/Cost	0.17	0.18	0.11	0.08	0.13	0.07	0.12	1.162609985
Socio-Cultural	0.25	0.18	0.32	0.23	0.19	0.30	0.24	1.04719097
Technical	0.22	0.18	0.32	0.47	0.38	0.44	0.34	0.880596922
Sensorial	0.25	0.37	0.21	0.12	0.13	0.15	0.20	1.377336489
Total	1.00	1.00	1.00	1.00	1.00	1.00	1.00	6.52
						Matrix Size	6	
						RI	1.24	
						CI	0.103	
						CR	0.083064516	

Figure 12: Relative priority scores of the parent factors.

Table 5: Derived priority scores of the parent factors

Factor/Criterion	Relative Priority
General/Site	0.030
Environmental/Health	0.070
Economic/Cost	0.120
Socio Cultural	0.240
Technical	0.340
Sensorial	0.200

Table 6: Random index values for $1 \leq n \leq 15$

n	1	2	3	4	5	6	7	8	9	10	11	12	13	14	15
RI	0	0	0.58	0.9	1.12	1.24	1.32	1.41	1.45	1.49	1.51	1.54	1.55	1.57	1.58

According to the AHP model, a matrix is considered as being consistent when the CR is less than 10%. With a Consistency Ratio (CR) of 0.083, the matrix was considered consistent since it was less than 0.1.

Pair-Wise Analysis of Sub-Factors

The results of the next pair-wise comparison matrices amongst the relative sub-factors are shown from Figures 13-24. The same calculations done for the principal matrices of the parent factors were also done for the matrices of the sub-factors. The local priority vector and the consistency ratio for each sub-criterion matrix were also computed and displayed on each corresponding table as fully displayed below. After comparing each sub-factor according to the user's system of value over other sub-factors, the weightings were obtained to establish each priority weightings in the context of the overall goal: selecting the most sustainable low-cost green floor material. The criteria matrices of each sub-factor were then normalised (by dividing a cell value by the sum of each column) and then checked for consistency as shown in Figures 13-24.

Determining the Weightings of Sub-Factors

The next stage of the assessment process was to find the final weightings of both the parent and sub-factors that will be used subsequently to evaluate the material attributes for sustainable building material selection. To determine the final weightings of the selected factors, the priority vectors (1) of the parent factors are multiplied by the corresponding relative priority vectors of each sub-criterion weighting vectors (2) to obtain the (final) weighting (3) as shown in Table 7. The main/parent factor weighting is derived from users' judgment with respect to a single main criterion. The resultant value of the comparison of each parent factor serves as the priority vector of the main criteria needed for evaluating material attributes. The selected value for each parent factor as shown in Table 7 include: GS = 0.026, EH = 0.068, C = 0.122, SC = 0.245, T = 0.335 and SN = 0.203.

Table 7: Derived final weightings for G-site factors

Criteria	Parent factor/Criteria Weighting (1)				Sub-Factor/Criteria Weighting (2)				Final Weighting (3)
	User Value	Default	CR <0.1	Selected Value	Sub-Criteria	User Value	CR < 0.1	Selected Value	Total = 1.0000
General/Site	0.03	0.057	0.08	**0.026**	**GS1-Location (Mph)**	**0.197**	0.09	**0.197**	**0.0051**
					GS2-Material Availability	0.158		0.158	0.0041
					GS3-Distance to Market (km/h)	0.127		0.127	0.0033
					GS4-Building Certification code	0.115		0.115	0.0030
					GS6-Withstand site natural disaster	0.083		0.083	0.0022
					GS8-Conforms to site geometry	0.114		0.114	0.0030
					GS9-Conforms to spatial structure	0.069		0.069	0.0018
					GS10-Conforms to all spatial activities	0.053		0.053	0.0014
					GS11-Conforms to design geometry	0.044		0.044	0.0012
					GS12-Mat. Spatial scale/Size (sq./m)	0.040		0.040	0.0010

The sub-factor weighting is derived from user's judgment with respect to each sub-factor. Some of the selected values that serve as the corresponding relative priority vectors of the general/site variable include: 0.197, 0.158, 0.127, 0.115, 0.083, 0.114, 0.069, 0.053, 0.044, and 0.040 as shown in Table 7. Final weighting is derived from multiplying the selected value of the main criteria-weighting or priority vector by the selected value of the sub-factor priority vector. This entry is obtained as follows: 0.026 × 0.197= 0.005122 (as highlighted in Table 7). The same process was applied to the other parent factors of the respective categories. The following steps describe the ways by which the various weighting vectors of each criterion are derived.

Pair-Wise Comparison of the Selected Material Alternatives against Each Sub-Factor

The final step of the exercise was for the respondents to compare each pair of low-cost green material alternatives with respect to each sub-factor. Here the user evaluates the criteria/factors and material alternatives by comparing them through direct rating, to know which factor is more important; how many times; and which material alternative is better in the context of each factor. The corresponding weightings were based on the importance that the evaluators attached to the dominance of each material alternative relative to all other alternatives under each sub-criterion. These matrices were also normalized and checked for consistency as shown in Figures 25-38.

Figures 25-38 present some results of the analyses, which explain the pair-wise matrix priority weightings and normalisation of the various materials with respect to each sub-criterion.

Determining the Weightings of Sub-Factors

The next phase, after analysing the pair-wise matrices of the sub-factors against the various low-cost green floor material alternatives was to normalize the priority weights for each pair-wise comparison judgment matrices. Once the normalised matrices of the floor material alternatives and various sub-factors were obtained, the values derived from the analysis were multiplied and summed to obtain the final composite priority weights of all material alternatives, focusing particularly on the three floor materials used in the fourth level of the AHP model of decision-making shown in Figure 6.

In this case, the final weighting scores (obtained from multiplying the priorities vectors of the parent criteria with that of individual sub-factors), is further multiplied by the priority vector of each material alternative after the pair-wise comparison against each sub-factor (as shown in Figure 38). This resulted in a final composite priority/weighting score of each sub-factor for the three floor material alternatives.

	Score	GS1	GS2	GS3	GS4	GS6	GS8	GS9	GS10	GS11	GS12
GS1-Location (Mph)	0.197	1.00	2.00	3.00	2.00	4.00	2.00	2.00	2.00	3.00	3.00
GS2-Material Availability	0.158	0.50	1.00	2.00	2.00	2.00	3.00	3.00	3.00	2.00	3.00
GS3-Distance to Market (km/h)	0.127	0.33	0.50	1.00	2.00	2.00	2.00	3.00	3.00	3.00	2.00
GS4-Building Certification code	0.115	0.50	0.50	0.50	1.00	2.00	2.00	3.00	2.00	4.00	2.00
GS6-Withstand site natural disaster	0.083	0.25	0.50	0.50	0.50	1.00	2.00	2.00	2.00	2.00	2.00
GS8-Conforms to site geometry	0.114	0.50	0.33	0.50	0.50	0.50	1.00	3.00	7.00	3.00	4.00
GS9-Conforms to spatial structure	0.069	0.50	0.33	0.33	0.33	0.50	0.33	1.00	3.00	3.00	2.00
GS10-Conforms to all spatial activities	0.053	0.50	0.33	0.33	0.50	0.50	0.14	0.33	1.00	2.00	2.00
GS11-Conforms to design geometry	0.044	0.33	0.50	0.33	0.25	0.50	0.33	0.33	0.50	1.00	2.00
GS12-Mat. Spatial scale/Size (sq./m)	0.040	0.33	0.33	0.50	0.50	0.50	0.25	0.50	0.50	0.50	1.00
CR	0.09										

Figure 13: Pair-wise matrix for general/site factors.

Normalised Matrix										λ_{MAX}	λ_{MAX}	11
0.210	0.315	0.333	0.208	0.296	0.153	0.110	0.083	0.127	0.130	0.935	Matrix Size	10
0.105	0.157	0.222	0.208	0.148	0.229	0.165	0.125	0.085	0.130	0.999	CI	0.14
0.070	0.078	0.111	0.208	0.148	0.153	0.165	0.125	0.127	0.086	1.147	RI	1.49
0.105	0.078	0.055	0.104	0.148	0.153	0.165	0.083	0.170	0.086	1.103	CR	0.09
0.052	0.078	0.055	0.052	0.074	0.153	0.110	0.083	0.085	0.086	1.123		
0.105	0.052	0.055	0.052	0.037	0.076	0.165	0.291	0.127	0.173	1.486		
0.105	0.052	0.037	0.034	0.037	0.025	0.055	0.12	0.127	0.086	1.248		
0.105	0.052	0.037	0.052	0.037	0.010	0.018	0.041	0.085	0.086	1.265		
0.070	0.078	0.037	0.026	0.037	0.025	0.018	0.020	0.042	0.086	1.042		
0.070	0.052	0.055	0.052	0.037	0.019	0.027	0.020	0.021	0.043	0.920		

Figure 14: Normalised matrix for general/site factors.

	Score	EH1	EH2	EH3	EH4	EH5	EH6	EH7	EH8	EH9	EH10
EH1-Env. Statutory Compliance	0.202	1.00	4.00	3.00	2.00	2.00	3.00	3.00	2.00	2.00	2.00
EH2-Embodied CO_2 Emission ($KgCO_2/m^2$)	0.124	0.25	1.00	2.00	3.00	2.00	2.00	2.00	2.00	3.00	0.50
EH3-Human Toxicity-Users Safety level	0.113	0.33	0.50	1.00	2.00	2.00	2.00	3.00	3.00	3.00	0.50
EH4-Ozone depletion rate	0.086	0.50	0.33	0.50	1.00	2.00	2.00	2.00	2.00	2.00	0.33
EH5-Amt. of Pesticide Treatment (l/m^2)	0.078	0.50	0.50	0.50	0.50	1.00	2.00	3.00	2.00	0.33	0.50
EH6-Complies with the Climate of the region	0.067	0.33	0.50	0.50	0.50	0.50	1.00	2.00	2.00	2.00	0.50
EH7-Env. Toxicity (land, water, Animals)	0.053	0.33	0.50	0.33	0.50	0.33	0.50	1.00	2.00	2.00	0.33
EH8-Fossil fuel/Habitat depletion	0.058	0.50	0.50	0.33	0.50	0.50	0.50	0.50	1.00	4.00	0.25
EH9-Nuclear waste rate	0.057	0.50	0.33	0.33	0.50	3.00	0.50	0.50	0.25	1.00	0.33
EH10-Waste Disposal rate	0.162	0.50	2.00	2.00	3.00	2.00	2.00	3.00	4.00	3.00	1.00
CR	0.10										

Figure 15: Pair-wise matrix for environmental factors.

Normalised Matrix										λ_{MAX}	λ_{MAX}	11
0.210	0.393	0.285	0.148	0.130	0.193	0.15	0.098	0.089	0.32	0.960	Matrix Size	10
0.052	0.098	0.150	0.222	0.130	0.129	0.1	0.098	0.134	0.08	1.257	CI	0.15
0.070	0.049	0.095	0.148	0.130	0.129	0.15	0.148	0.134	0.08	1.191	RI	1.49
0.105	0.032	0.047	0.074	0.130	0.129	0.1	0.098	0.089	0.05	1.162	CR	0.10
0.105	0.049	0.047	0.037	0.065	0.129	0.15	0.098	0.014	0.08	1.191		
0.070	0.049	0.047	0.037	0.032	0.064	0.1	0.098	0.089	0.08	1.038		
0.070	0.049	0.031	0.037	0.020	0.032	0.05	0.098	0.089	0.05	1.068		
0.105	0.049	0.031	0.037	0.032	0.032	0.025	0.049	0.179	0.04	1.178		
0.105	0.032	0.031	0.037	0.195	0.032	0.025	0.012	0.044	0.05	1.273		
0.105	0.196	0.150	0.222	0.130	0.129	0.15	0.197	0.134	0.16	1.010		

Figure 16: Normalised matrix for environmental factors.

	Score	C1	C2	C3	C4	C5	C6
C1-Total life-cycle cost ($)	0.347	1.00	2.00	2.00	3.00	5.00	9.00
C2-Material embodied energy cost ($)	0.247	0.50	1.00	2.00	4.00	4.00	3.00
C3-Material capital cost ($)	0.186	0.50	0.50	1.00	2.00	4.00	6.00
C4-Labour/Installation cost ($/sqft)	0.120	0.33	0.25	0.50	1.00	3.00	5.00
C5-Material replacement cost ($)	0.063	0.20	0.25	0.25	0.33	1.00	3.00
C6-Material Maintenance cost ($)	0.037	0.11	0.33	0.17	0.20	0.33	1.00
CR	0.07						

Figure 17: Pair-wise matrix for economic/cost factors.

Normalised Matrix						λ_{MAX}	λ_{MAX}	6
0.378	0.461	0.338	0.284	0.288	0.333	0.919	Matrix Size	6
0.18	0.230	0.338	0.379	0.230	0.111	1.069	CI	0.09
0.18	0.115	0.169	0.189	0.230	0.222	1.101	RI	1.24
0.12	0.057	0.084	0.094	0.173	0.185	1.267	CR	0.07
0.075	0.057	0.042	0.031	0.057	0.111	1.086		
0.042	0.076	0.028	0.018	0.019	0.037	1.001		

Figure 18: Normalised matrix for economic/cost factors.

	Score	SC1	SC2	SC3	SC4	SC5
SC1-Material compatibility with traditions	0.164	1.00	2.00	0.33	0.50	2.00
SC2-Material compatibility with region	0.102	0.50	1.00	0.50	0.50	0.33
SC3-Cultural restriction on usury	0.362	3.00	2.00	1.00	2.00	3.00
SC4-Client's preference rating	0.227	2.00	2.00	0.50	1.00	2.00
SC5-Conforms to Knowledge of custom	0.146	0.50	3.00	0.33	0.50	1.00
CR	0.08					

Figure 19: Pair-wise matrix for socio-cultural factors.

Normalised Matrix					λ_{MAX}	λ_{MAX}	5
0.142	0.2	0.125	0.111	0.24	1.147	Matrix Size	5
0.071	0.1	0.187	0.111	0.04	1.020	CI	0.09
0.428	0.2	0.375	0.444	0.36	0.964	RI	1.12
0.285	0.2	0.1875	0.222	0.24	1.022	CR	0.08
0.071	0.3	0.125	0.111	0.12	1.213		

Figure 20: Normalised matrix for socio-cultural factors.

	Score	T1	T2	T3	T4	T5	T6	T7	T8	T9	T10	T11	T12	T13	T14	T15	T16	T17	T17
T1-Recyclable	0.09	1.00	2.00	2.00	3.00	0.50	2.00	2.00	0.50	0.50	2.00	3.00	2.00	2.00	2.00	3.00	0.50	0.33	0.50
T2-Ease to remove	0.10	0.50	1.00	0.33	0.33	0.33	3.00	2.00	3.00	0.50	2.00	3.00	2.00	2.00	3.00	2.00	3.00	2.00	2.00
T3- Maintenance level	0.06	0.50	3.00	1.00	1.00	1.00	1.00	1.00	1.00	1.00	1.00	1.00	1.00	1.00	1.00	1.00	1.00	1.00	1.00
T4-Expansion Tolerance	0.06	0.33	3.00	1.00	1.00	1.00	1.00	1.00	1.00	1.00	1.00	1.00	1.00	1.00	1.00	1.00	1.00	1.00	1.00
T5-Conforms to skills	0.06	2.00	3.00	1.00	1.00	1.00	1.00	1.00	1.00	1.00	1.00	1.00	1.00	1.00	1.00	1.00	1.00	1.00	1.00
T6-Ease of fixing	0.05	0.50	0.33	1.00	1.00	1.00	1.00	1.00	1.00	1.00	1.00	1.00	1.00	1.00	1.00	1.00	1.00	1.00	1.00
T7-Fire resistance	0.04	0.50	0.50	1.00	1.00	1.00	1.00	1.00	1.00	1.00	1.00	1.00	1.00	0.11	1.00	0.14	1.00	1.00	1.00
T8-Thermal resistance	0.05	2.00	0.33	1.00	1.00	1.00	1.00	1.00	1.00	1.00	1.00	1.00	1.00	1.00	1.00	1.00	0.14	1.00	1.00
T9-Moisture resistance	0.06	2.00	2.00	1.00	1.00	1.00	1.00	1.00	1.00	1.00	1.00	1.00	1.00	1.00	1.00	1.00	1.00	1.00	1.00
T10-Scratch resistance	0.05	0.50	0.50	1.00	1.00	1.00	1.00	1.00	1.00	1.00	1.00	1.00	1.00	1.00	1.00	1.00	1.00	1.00	1.00
T11-Weather resistance	0.05	0.33	0.33	1.00	1.00	1.00	1.00	1.00	1.00	1.00	1.00	1.00	1.00	1.00	1.00	1.00	1.00	1.00	1.00
T12-Chemical resistance	0.05	0.50	0.50	1.00	1.00	1.00	1.00	1.00	1.00	1.00	1.00	1.00	1.00	1.00	1.00	1.00	1.00	1.00	1.00
T13-Resistance to decay	0.07	0.50	0.50	1.00	1.00	1.00	1.00	9.00	1.00	1.00	1.00	1.00	1.00	1.00	1.00	1.00	1.00	1.00	1.00
T14-Weight of material	0.05	0.50	0.33	1.00	1.00	1.00	1.00	1.00	1.00	1.00	1.00	1.00	1.00	1.00	1.00	1.00	1.00	1.00	1.00
T15-Life expectancy	0.07	0.33	0.50	1.00	1.00	1.00	1.00	7.00	1.00	1.00	1.00	1.00	1.00	1.00	1.00	0.25	1.00	1.00	
T16-Biodegradable	0.08	2.00	0.33	1.00	1.00	1.00	1.00	1.00	7.00	1.00	1.00	1.00	1.00	1.00	1.00	4.00	1.00	1.00	
T17-UV Resistance	0.05	3.00	0.50	1.00	1.00	1.00	1.00	1.00	1.00	1.00	1.00	1.00	1.00	1.00	1.00	1.00	1.00	1.00	
T18-Compatibility	0.05	0.50	1.00	1.00	1.00	1.00	1.00	1.00	1.00	1.00	1.00	1.00	1.00	1.00	1.00	1.00	1.00	1.00	1.00
CR	0.09																		

Figure 21: Pair-wise matrix for technical factors.

Normalised Matrix

																			λ_{MAX}	λ_{MAX}	21
0.05	0.11	0.11	0.17	0.02	0.11	0.11	0.02	0.02	0.11	0.17	0.11	0.11	0.11	0.17	0.02	0.01	0.02	1.602	**Size**	18
0.02	0.05	0.01	0.01	0.01	0.17	0.11	0.17	0.02	0.11	0.17	0.11	0.11	0.17	0.11	0.17	0.11	0.11	1.778	**CI**	0.15
0.02	0.17	0.05	0.05	0.05	0.05	0.05	0.05	0.05	0.05	0.05	0.05	0.05	0.05	0.05	0.05	0.05	0.05	1.083	**RI**	1.69
0.01	0.17	0.05	0.05	0.05	0.05	0.05	0.05	0.05	0.05	0.05	0.05	0.05	0.05	0.05	0.05	0.05	0.05	1.074	**CR**	0.09
0.11	0.17	0.05	0.05	0.05	0.05	0.05	0.05	0.05	0.05	0.05	0.05	0.05	0.05	0.05	0.05	0.05	0.05	1.167		
0.02	0.01	0.05	0.05	0.05	0.05	0.05	0.05	0.05	0.05	0.05	0.05	0.05	0.05	0.05	0.05	0.05	0.05	0.935		
0.02	0.02	0.05	0.05	0.05	0.05	0.05	0.05	0.05	0.05	0.05	0.05	0.00	0.05	0.01	0.05	0.05	0.05	0.847		
0.11	0.01	0.05	0.05	0.05	0.05	0.05	0.05	0.05	0.05	0.05	0.05	0.05	0.05	0.01	0.05	0.05	0.05	0.971		
0.11	0.11	0.05	0.05	0.05	0.05	0.05	0.05	0.05	0.05	0.05	0.05	0.05	0.05	0.05	0.05	0.05	0.05	1.111		
0.02	0.02	0.05	0.05	0.05	0.05	0.05	0.05	0.05	0.05	0.05	0.05	0.05	0.05	0.05	0.05	0.05	0.05	0.944		
0.01	0.01	0.05	0.05	0.05	0.05	0.05	0.05	0.05	0.05	0.05	0.05	0.05	0.05	0.05	0.05	0.05	0.05	0.926		
0.02	0.02	0.05	0.05	0.05	0.05	0.05	0.05	0.05	0.05	0.05	0.05	0.05	0.05	0.05	0.05	0.05	0.05	0.944		
0.02	0.02	0.05	0.05	0.05	0.05	0.51	0.05	0.05	0.05	0.05	0.05	0.05	0.05	0.05	0.05	0.05	0.05	1.389		
0.02	0.01	0.05	0.05	0.05	0.05	0.05	0.05	0.05	0.05	0.05	0.05	0.05	0.05	0.05	0.05	0.05	0.05	0.935		
0.01	0.02	0.05	0.05	0.05	0.05	0.4	0.05	0.05	0.05	0.05	0.05	0.05	0.05	0.01	0.05	0.05		1.227		
0.11	0.01	0.05	0.05	0.05	0.05	0.05	0.4	0.05	0.05	0.05	0.05	0.05	0.05	0.22	0.05	0.05	0.05	1.519		
0.17	0.02	0.05	0.05	0.05	0.05	0.05	0.05	0.05	0.05	0.05	0.05	0.05	0.05	0.05	0.05	0.05	0.05	1.083		
0.02	0.05	0.05	0.05	0.05	0.05	0.05	0.05	0.05	0.05	0.05	0.05	0.05	0.05	0.05	0.05	0.05	0.05	0.972		

Figure 22: Normalised matrix for technical factors.

	Score	SN1	SN2	SN3	SN4	SN5	SN6	SN7	SN8	SN9	SN10	SN11	SN12	SN13
SN1-Aesthetics	0.077	1.00	1	1	1	1	1	1	1	1	1	1	1	1
SN2-Texture	0.077	1.00	1.00	1	1	1	1	1	1	1	1	1	1	1
SN3-Colour	0.077	1.00	1.00	1.00	1	1	1	1	1	1	1	1	1	1
SN4-Temperature	0.077	1.00	1.00	1.00	1.00	1	1	1	1	1	1	1	1	1
SN5-Acoustics	0.106	1.00	1.00	1.00	1.00	1.00	2	0	4	0	2	0	2	2
SN6-Odour	0.087	1.00	1.00	1.00	1.00	0.50	1.00	2	1	0	2	1	2	2
SN7-Thickness/Thinness	0.107	1.00	1.00	1.00	1.00	3.00	0.50	1.00	2	2	2	3	0	0
SN8-Glossiness/fineness	0.075	1.00	1.00	1.00	1.00	0.25	2.00	0.50	1.00	1	1	1	1	1
SN9-Strength/Hardness	0.109	1.00	1.00	1.00	1.00	3.00	5.00	0.50	1.00	1.00	1	1	1	1
SN10-Lighting effect	0.068	1.00	1.00	1.00	1.00	0.50	0.50	0.50	1.00	1.00	1.00	1	1	1
SN11-Translucence	0.108	1.00	1.00	1.00	1.00	6.00	2.00	0.33	1.00	1.00	1.00	1.00	1	1
SN12-Structure	0.089	1.00	1.00	1.00	1.00	0.50	0.50	4.00	1.00	1.00	1.00	1.00	1.00	1
SN13-Thermal	0.083	1.00	1.00	1.00	1.00	0.50	0.50	3.00	1.00	1.00	1.00	1.00	1.00	1.00
CR	0.10													

Figure 23: Pair-wise matrix for sensorial factors.

Normalised Matrix													λ_{MAX}	λ_{MAX}	15
0.076	0.076	0.076	0.076	0.076	0.076	0.076	0.076	0.076	0.076	0.076	0.076	0.076	1.000	Matrix Size	13
0.076	0.076	0.076	0.076	0.076	0.076	0.076	0.076	0.076	0.076	0.076	0.076	0.076	1.000	CI	0.15
0.076	0.076	0.076	0.076	0.076	0.076	0.076	0.076	0.076	0.076	0.076	0.076	0.076	1.000	RI	1.5551
0.076	0.076	0.076	0.076	0.076	0.076	0.076	0.076	0.076	0.076	0.076	0.076	0.076	1.000	CR	0.10
0.076	0.076	0.076	0.076	0.076	0.153	0.025	0.307	0.025	0.153	0.012	0.153	0.153	1.372		
0.076	0.076	0.076	0.076	0.038	0.076	0.153	0.038	0.015	0.153	0.038	0.153	0.153	1.131		
0.076	0.076	0.076	0.076	0.230	0.038	0.076	0.153	0.153	0.153	0.230	0.019	0.025	1.391		
0.076	0.076	0.076	0.076	0.019	0.153	0.038	0.076	0.076	0.076	0.076	0.076	0.076	0.981		
0.076	0.076	0.076	0.076	0.230	0.384	0.038	0.076	0.076	0.076	0.076	0.076	0.076	1.423		
0.076	0.076	0.076	0.076	0.038	0.038	0.038	0.076	0.076	0.076	0.076	0.076	0.076	0.885		
0.076	0.076	0.076	0.076	0.461	0.153	0.025	0.076	0.076	0.076	0.076	0.076	0.076	1.410		
0.076	0.076	0.076	0.076	0.038	0.038	0.307	0.076	0.076	0.076	0.076	0.076	0.076	1.154		
0.076	0.076	0.076	0.076	0.038	0.038	0.230	0.076	0.076	0.076	0.076	0.076	0.076	1.077		

Figure 24: Normalised matrix for sensorial factors.

Using the priorities determined through these matrices, the weighted overall priority of each candidate material was determined. The amalgamation method yielded a single green utility index of alternative worth, which allowed the material options to be ranked according to their overall priorities. The material with the highest score then becomes the selected candidate material as shown in Figure 38. Looking at Figure 38, it is clear from the results of the analysis that Material option (A) turns out to be the most preferred material among the three material options identified in Table 4, with an overall priority or index score of 0.086. It is based on the concept of the higher the green utility index value, the better the option. The green utility index as calculated for each of the three material alternatives was M(C) = 0.086, M(A) = 0.072 and M(B) = 0.062 for material options C, A and B respectively, making Option C (fly-ash cement concrete floor slab) emerge as the best option amongst the other alternatives as shown in Figure 38. The above example has illustrated the application of the MSDSS in a material selection problem for a proposed 5-bedroom low-cost residential green building project in the London Borough of Sutton. From the illustrated example it can be deduced that the MSDSS model is able to provide rankings in low-cost green building material assessment combining site, economic, technical, social-cultural, sensorial and environmental criteria into a composite index system based on the AHP technique. This model is therefore, based on the presumption that decision makers, given full knowledge of all possible consequences of all possible alternatives and factors, will select the material with the highest-ranking score.

GSI-Location (km)	CSR	CP	RL	B.XL	FA	RT	FPH	SS	RPB	T&GJ	PB	T&GW	SC	SIT	SIS
Compressed Stabilized Rammed Earth blocks	1.0	2.0	2.0	4.0	2.0	5.0	8.0	8.0	4.0	4.00	4.0	4.00	7.0	2.00	4.0
Clay Products-Unfired Bricks	0.5	1.0	1.0	3.0	1.0	4.0	7.0	7.0	3.0	3.00	3.0	3.00	6.0	1.00	3.0
Reclaimed/Recycled laminated Wood Flooring and Panelling	0.5	1.0	1.0	3.0	1.0	4.0	7.0	7.0	3.0	3.00	3.0	3.00	6.0	1.00	3.0
Bamboo XL laminated Split Paneled Flooring	0.3	0.3	0.3	1.0	0.3	2.0	5.0	5.0	1.0	1.0	1.0	1.00	4.0	0.33	1.0
Fly Ash Sand Lime interlocking Paving Bricks/Block	0.5	1.0	1.0	3.0	1.0	4.00	7.00	7.00	3.0	3.00	3.0	3.0	6.0	1.0	3.0
Recycled timber clad Aluminium framed window unit	0.2	0.3	0.3	0.5	0.3	1.0	4.00	4.00	0.50	0.50	0.5	0.50	3.0	0.3	0.5
Four panel hardwood door finished with Alpilignum.	0.1	0.1	0.1	0.2	0.1	0.3	1.0	1.0	0.2	0.2	0.2	0.2	0.5	0.1	0.2
Stainless Steel Entry Door.	0.1	0.1	0.1	0.2	0.1	0.3	1.0	1.0	0.2	0.2	0.2	0.2	0.5	0.1	0.20
Reprocessed Particleboard wood chipboard to BS EN 312 Type P5,	0.3	0.3	0.3	1.0	0.3	2.0	5.0	5.0	1.0	1.00	1.0	1.00	4.0	0.3	1.00
Tongue & grooved Wooddeco Multiline ceiling tiles to BS EN 636-2|	0.3	0.3	0.3	1.0	0.3	2.0	5.0	5.0	1.0	1.00	1.0	1.00	4.0	0.33	1.00
Plasterboard on 70 mm steel studs with 50 mm 12.9 kg/m³ insulation,	0.3	0.3	0.3	1.0	0.3	2.0	5.0	5.0	1.0	1.00	1.0	1.00	4.00	0.33	1.00
Tongue & Grooved Laminated Wooden column bolted to steel plate on concrete base.	0.3	0.3	0.3	1.0	0.3	2.0	5.0	5.0	1.0	1.0	1.0	1.00	4.0	0.33	1.0
Steel Column UC	0.1	0.2	0.2	0.3	0.2	0.33	2.00	2.00	0.3	0.25	0.3	0.3	1.0	0.17	0.3
Structurally insulated timber panel system with OSB/3 each side, roofing underlay reclaimed clay tiles	0.5	1.0	1.0	3.0	1.0	4.0	7.0	7.0	3.0	3.0	3.0	3.0	6.0	1.0	3.0
Structurally insulated natural slate (temperate EN 636-2) decking each side|	0.3	0.3	0.3	1.0	0.3	2.0	5.00	5.00	1.00	1.00	1.0	1.00	4.0	0.3	1.0
Total	5.1	8.7	8.7	23.2	8.7	34.8	74.0	74.0	23.2	23.2	23.2	23.2	60.0	8.7	23.2

Figure 25: Pair-wise matrix: location.

CS	CP	RL	B.XL	FA	RT	FPH	SS	RP,	T&GJ	PB	T&GW.	SC	SIT	SIS
0.2	0.2	0.2	0.2	0.23	0.14	0.11	0.11	0.17	0.17	0.17	0.17	0.12	0.23	0.17
0.1	0.1	0.1	0.1	0.11	0.11	0.09	0.09	0.13	0.13	0.13	0.13	0.10	0.11	0.13
0.1	0.1	0.1	0.1	0.11	0.11	0.09	0.09	0.13	0.13	0.13	0.13	0.10	0.11	0.13
0.0	0.0	0.0	0.0	0.04	0.1	0.07	0.07	0.04	0.04	0.04	0.04	0.07	0.04	0.04
0.10	0.11	0.11	0.13	0.11	0.11	0.09	9.46E-02	0.13	0.13	0.13	0.13	0.10	0.11	0.13
0.0	0.0	0.03	0.02	0.03	0.03	0.05	0.05	0.02	0.02	0.02	0.02	0.05	0.03	0.02
0.0	0.0	0.0	0.0	0.02	0.0	0.01	0.0135134	0.01	0.01	0.01	0.01	0.01	0.02	0.01
0.0	0.0	0.0	0.0	0.02	0.0	0.01	4	0.01	0.01	0.01	0.01	0.01	0.02	0.01
0.0	0.0	0.0	0.0	0.04	0.1	0.07	0.067567568	0.04	0.04	0.04	0.04	0.07	0.04	0.04
0.0	0.0	0.0	0.0	0.0	0.1	0.1	0.067567568	0.04	0.04	0.04	0.04	0.07	0.04	0.04
0.05	0.04	0.04	0.0	0.04	0.1	0.1	0.067567568	0.04	0.04	0.04	0.04	0.07	0.04	0.04
0.0	0.0	0.0	0.0	0.0	0.1	0.1	0.067567568	0.04	0.04	0.04	0.04	0.07	0.04	0.04
0.0	0.0	0.0	0.0	0.0	0.0	0.0	0.027027027	0.01	0.01	0.01	0.01	0.02	0.02	0.01
0.1	0.1	0.1	0.1	0.1	0.1	0.1	0.094594595	0.13	0.13	0.13	0.13	0.10	0.11	0.13
0.0	0.0	0.0	0.0	0.04	0.1	0.07	0.067567568	0.04	0.04	0.04	0.04	0.07	0.04	0.04
1.00	1.00	1.00	1.00	1.00	1.00	1.00	1.00	1.00	1.00	1.00	1.00	1.00	1.00	1.00

Figure 26: Normalised matrix: location.

EH2-Embodied CO_2 Emission ($kgCO_2/m^2$)	Compressed Stabilized Rammed Earth blocks	Clay Products- Unfired Bricks	Reclaimed/Recycled laminated Wood Flooring and Panelling	Bamboo XL laminated Split Paneled Flooring	Fly Ash Sand Lime interlocking Paving Bricks/Block	Recycled timber clad Aluminium framed window unit	Four panel hardwood door finished with Alpilignum.	Stainless Steel Entry Door.	Reprocessed Particleboard wood chipboard to BS EN 312 Type P5.	Tongue & grooved Wooddeco Multiline ceiling tiles to BS EN 636–2]	Plasterboard on 70 mm steel studs with 50 mm 12.9 kg/m³ insulation,	Tongue & Grooved Laminated Wooden column bolted to steel plate on concrete base	Steel Column UC	Structurally insulated timber panel system with OSB/3 each side, roofing underlay reclaimed clay tiles	Structurally insulated natural slate (temperate EN 636–2) decking each side]
Compressed Stabilized Rammed Earth blocks	1.0	1.0	5.0	1.0	1.0	5.0	5.0	8.0	2.0	1.0	4.0	5.00	6.00	5.00	1.00
Clay Products—Unfired Bricks	1.0	1.0	5.0	1.0	1.0	5.0	5.0	8.0	2.0	1.0	4.0	5.0	6.0	5.0	1.0
Reclaimed/Recycled laminated Wood Flooring and Panelling	0.2	0.2	1.0	0.2	0.2	1.0	1.0	4.0	0.3	0.2	0.5	1.0	2.0	1.0	0.2
Bamboo XL laminated Split Paneled Flooring	1.0	1.0	5.0	1.0	1.0	5.0	5.0	8.0	2.0	1.0	4.0	5.0	6.0	5.0	1.0
Fly Ash Sand Lime interlocking Paving Bricks/Block	1.0	1.0	5.0	1.0	1.0	5.0	5.0	8.0	2.0	1.0	4.0	5.0	6.0	5.0	1.0
Recycled timber clad Aluminium framed window unit	0.2	0.2	1.0	0.2	0.2	1.0	1.0	4.0	0.3	0.2	0.5	1.0	2.0	1.0	0.2
Four panel hardwood door finished with Alpilignum.	0.2	0.2	1.0	0.2	0.2	1.0	1.0	4.0	0.3	0.2	0.5	1.0	2.0	1.0	0.2
Stainless Steel Entry Door.	0.125	0.125	0.25	0.125	0.125	0.25	0.25	1	0.14	0.125	0.2	0.25	0.3	0.25	0.125
Reprocessed Particleboard wood chipboard to BS EN 312 Type P5,	0.5	0.5	4.0	0.5	0.5	4.0	4.0	7.0	1.0	0.5	3.0	4.0	5.0	4.0	0.5
Tongue & grooved Wooddeco Multiline ceiling tiles to BS EN 636–2]	1	1	5	1	1	5	5	8	2	1	4	5	6	5	1
Plasterboard on 70 mm steel studs with 50 mm 12.9 kg/m³ insulation,	0.25	0.25	2	0.25	0.25	2	2	5	0.3	0.25	1	2	3	2	0.25
Tongue & Grooved Laminated Wooden column bolted to steel plate on concrete base.	0.20	0.20	1.00	0.20	0.20	1.00	1.00	4.00	0.25	0.20	0.50	1.00	2.00	1.00	0.20
Steel Column UC	0.2	0.2	0.5	0.2	0.2	0.5	0.5	3.0	0.2	0.2	0.3	0.5	1.0	0.5	0.2
Structurally insulated timber panel system with OSB/3 each side, roofing underlay reclaimed clay tiles	0.2	0.2	1.0	0.2	0.2	1.0	1.0	4.0	0.3	0.2	0.5	1.0	2.0	1.0	0.2
Structurally insulated natural slate (temperate EN 636-2) decking each side]	1.0	1.0	5.0	1.0	1.0	5.0	5.0	8.0	2.0	1.0	4.0	5.00	6.00	5.00	1.00
Total			8.0		8.0	41.8	8.0	8.0	41.8	41.8	84.0	14.9		8.0	

Figure 27: Pair-wise matrix: embodied CO_2 emissions.

Compressed Stabilized Rammed Earth blocks	Clay Products- Unfired Bricks	Reclaimed/Recycled laminated Wood Flooring and Panelling	Bamboo XL laminated Split Paneled Flooring	Fly Ash Sand Lime interlocking Paving Bricks/Block	Recycled timber clad Aluminium framed window unit	Four panel hardwood door finished with Alphlgmum	Stainless Steel Entry Door	Reprocessed Particleboard wood chipboard to BS EN 312 Type P5.	Tongue & grooved Wooddeco Multiline ceiling tiles to BS EN 636-2]	Plasterboard on 70 mm steel studs with 50 mm 12.9 kg/m³ insulation,	Tongue & Grooved Laminated Wooden column bolted to steel plate on concrete base.	Steel Column UC	Structurally insulated timber panel system with OSB/3 each side, roofing underlay reclaimed clay tiles	Structurally insulated natural slate (temperate EN 636-2) decking each side]	CI 0.03	
0.12	0.12	0.12	0.1	0.12	0.1	0.1	0.1	0.12	0.12	0.13	0.12	0.11	0.12	0.12	0.12	0.97 RI 1.58
0.1	0.1	0.1	0.1	0.1	0.1	0.1	0.1	0.1	0.12	0.13	0.12	0.11	0.12	0.12	0.12	0.97 CR 0.02
0.0	0.0	0.0	0.0	0.0	0.0	0.0	0.05	0.01	0.02	0.02	0.02	0.04	0.02	0.02	0.03	1.07
0.1	0.1	0.1	0.1	0.1	0.1	0.1	0.1	0.12	0.12	0.13	0.12	0.11	0.12	0.12	0.12	0.97
0.1	0.1	0.1	0.1	0.12	0.1	0.12	0.1	0.12	0.12	0.13	0.12	0.11	0.12	0.12	0.12	0.97
0.0	0.0	0.0	0.0	0.02	0.0	0.02	0.05	0.01	0.02	0.02	0.02	0.04	0.02	0.02	0.03	1.07
0.0	0.0	0.0	0.0	0.02	0.0	0.02	0.05	0.01	0.02	0.02	0.02	0.04	0.02	0.02	0.03	1.07
0.015544041	0.015544041	0.005988024	0.01	0.02	0.004	0.01	0.01	0.009	0.015544041	0.03	0.005	0.006	0.005	0.015	0.01	0.88
0.1	0.1	0.1	0.1	0.06	0.1	0.10	0.1	0.06	0.06	0.10	0.10	0.09	0.10	0.06	0.08	1.18
0.124352332	0.124352332	0.119760479	0.12	0.12	0.11	0.12	0.1	0.1	0.124352332	0.12	0.11	0.105	0.11	0.122	0.12	0.97
0.031088083	0.031088083	0.047904192	0.03	0.03	0.04	0.05	0.1	0.021	0.031088083	0.03	0.047	0.057	0.047	0.03	0.04	1.23
0.02	0.02	0.02	0.02	0.02	0.02	0.02	0.1	0.01	0.024870466	0.01	0.02	0.038	0.02	0.026	0.03	1.07
0.0	0.0	0.0	0.0	0.02	0.0	0.01	0.01	0.01	0.02	0.01	0.01	0.02	0.01	0.02	0.02	0.97
0.0	0.0	0.0	0.0	0.0	0.0	0.0	0.05	0.01	0.02	0.02	0.02	0.04	0.02	0.02	0.03	1.07
0.12	0.12	0.12	0.1	0.12	0.1	0.1	01	0.1	0.12	0.13	0.12	0.11	0.12	0.12	0.12	0.97
1.00	1.00	1.00	1.00	1.00	1.00	1.00	1.00	1.00	1.00	1.00	1.00	1.00	1.00	1.00	1.00	15.5

Figure 28: Normalised matrix: embodied CO_2 emissions.

C1- Total life-cycle cost ($)	Compressed Stabilized Rammed Earth blocks	Clay Products- Unfired Bricks	Reclaimed/Recycled laminated Wood Flooring and Panelling	Bamboo XL laminated Split Paneled Flooring	Fly Ash Sand Lime interlocking Paving Bricks/Block	Recycled timber clad Aluminium framed window unit	Four panel hardwood door finished with Alpilgnum.	Stainless Steel Entry Door	Reprocessed Particleboard wood chipboard to BS EN 312 Type P5,	Tongue & grooved Wooddeco Multiline ceiling tiles to BS EN 636-2]	Plasterboard on 70 mm steel studs with 50 mm 12.9 kg/m³ insulation,	Tongue & Grooved Laminated Wooden column bolted to steel plate on concrete base	Steel Column UC	Structurally insulated timber panel system with OSB/3 each side, roofing underlay reclaimed clay tiles	Structurally insulated natural slate (temperate EN 636-2) decking each side]
Compressed Stabilized Rammed Earth blocks	1.0	0.5	3.0	0.5	2.0	7.0	8.0	7.0	7.0	8.0	8.0	8.0	7.0	7.0	7.0
Clay Products- Unfired Bricks	2	1	4	1	3	8	9	8	8	9	9	9	8	8	8
Reclaimed/Recycled laminated Wood Flooring and Panelling	0.3	0.3	1.0	0.3	0.5	5.0	6.0	5.0	5.0	6.0	6.0	6.00	5.00	5.00	5.00
Bamboo XL laminated Split Paneled Flooring	2	1	4	1	3	8	9	8	8	9	9	9	8	8	8
Fly Ash Sand Lime interlocking Paving Bricks/Block	0.5	0.3	2	0.3	1	6	7	6	6	7	7	7	6	6	6
Recycled timber clad Aluminium framed window unit	0.14	0.13	0.20	0.13	0.17	1.00	2.00	1.00	1.00	2.00	2.00	2.00	1.00	1.00	1.00
Four panel hardwood door finished with Alpilignum.	0.1	0.1	0.2	0.1	0.1	0.5	1.0	0.5	0.5	1.0	1.0	1.0	0.5	0.5	0.5
Stainless Steel Entry Door.	0.1	0.1	0.2	0.1	0.2	1.0	2.0	1.0	1.0	2.0	2.0	2.0	1.0	1.0	1.0
Reprocessed Particleboard wood chipboard to BS EN 312 Type P5,	0.1	0.1	0.2	0.1	0.2	1.0	2.0	1.0	1.0	2.0	2.0	2.0	1.0	1.0	1.0
Tongue & grooved Wooddeco Multiline ceiling tiles to BS EN 636-2]	0.1	0.1	0.2	0.1	0.1	0.5	1.0	0.5	0.5	1.0	1.0	1.0	0.5	0.5	0.5
Plasterboard on 70 mm steel studs with 50 mm 12.9 kg/m³ insulation,	0.1	0.1	0.2	0.1	0.1	0.5	1.0	0.5	0.5	1.0	1.0	1.0	0.5	0.5	0.5
Tongue & Grooved Laminated Wooden column bolted to steel plate on concrete base.	0.1	0.1	0.2	0.1	0.1	0.5	1.0	0.5	0.5	1.0	1.0	1.0	0.5	0.5	0.5
Steel Column UC	0.1	0.1	0.2	0.1	0.2	1.0	2.0	1.0	1.0	2.0	2.0	2.0	1.0	1.0	1.0
Structurally insulated timber panel system with OSB/3 each side, roofing underlay reclaimed clay tiles	0.1	0.1	0.2	0.1	0.2	1.0	2.0	1.0	1.0	2.0	2.0	2.0	1.0	1.0	1.0
Structurally insulated natural slate (temperate EN 636-2) decking each side]	0.1	0.1	0.2	0.1	0.2	1.0	2.0	1.0	1.0	2.0	2.0	2.0	1.0	1.0	1.0
Total	7.2	4.3	15.9	4.3	11.1	42.0	55.0	42.0	42.0	55.0	55.0	55.0	42.0	42.0	42.0

Figure 29: Pair-wise matrix: total life-cycle cost.

Compressed Stabilized Rammed Earth blocks	Clay Products- Unfired Bricks	Reclaimed/Recycled laminated Wood Flooring and Panelling	Bamboo XL laminated Split Paneled Flooring	Fly Ash Sand Lime interlocking Paving Bricks/Block	Recycled timber clad Aluminium framed window unit	Four panel hardwood door finished with Alpiligraum.	Stainless Steel Entry Door.	Reprocessed Particleboard wood chipboard to BS EN 312 Type P5,	Tongue & grooved Wooddeco Multiline ceiling tiles to BS EN 636-2]	Plasterboard on 70 mm steel studs with 50 mm 12.9 kg/m³ insulation,	Tongue & Grooved Laminated Wooden column bolted to steel plate on concrete base.	Steel Column UC	Structurally insulated timber panel system with OSB/3 each side, roofing underlay reclaimed clay tiles	Structurally insulated natural slate (temperate EN 636-2) decking each side]	Average	Lambda Max CI 0.06
0.1	0.1	0.2	0.1	0.18	0.2	0.15	0.17	0.1667	0.15	0.15	0.15	0.17	0.17	0.17	0.15	1.11 RI 1.58
0.2	0.2	0.2	0.2	0.2	0.19	0.1	0.19	0.190	0.163636364	0.163636364	0.163636364	0.19047619	0.19047619	0.19047619	0.20	0.87 CR 0.04
0.05	0.06	0.06	0.1	0.05	0.1	0.1	0.19	0.119	0.11	0.11	0.11	0.12	0.12	0.12	0.09	1.50
0.2	0.24	0.25	0.2	0.2	0.19	0.16	0.1909	0.190	0.163636364	0.163636364	0.163636364	0.19047619	0.19047619	0.19047619	0.20	
0.0	0.08	0.12	0.07	0.09	0.14	0.12	3	0.14	0.127272727	0.127272727	0.127272727	0.142857143	0.142857143	0.142857143	0.12	
0.02	0.03	0.01	0.03	0.02	0.02	0.04	0.024	0.02	0.04	0.036363636	0.036363636	0.023809524	0.023809524	0.023809524	0.03	
0.0	0.0	0.0	0.0	0.01	0.0	0.02	0.0162	0.01	0.02	0.02	0.02	0.01	0.01	0.01	0.02	

Figure 30: Normalised matrix: total life-cycle cost.

SC3- Cultural restriction on usury	Compressed Stabilized Rammed Earth blocks	Clay Products- Unfired Bricks	Reclaimed/Recycled laminated Wood Flooring and Panelling	Bamboo XL laminated Split Paneled Flooring	Fly Ash Sand Lime interlocking Paving Bricks/Block	Recycled timber clad Aluminium framed window unit	Four panel hardwood door finished with Alpiligraum.	Stainless Steel Entry Door.	Reprocessed Particleboard wood chipboard to BS EN 312 Type P5,	Tongue & grooved Wooddeco Multiline ceiling tiles to BS EN 636-2]	Plasterboard on 70 mm steel studs with 50 mm 12.9 kg/m³ insulation,	Tongue & Grooved Laminated Wooden column bolted to steel plate on concrete base.	Steel Column UC	Structurally insulated timber panel system with OSB/3 each side, roofing underlay reclaimed clay tiles	Structurally insulated natural slate (temperate EN 636-2) decking each side]
Compressed Stabilized Rammed Earth blocks	1.0	1.0	1.0	1.0	1.0	0.3	0.3	0.2	1.0	1.0	0.3	1.0	0.1	1.0	1.0
Clay Products—Unfired Bricks	1.0	1.0	1.0	1.0	1.0	0.3	0.3	0.2	1.0	1.0	0.3	1.0	0.1	1.0	1.0
Reclaimed/Recycled laminated Wood Flooring and Panelling	1.0	1.0	1.0	1.0	1.0	0.3	0.3	0.2	1.0	1.0	0.3	1.0	0.1	1.0	1.0
Bamboo XL laminated Split Paneled Flooring	1.0	1.0	1.0	1.0	1.0	0.3	0.3	0.2	1.0	1.0	0.3	1.0	0.1	1.0	1.0
Fly Ash Sand Lime interlocking Paving Bricks/Block	1.0	1.0	1.0	1.0	1.0	0.3	0.3	0.2	1.0	1.0	0.3	1.0	0.1	1.0	1.0

Recycled timber clad Aluminium framed window unit	3.0	3.0	3.0	3.0	3.0	1.0	1.0	0.3	3.0	3.0	1.0	3.0	0.2	3.0	3.0
Four panel hardwood door finished with Alpilignum.	3.0	3.0	3.0	3.0	3.0	1.0	1.0	0.3	3.0	3.0	1.0	3.0	0.2	3.0	3.0
Stainless Steel Entry Door.	5.0	5.0	5.0	5.0	5.0	3.0	3.0	1.0	5.0	5.0	3.0	5.0	0.3	5.0	5.0
Reprocessed Particleboard wood chipboard to BS EN 312 Type P5,	1.0	1.0	1.0	1.0	1.0	0.3	0.3	0.2	1.0	1.0	0.3	1.0	0.1	1.0	1.0
Tongue & grooved Wooddeco Multiline ceiling tiles to BS EN 636–2]	1.0	1.0	1.0	1.0	1.0	0.3	0.3	0.2	1.0	1.0	0.3	1.0	0.1	1.0	1.0
Plasterboard on 70 mm steel studs with 50 mm 12.9 kg/m³ insulation,	3.0	3.0	3.0	3.0	3.0	1.0	1.0	0.3	3.0	3.0	1.0	3.00	0.20	3.00	3.00
Tongue & Grooved Laminated Wooden column bolted to steel plate on concrete base.	1.0	1.0	1.0	1.0	1.0	0.3	0.3	0.2	1.0	1.0	0.3	1.0	0.1	1.0	1.0
Steel Column UC	7.0	7.0	7.0	7.0	7.0	5.0	5.0	3.0	7.0	7.0	5.0	7.0	1.0	7.0	7.0
Structurally insulated timber panel system with OSB/3 each side, roofing underlay reclaimed clay tiles	1.0	1.0	1.0	1.0	1.0	0.3	0.3	0.2	1.0	1.0	0.3	1.0	0.1	1.0	1.0
Structurally insulated natural slate (temperate EN 636-2) decking each side]	1.0	1.0	1.0	1.0	1.0	0.3	0.3	0.2	1.0	1.0	0.3	1.0	0.1	1.0	1.0
Total	31.0	31.0	31.0	31.0	31.0	14.3	14.3	7.0	31.0	31.0	14.3	31.0	3.4	31.0	31.0

Figure 31: Pair-wise matrix: cultural restriction on usury.

Compressed Stabilized Rammed Earth blocks	Clay Products- Unfired Bricks	Reclaimed/Recycled laminated Wood Flooring and Paneling	Bamboo XL laminated Split Paneled Flooring	Fly Ash Sand Lime interlocking Paving Bricks/Block	Recycled timber clad Aluminum framed window unit	Four panel hardwood door finished with Alpilignum.	Stainless Steel Entry Door.	Reprocessed Particleboard wood chipboard to BS EN 312 Type P5,	Tongue & grooved Wooddeco Multiline ceiling tiles to BS EN 636-2]	Plasterboard on 70 mm steel studs with 50 mm 12.9 kg/m³ insulation,	Tongue & Grooved Laminated Wooden column bolted to steel plate on concrete base.	Steel Column UC	Structurally insulated timber panel system with OSB/3 each side, roofing underlay reclaimed clay tiles	Structurally insulated natural slate (temperate EN 636-2) decking each side]	Average	Lambda Max	CI	0.02
0.0	0.0	0.0	0.0	0.03	0.0	0.02	0.02	0.03	0.03	0.02	0.03	0.04	0.03	0.03	0.03	0.96	RI	1.58
0.0	0.0	0.0	0.0	0.03	0.0	0.02	0.02	0.03	0.03	0.02	0.03	0.04	0.03	0.03	0.03	0.96	CR	0.01
0.0	0.0	0.0	0.0	0.03	0.0	0.02	0.02	0.03	0.03	0.02	0.03	0.04	0.03	0.03	0.03	0.96		
0.0	0.0	0.0	0.0	0.03	0.0	0.02	0.02	0.03	0.03	0.02	0.03	0.04	0.03	0.03	0.03	0.96		
0.0	0.0	0.0	0.0	0.03	0.0	0.02	0.02	0.03	0.03	0.02	0.03	0.04	0.03	0.03	0.03	0.96		
0.1	0.1	0.1	0.1	0.10	0.1	0.07	0.04	0.09	0.10	0.07	0.10	0.06	0.10	0.10	0.09	1.23		
0.1	0.1	0.1	0.1	0.10	0.1	0.07	0.04	0.09	0.10	0.07	0.10	0.06	0.10	0.10	0.09	1.23		
0.2	0.2	0.2	0.2	0.16	0.2	0.21	0.14	0.16	0.16	0.21	0.16	0.10	0.16	0.16	0.17	1.16		
0.0	0.0	0.0	0.0	0.03	0.0	0.02	0.02	0.03	0.03	0.02	0.03	0.04	0.03	0.03	0.03	0.96		
0.0	0.0	0.0	0.0	0.0	0.0	0.0	0.02	0.03	0.03	0.02	0.03	0.04	0.03	0.03	0.03	0.96		
0.10	0.10	0.10	0.1	0.10	0.1	0.1	0.04	0.09	0.10	0.07	0.10	0.06	0.10	0.10	0.09	1.23		
0.0	0.0	0.0	0.0	0.0	0.0	0.0	0.028	0.03	0.03	0.02	0.03	0.04	0.03	0.03	0.03	0.96		

0.2	0.2	0.2	0.2	0.2	0.3	0.3	0.42	0.22	0.23	0.35	0.23	0.30	0.23	0.23	0.27	0.90
0.0	0.0	0.0	0.0	0.0	0.0	0.0	0.02	0.03	0.03	0.02	0.03	0.04	0.03	0.03	0.03	0.96
0.0	0.0	0.0	0.0	0.03	0.0	0.02	0.02	0.03	0.03	0.02	0.03	0.04	0.03	0.03	0.03	0.96
1.00	1.00	1.00	1.00	1.00	1.00	1.00	1.00	1.00	1.00	1.00	1.00	1.00	1.00	1.00	1.00	15.3

Figure 32: Normalised matrix: cultural restriction on usury.

T2-Ease to remove/reaffix/replace	Compressed Stabilized Rammed Earth blocks	Clay Products- Unfired Bricks	Reclaimed/Recycled laminated Wood Flooring and Panelling	Bamboo XL laminated Split Paneled Flooring	Fly Ash Sand Lime interlocking Paving Bricks/Block	Recycled timber clad Aluminium framed window unit	Four panel hardwood door finished with Alpilignum.	Stainless Steel Entry Door.	Reprocessed Particleboard wood chipboard to BS EN 312 Type P5,	Tongue & grooved Wooddeco Multiline ceiling tiles to BS EN 636-2]	Plasterboard on 70 mm steel studs with 50 mm 12.9 kg/m³ insulation,	Tongue & Grooved Laminated Wooden column bolted to steel plate on concrete base.	Steel Column UC	Structurally insulated timber panel system with OSB/3 each side, roofing underlay reclaimed clay tiles	Structurally insulated natural slate (temperate EN 636-2) decking each side]
Compressed Stabilized Rammed Earth blocks	1.0	0.3	0.2	0.2	0.3	0.2	0.2	0.3	0.2	0.2	0.3	0.20	0.33	0.20	0.20
Clay Products—Unfired Bricks	3.0	1.0	0.3	0.3	0.5	0.3	0.3	1.0	0.3	0.3	1.0	0.3	1.0	0.3	0.3
Reclaimed/Recycled laminated Wood Flooring and Panelling	5.0	3.0	1.0	1.0	2.0	1.0	1.0	3.0	1.0	1.0	3.0	1.0	3.0	1.0	1.0
Bamboo XL laminated Split Paneled Flooring	5.0	3.0	1.0	1.0	2.0	1.0	1.0	3.0	1.0	1.0	3.0	1.0	3.0	1.0	1.0
Fly Ash Sand Lime interlocking Paving Bricks/Block	4.0	2.0	0.5	0.5	1.0	0.5	0.5	2.0	0.5	0.5	2.0	0.5	2.0	0.5	0.5
Recycled timber clad Aluminium framed window unit	5.0	3.0	1.0	1.0	2.0	1.0	1.0	3.0	1.0	1.0	3.0	1.0	3.0	1.0	1.0
Four panel hardwood door finished with Alpilignum.	5.0	3.0	1.0	1.0	2.0	1.0	1.0	3.0	1.0	1.0	3.0	1.0	3.0	1.0	1.0
Stainless Steel Entry Door.	3.0	1.0	0.3	0.3	0.5	0.3	0.3	1.0	0.3	0.3	1.0	0.3	1.0	0.3	0.3
Reprocessed Particleboard wood chipboard to BS EN 312 Type P5,	5.0	3.0	1.0	1.0	2.0	1.0	1.0	3.0	1.0	1.0	3.0	1.0	3.0	1.0	1.0
Tongue & grooved Wooddeco Multiline ceiling tiles to BS EN 636-2]	5.0	3.0	1.0	1.0	2.0	1.0	1.0	3.0	1.0	1.0	3.0	1.0	3.0	1.0	1.0
Plasterboard on 70 mm steel studs with 50 mm 12.9kg/m³ insulation,	3.0	1.0	0.3	0.3	0.5	0.3	0.3	1.0	0.3	0.3	1.0	0.3	1.0	0.3	0.3
Tongue & Grooved Laminated Wooden column bolted to steel plate on concrete base.	5.0	3.0	1.0	1.0	2.0	1.0	1.0	3.0	1.0	1.0	3.0	1.0	3.0	1.0	1.0
Steel Column UC	3.0	1.0	0.3	0.3	0.5	0.3	0.3	1.0	0.3	0.3	1.0	0.3	1.0	0.3	0.3
Structurally insulated timber panel system with OSB/3 each side, roofing underlay reclaimed clay tiles	5.0	3.0	1.0	1.0	2.0	1.0	1.0	3.0	1.0	1.0	3.0	1.0	3.0	1.0	1.0
Structurally insulated natural slate (temperate EN 636-2) decking each side]	5.0	3.0	1.0	1.0	2.0	1.0	1.0	3.0	1.0	1.0	3.0	1.0	3.00	1.00	1.00
Total	62.0	33.3	11.0	11.0	21.3	11.0	11.0	33.3	11.0	11.0	33.3	11.0	33.3	11.0	11.0

Figure 33: Pair-wise matrix: ease to remove/affix/replace.

Compressed Stabilized Rammed Earth blocks	Clay Products- Unfired Bricks	Reclaimed/Recycled laminated Wood Flooring and Panelling	Bamboo XL laminated Split Paneled Flooring	Fly Ash Sand Lime interlocking Paving Bricks/Block	Recycled timber clad Aluminium framed window unit	Four panel hardwood door finished with Alpilgnum.	Stainless Steel Entry Door.	Reprocessed Particleboard wood chipboard to BS EN 312 Type P5,	Tongue & grooved Wooddeco Multiine ceiling tiles to BS EN 636-2]	Plasterboard on 70 mm steel studs with 50 mm 12.9 kg/m³ insulation,	Tongue & Grooved Laminated Wooden column bolted to steel plate on concrete base.	Steel Column UC	Structurally insulated timber panel system with OSB/3 each side, roofing underlay reclaimed clay tiles	Structurally insulated natural slate (temperate EN 636-2) decking each side]	Average	Lambda Max	CI	0.01
0.02	0.01	0.02	0.0	0.01	0.0	0.0	0.01	0.01	0.02	0.01	0.02	0.01	0.02	0.02	0.02	0.95	RI	
0.0	0.0	0.0	0.0	0.0	0.0	0.0	0.03	0.03	0.03	0.03	0.03	0.03	0.03	0.03	0.03	1.03	CR	
0.1	0.1	0.1	0.1	0.1	0.1	0.1	0.09	0.09	0.09	0.09	0.09	0.09	0.09	0.09	0.09	0.99		
0.1	0.1	0.1	0.1	0.1	0.1	0.1	0.09	0.09	0.09	0.09	0.09	0.09	0.09	0.09	0.09	0.99		
0.1	0.1	0.0	0.1	0.05	0.0	0.05	0.06	0.04	0.05	0.06	0.05	0.06	0.05	0.05	0.05	1.08		
0.1	0.1	0.1	0.1	0.09	0.1	0.09	0.09	0.09	0.09	0.09	0.09	0.09	0.09	0.09	0.09	0.99		
0.1	0.1	0.1	0.1	0.09	0.1	0.09	0.09	0.09	0.09	0.09	0.09	0.09	0.09	0.09	0.09	0.99		
0.0	0.0	0.0	0.0	0.02	0.0	0.03	0.03	0.03	0.03	0.03	0.03	0.03	0.03	0.03	0.03	1.03		
0.1	0.1	0.1	0.1	0.09	0.1	0.09	0.09	0.09	0.09	0.09	0.09	0.09	0.09	0.09	0.09	0.99		
0.0	0.0	0.0	0.0	0.02	0.0	0.03	0.03	0.03	0.03	0.03	0.03	0.03	0.03	0.03	0.03	1.03		
0.1	0.1	0.1	0.1	0.09	0.1	0.09	0.09	0.09	0.09	0.09	0.09	0.09	0.09	0.09	0.09	0.99		
0.0	0.0	0.0	0.0	0.02	0.0	0.03	0.03	0.03	0.03	0.03	0.03	0.03	0.03	0.03	0.03	1.03		
0.1	0.1	0.1	0.1	0.1	0.1	0.1	0.09	0.09	0.09	0.09	0.09	0.09	0.09	0.09	0.09	0.99		
0.08	0.09	0.09	0.1	0.09	0.1	0.1	0.09	0.09	0.09	0.09	0.09	0.09	0.09	0.09	0.09	0.99		
1.00	1.00	1.00	1.00	1.00	1.00	1.00	1.00	1.00	1.00	1.00	1.00	1.00	1.00	1.00	1.00	15.1		

Figure 34: Normalised matrix: ease to remove/affix/replace.

SN5- Acoustics Performance	Compressed Stabilized Rammed Earth blocks	Clay Products- Unfired Bricks	Reclaimed/Recycled laminated Wood Flooring and Panelling	Bamboo XL laminated Split Paneled Flooring	Fly Ash Sand Lime interlocking Paving Bricks/Block	Recycled timber clad Aluminium framed window unit	Four panel hardwood door finished with Alpilgnum.	Stainless Steel Entry Door.	Reprocessed Particleboard wood chipboard to BS EN 312 Type P5,	Tongue & grooved Wooddeco Multiine ceiling tiles to BS EN 636-2]	Plasterboard on 70 mm steel studs with 50 mm 12.9 kg/m³ insulation,	Tongue & Grooved Laminated Wooden column bolted to steel plate on concrete base.	Steel Column UC	Structurally insulated timber panel system with OSB/3 each side, roofing underlay reclaimed clay tiles	Structurally insulated natural slate (temperate EN 636-2) decking each side]
Compressed Stabilized Rammed Earth blocks	1.0	0.2	0.3	0.2	0.2	0.3	0.2	1.0	0.3	0.3	0.3	0.3	0.3	0.3	1.0
Clay Products—Unfired Bricks	5.0	1.0	2.0	1.0	1.0	3.0	1.0	5.0	2.0	2.0	2.0	2.0	2.0	2.0	5.0

Material															
Reclaimed/Recycled laminated Wood Flooring and Panelling	4.0	0.5	1.0	0.5	0.5	2.0	0.5	4.0	1.0	1.0	1.0	1.0	1.0	1.0	4.0
Bamboo XL laminated Split Paneled Flooring	5.0	1.0	2.0	1.0	1.0	3.0	1.0	5.0	2.0	2.0	2.0	2.0	2.0	2.0	5.0
Fly Ash Sand Lime interlocking Paving Bricks/Block	5.0	1.0	2.0	1.0	1.0	3.0	1.0	5.0	2.0	2.0	2.0	2.0	2.0	2.0	5.0
Recycled timber clad Aluminium framed window unit	3.0	0.3	0.5	0.3	0.3	1.0	0.3	3.0	0.5	0.5	0.5	0.5	0.5	0.5	3.0
Four panel hardwood door finished with Alpilignum.	5.0	1.0	2.0	1.0	1.0	3.0	1.0	5.0	2.0	2.0	2.0	2.0	2.0	2.0	5.0
Stainless Steel Entry Door.	1.0	0.2	0.3	0.2	0.2	0.3	0.2	1.0	0.3	0.3	0.3	0.3	0.3	0.3	1.0
Reprocessed Particleboard wood chipboard to BS EN 312 Type P5,	4	0.5	1	0.5	0.5	2	0.5	4	1	1	1	1	1	1	4
Tongue & grooved Wooddeco Multiline ceiling tiles to BS EN 636–2]	4.0	0.5	1.0	0.5	0.5	2.0	0.5	4.0	1.0	1.0	1.0	1.0	1.0	1.0	4.0
Plasterboard on 70 mm steel studs with 50 mm 12.9 kg/m² insulation,	4	0.5	1	0.5	0.5	2	0.5	4	1	1	1	1	1	1	4
Tongue & Grooved Laminated Wooden column bolted to steel plate on concrete base.	4	0.5	1	0.5	0.5	2	0.5	4	1	1	1	1	1	1	4
Steel Column UC	4.00	0.50	1.00	0.50	0.50	2.00	0.50	4.00	1.00	1.00	1.00	1.00	1.00	1.00	4.00
Structurally insulated timber panel system with OSB/3 each side, roofing underlay reclaimed clay tiles	4.0	0.5	1.0	0.5	0.5	2.0	0.5	4.0	1.0	1.0	1.0	1.0	1.0	1.0	4.0
Structurally insulated natural slate (temperate EN 636-2) decking each side]	1.0	0.2	0.3	0.2	0.2	0.3	0.2	1.0	0.3	0.3	0.3	0.3	0.3	0.3	1.0
Total	54.0	8.4	16.3	8.4	8.4	28.0	8.4	54.0	16.3	16.3	16.3	16.3	16.3	16.3	54.0

Figure 35: Pair-wise matrix: acoustics performance.

Compressed Stabilized Rammed Earth blocks	Clay Products- Unfired Bricks	Reclaimed/Recycled laminated Wood Flooring and Panelling	Bamboo XL laminated Split Paneled Flooring	Fly Ash Sand Lime interlocking Paving Bricks/Block	Recycled timber clad Aluminium framed window unit	Four panel hardwood door finished with Alpilignum.	Stainless Steel Entry Door.	Reprocessed Particleboard wood chipboard to BS EN 312 Type P5,	Tongue & grooved Wooddeco Multiline ceiling tiles to BS EN 636–2]	Plasterboard on 70 mm steel studs with 50 mm 12.9 kg/m³ insulation,	Tongue & Grooved Laminated Wooden column bolted to steel plate on concrete base.	Steel Column UC	Structurally insulated timber panel system with OSB/3 each side, roofing underlay reclaimed clay tiles	Structurally insulated natural slate (temperate EN 636-2) decking each side]	Average	Lambda Max	CI	0.01
0.0	0.0	0.0	0.0	0.0	0.0	0.0	0.01	0.01	0.02	0.02	0.02	0.02	0.02	0.02	0.02	0.97	RI	1.58
0.1	0.1	0.1	0.1	0.1	0.1	0.1	0.09	0.12	0.12	0.12	0.12	0.12	0.12	0.09	0.11	0.97	CR	0.01
0.1	0.1	0.1	0.1	0.1	0.1	0.1	0.07	0.061	0.06	0.06	0.06	0.06	0.06	0.07	0.06	1.04		
0.1	0.1	0.1	0.1	0.1	0.1	0.1	0.09	0.123	0.12	0.12	0.12	0.12	0.12	0.09	0.11	0.97		
0.1	0.1	0.1	0.1	0.1	0.1	0.1	0.09	0.123	0.12	0.12	0.12	0.12	0.12	0.09	0.11	0.97		
0.1	0.0	0.0	0.0	0.0	0.0	0.0	0.05	0.030	0.03	0.03	0.03	0.03	0.03	0.06	0.04	1.07		
0.1	0.1	0.1	0.1	0.1	0.1	0.1	0.09	0.123	0.12	0.12	0.12	0.12	0.12	0.09	0.11	0.97		
0.0	0.0	0.0	0.0	0.0	0.0	0.0	0.018	0.01	0.02	0.02	0.02	0.02	0.02	0.02	0.02	0.97		
0.074	0.059	0.06	0.059	0.059	0.071	0.059	0.074	0.061	0.061	0.061	0.061538462	0.62	0.061538462	0.074074074	0.06	1.04		
0.1	0.1	0.1	0.1	0.1	0.1	0.1	0.074	0.061	0.06	0.06	0.06	0.06	0.06	0.07	0.06	1.04		
0.07	0.059	0.06	0.059	0.059	0.071	0.059	0.074	0.061	0.061	0.061	0.061538462	0462	0.061538462	0.074074074	0.06	1.04		
0.07	0.059	0.06	0.059	0.059	0.071	0.059	0.074	0.061	0.061	0.061	0.061538462	0.0	0.061538462	0.074074074	0.06	1.04		
0.07	0.06	0.06	0.06	0.06	0.07	0.06	0.074	0.06	0.061	0.061	0.061538462	62	0.061538462	0.074074074	0.06	1.04		
0.1	0.1	0.1	0.1	0.1	0.1	0.1	0.074	0.061	0.06	0.06	0.06	0.06	0.06	0.07	0.06	1.04		
0.0	0.0	0.0	0.0	0.0	0.0	0.0	0.018518519	0.015384615	0.02	0.02	0.02	0.02	0.02	0.02	0.02	0.97		
1.00	1.00	1.00	1.00	1.00	1.00	1.00	1.00	1.00	1.00	1.00	1.00	1.00	1.00	1.00	1.00	15.2		

Figure 36: Normalised matrix: acoustics performance.

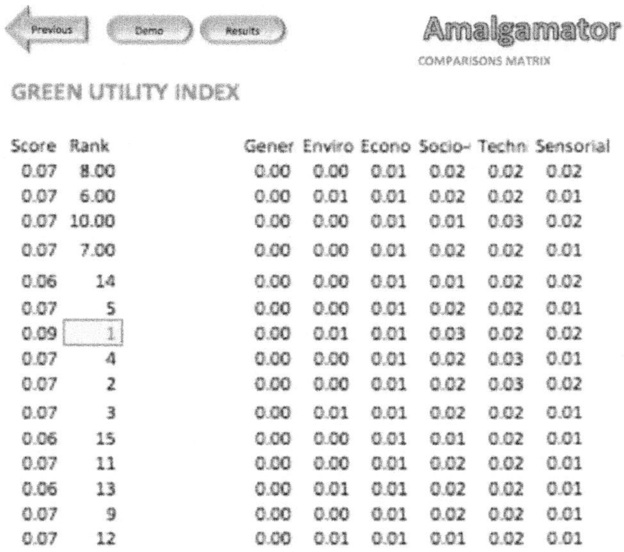

Previous   Demo   Results

Amalgamator
COMPARISONS MATRIX

GREEN UTILITY INDEX

Score	Rank		Gener	Enviro	Econo	Socio-	Techn	Sensorial
0.07	8.00		0.00	0.00	0.01	0.02	0.02	0.02
0.07	6.00		0.00	0.01	0.01	0.02	0.02	0.01
0.07	10.00		0.00	0.00	0.01	0.01	0.03	0.02
0.07	7.00		0.00	0.00	0.01	0.02	0.02	0.01
0.06	14		0.00	0.00	0.01	0.01	0.02	0.02
0.07	5		0.00	0.00	0.01	0.02	0.02	0.01
0.09	1		0.00	0.01	0.01	0.03	0.02	0.02
0.07	4		0.00	0.00	0.01	0.02	0.03	0.01
0.07	2		0.00	0.00	0.01	0.02	0.03	0.02
0.07	3		0.00	0.01	0.01	0.02	0.02	0.01
0.06	15		0.00	0.00	0.01	0.01	0.02	0.01
0.07	11		0.00	0.00	0.01	0.02	0.02	0.01
0.06	13		0.00	0.01	0.01	0.02	0.02	0.01
0.07	9		0.00	0.00	0.01	0.02	0.02	0.01
0.07	12		0.00	0.01	0.01	0.01	0.02	0.01

Figure 37: Green utility indices of the selected materials.

POTENTIAL BENEFITS OF THE MSDSS MODEL

The following are the benefits expected from the application of the MSDSS Model. However the model developed for this research differs from that of the previous works in the following ways:

- The main point of difference from the off-the-shelf assessment tools is that they only trade-off numerical values based on the single-attributes. These singleattribute claims ignore the possibility of what other variables can yield. MSDSS supports trade-off with and without tangible variables, such as a client's preference, environmental statutory compliance, and cultural restriction on usury. This feature is important as decision making in reality engages with solid, verbal and subjective elements.

- In terms of cost, it provides an opportunity for designers to be able to advise their clients as to what the probable financial estimate of the project may be. This helps clients to decide how much they are prepared to spend on different variables of construction.

- A separate set of contextual considerations was included as a heuristics base to facilitate site-specific feasibility and appropriateness testing of each material choice. Boundaries of sustainability inform of knowledge base rules as contained in the MSDSS model could help reduce bias that is often associated with the material selection process.

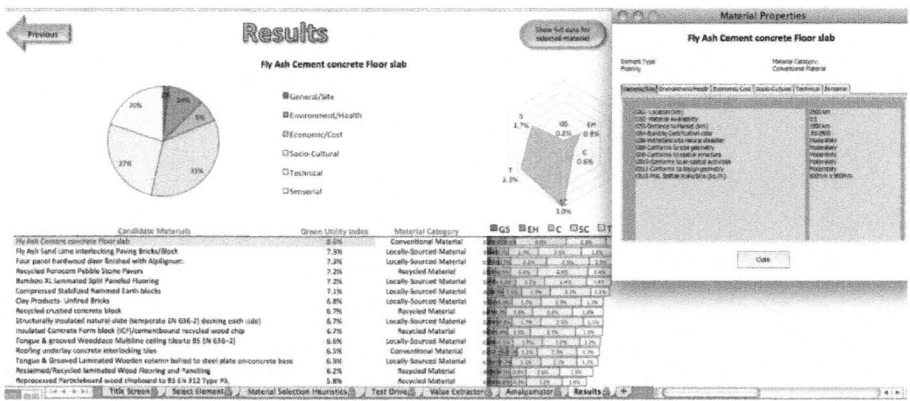

Figure 38: Corresponding indices of the ranked materials.

- Available material assessment tools are particularity ill-adapted for the early stages of the design process and are generally labour intensive. The MSDSS model consists of a resource for relatively small information input to produce quick and fairly accurate or approximate output of results with little or no training on the part of experienced users. This means that users that may require little training are inexperienced users but not as extensive as obtainable in previous tools.

- There are still significant numbers of smaller firms who cannot afford most material assessment tools because they are extremely expensive. This tool is more or less open source software recommended to provide solution to this challenge.

- Context is a critical consideration for all project decision-making, since even projects located on neighbouring sites will have different end users, and different specific site characteristics. This tool could be applied to other regions with minimal or no changes, and therefore has the ability to adapt to any situation, or change in design according to users' needs or different material alternatives.

- Unlike in the previous models, this tool contains tutorials and help menu as well as video guidance on how to use the software. This provides adequate help to beginners or inexperienced designers.

- For the visual aspect, the MSDSS model has the ability to produce a picture representative of data input rather than abstract. It is able to transfer data from it to other software, applicable to building material selection, and present the properties of each material in a successive window.

- User weightings have been included in the selection methodology to supplement, and not supplant human judgment in the decision-making process. By incorporating user weightings into the selection process, the methodology gains greater acceptability to the user who supplies the weightings.

- Materials change in their innovation, composition, price and availability and most tools find it challenging to update information relating to products. In this MSDSS model, the materials and the corresponding performance of the selected products is updated through a link to the manufacturers web page on the internet, and the users may access more information regarding the selected material or technology through internet from the supplier's web pages.

- The system has been designed to produce an artistic output, accurate, detailed representation and close to reality as much as it can be, without attempt to conceal any feature whether attractive or not;

- Provision of only a limited set of operations or criteria restricts the techniques and solutions that can be applied and consequently restricts the decision-making process. On the other hand, the inclusion of many objectives and the permitting of user specification of input data, system parameters and models, generally increases system flexibility and increases decision support freedom;

- In most tools, AHP technique at the pare-wise comparison stage, tend to be quite cumbersome and often takes a lot of time to maintain the consistency of the response. To eliminate this challenge MSDSS automatically debugs the system at every stage of the evaluation and selection process.

- The system has been thoroughly debugged to be less error prone, so that practitioners can integrate the decisions made by the tools more smoothly into practice, and that it takes less than few seconds to respond to users inputs;

- Responses/feedback from system programmers and accredited green building experts have also been included in the study to prove the ease of use, applicability and usability of the MSDSS model (see appendix A). As a result, some features have been adjusted based on expert feedbacks to support more reliable and expedient, timelier feedback to different design alternatives or changes.

REFLECTIVE SUMMARY

This paper discussed the process of developing a decision-support system to support choices in low-cost green building materials. The research presented in this paper acknowledged the lack of a reliable database model that decision makers can readily use to aid informed decision-making when selecting low-cost green materials for low-cost green residential housing development. The findings from the reviewed literature and the results of the surveyed questionnaire further underscored the need for improving understanding of relevant data associated with the use of such building materials and components, with the goal to change and positively influence the current mental models, attitudes and priorities of multiple stakeholders involved in the production of the built environment, so as to encourage their wider-scale use in mainstream housing. Based on the data obtained from selected expert builder/developer companies, a prototype MSDSS model was developed to aid designers in making informed decisions regarding their choice of materials for low-cost green residential housing projects. This model was consolidated in to an excel-based decision tool that allows designers to select low-cost green building products from a range of possibilities, and view the resulting impacts and difference in the cost, durability and performance of a range of alternatives. An analysis using the Analytical Hierarchy Process (AHP), based on the results of the participants was performed to show how optimal choices could change with changing user weightings and variables. The participants gained views from participating in the evaluation exercise for a real-life project, including the difficulties in choosing preference scores. This study thus, indicates that perhaps the development of a DSS model associated with the impacts of lowcost green building materials is useful in that it gives designers a new approach of going through the process of value elicitation, which allows them to explicitly and transparently test the impacts of their elicited values. Providing a visual representation, allowing designers or specifiers to compare multiple alternatives across multiple criteria, was a particularly useful aspect of this study.

CONCLUSIONS

This report has demonstrated how a DSS model can be used to support multi-stakeholder involvement in the selection of low-cost green construction materials in ways that enable building energy performance and lifecycle cost to be considered at the early stage of residential housing design. The study further reinforced the significance in taking a multi-attribute approach to assessing a building product's sustainable performance. To achieve this goal, the AHP model of decision-making

[57-60]. was adopted to deal with the ambiguities involved in the assessment of material alternatives and relative importance weightings of multiple factors, given its ability to solve multi-criteria decision-making (MCDM) between finite alternatives. To prove the validity of the model and the feasibility of the proposed selection methodology, a real-life but hypothetical application scenario was used to further illustrate the application of the MSDSS model in selecting the most appropriate floor material for a single 5-bedroom residential housing project located in the Sutton County of London.

The results demonstrated the capabilities of the system, and exposed the way in which the system transparently demonstrates the implications of each step of the analysis. It also proved the practicality of using the MSDSS model, as it combines multiple factors into a single performance value that is easily interpreted. Since the purpose of this research study was to develop an innovative concept to demonstrate a step-bystep methodology for selecting low-cost green materials with reasonable accuracy and in real time, as opposed to developing a fully-equipped commercial software, macro-in-excel database management technique was used in the back-end of the system to integrate the large volumes of data obtained from multiple sources.

Excel was adopted as the database management system since it has the capabilities to perform all necessary calculations and is common enough that most people are familiar with it. The process followed to develop the prototype MSDSS model in this research demonstrates that, depending on the domain and scope of the problem at hand, a DSS can be built fairly quickly and can be used effectively to help designers quantify how they compare materials that are yet to be certified under the standard specifications and codes of practice, and that which are already permitted under existing codes. However further work is required to fully validate the MSDSS and the methodology presented. To do so, this research intends to run further case studies ideally using "live" building design projects, by comparing the outputs from the algorithms of the MSDSS system to monitored data from the completed case study building, in order to review the potential savings of the new materials or components proposed by the MSDSS model.

Contributions to Research and Industry

Insights identified from addressing the research objectives in Section 3 represent part of the original contribution to knowledge made by this study. The following are itemised as key contributions of the study to research and practice:

- The contribution of this research includes the consideration of a holistic approach to low-cost green building product selection based on socio-cultural, technical, emotive, site, cost and environmental performance. Pre-design estimators and pre-construction managers could improve their estimating and product selection practices using the proposed MSDSS tool.

- Material suppliers can also benefit from this approach, as they can use it to enhance their pricing strategies, marketing plans, and overall product competitiveness.

- Decision problems about a product's choice are usually unstructured and ill-defined. By suggesting an alternative means of integrating the available resources associated with the informed selection of low-cost green building materials, it is hoped that the model will help decision makers to further refine their material selection criteria thus, encourage effective decision-making.

- The material selection process is characterized by competitive objectives, involving multiple stakeholders and key actors, dynamic and uncertain procedures and limited timeframes to make significant decisions. The decision makers within this domain: the designers, specifiers and other stakeholders are often confronted with conflicting subjective preferences and fragmented expertise; hence resulting in decision-making failures. The capacity of the system to compare materials using multiple factors with user-specified weightings, will therefore, encourage decision-makers to explicitly consider the effects of their previously-implicit judgments on the outcome of the project, and thus make choices that are timely, and result in more sustainable residential housing project design and implementation.

- The ability to quickly quantify and qualify the suitability outcomes of alternative materials may encourage greater industry acceptance of innovative technology for materials that are yet to be certified under the standard specifications and codes of practice.

- The overall approach used here could be tested in other contexts to determine its generalizability and applicability. In other words, the system could be extended to select materials for commercial development or for any other purpose.

- The material selection factors identified in the prototype model of the MSDSS, provides a unique insight into sustainability and environmental design information requirements for low-cost green housing.

- The adopted research methodology (see Table 1) employed to address the research objectives in Section 3 represents part of the original contribution to knowledge made by this study.

- The number of academic publications on the impacts of low-cost green materials was found to be low; hence makes a crucial contribution.

- In the short term, the model could be used in the housing sector as a catalogue of materials to support decision-making in low-cost green housing designs.

- As low-cost green building materials and components become well understood by design and building professionals, there is a likelihood of reducing over-dependency on conventional construction materials in the housing industry.

- The outcome of this study could aid top executives within the housing sector to consider low-cost green materials as part of existing regulatory frameworks and building codes of the Construction Standards Institute (CSI) in capital projects. By doing so, such an approach may create a potential market for local manufacturing and processing of such materials.

Setbacks, Challenges and Probable Solutions

There were few possible limitations that this research faced during the cause of the study. The limitations are hereby listed for future consideration.

- The process of developing the selection methodology was faced with critical issues that led to several changes in the research methodology and its objectives so many times, in order to achieve the aim of this research.

- Citing prior research studies formed the basis of the literature review and helped lay the foundation for understanding the research problem investigated in this study. However, there were reservations regarding the currency and scope of the research topic, as there was no compelling evidence of prior research on the topic. As literature on DSS for low-cost green housing design is still relatively low, the study therefore had to rely on the most current reports, interviews, and observations from the different and various organisations, and building professionals for its information.

- It remains true that sample sizes that are too small cannot adequately support claims of having achieved valid conclusions and sample sizes that are too large do not permit the deep, naturalistic, and inductive analysis that defines qualitative inquiry [47]. Yin [47]. Noted that determining adequate sample size in qualitative research is ultimately a matter of judgment and experience in evaluating the quality. Hair et al. [61]. warned that it is important to consider not only the statistical significance, but also the quality and practical significance of the results for managerial applications, when analysing data. They noted that unequal or uneven sample sizes amongst different professional groups could also bias or influence the results as getting equal sample sizes from different groups of respondents was unrealistic and demanding. To address this issue the study adopted a sampling strategy using the stratified random sampling approach where each group of the sample population had reasonable number of randomly selected participants, which helped to achieve sampling equivalence between the researcher and professionals of the various building professions both in higher institutions and practicing building design and housing construction firms.

- Giving that most respondents were practicing professionals, getting a list of the sample population for the study was very discouraging. Having access to people, and organizations, was otherwise limited, giving the time differences and tight-scheduled activities. However the use of progressive approach of reminding the subjects using any available means either through e-mails, LinkedIn, Facebook, Twitter or through phone calls helped to address this problem.

- Very few of the participants had little exposure to AHP quantitative-based decision-making process. Though they found the process a bit daunting, they were somewhat comfortable with the idea of ranking preferences, as they were used to considering the choice for alternatives based on unquantified methods, but without assigning personal values to criteria. Prior help manual sent to participants before embarking on expert evaluation survey helped to reduce the complexities associated with the MCDM technique adopted.

Potential Areas for Further Studies

Several areas were identified as potential areas for further research as itemised below:

- Although not demonstrated in this system but it is also possible that potential researchers can redesign or customize the database to best fit the needs of any particular region or could be extended to select materials for commercial development;

- While the findings of this research focused specifically on a subset of design and building professionals involved with public residential housing sector projects, the overall approach used here could be tested in other contexts to determine its generalizability and applicability.

ACKNOWLEDGEMENTS

This work was made possible through private funding, and was partially supported by a discount from the Journal of Building Construction and Planning Research Doctoral Research Student Discount Scheme program.

APPENDIX A: Feedbacks from Evaluators

The following are feedbacks and suggestions retrieved from users on the MSDS tool. The names of the participants were undisclosed to respect their anonymity. "The system relates to issues concerned with local knowledge, local materials data, local climate know-how, local experts needed to operate system, which are hardly considered in other systems". I think it shows great promise and the mechanics are very well-developed and user-friendly, "Material costs vary from location to location (especially in the USA where material costs vary not just from state to state but also from city to city". Perhaps when the material selection is sorted by the element choice, this will seem more useful". "It depends on what resources you are referring to; if referring to the underlying database, those are considerable. If referring to the resource needs of the organization that would use the model, not too costly to operate". "The interface is very well-designed and easy to navigate. However, there is a need for more explanatory material to allow the user to understand what s/he is actually doing, and how to operate some parts of the model appropriately". "In terms of its operation, interoperability, flexibility, usability and applicability, per se, it is very clear and straightforward; it's the underlying premise and data that needs little clarification in order for the user to operate the model effectively.

REFERENCES

1. IEA (International Energy Agency), "Energy Efficiency Requirements in Building Codes, Energy Efficiency Policies for New Buildings," OECD/IEA, Paris, 2008.

2. IEA (International Energy Agency), "IEA Net Zero Energy," Montreal, 2009.

3. World Bank, "Nigeria: State Building, Sustaining Growth, and Reducing Poverty. A Country Economic Report," Report 29551-NG, Poverty Reduction and Economic Management Sector Unit, West Africa Region, Washington DC, 2010.

4. UN-HABITAT, "Global Campaign on Urban Governance," Oxford University Press, New York, 2011. http://www.unhabitat.org.

5. United Nations Development Plan (UNDP), "African Economic Outlook 2011: Africa and Its Emerging Partners," African Development Bank, OECD, UNDP and UNECA, 2011.

6. United States Department of Energy (USDOE), "Energy Efficiency and Renewable Energy," Federal Energy Management Program, 2010, pp. 1-34.

7. United States Department of Energy, "About the Weatherization Assistance Program," Washington DC, 2010. http://www1.eere.energy.gov/wip/wap.html

8. J. Kennedy, "Building without Borders: Sustainable Construction for the Global Village," New Society Publishers, Gabriola, 2004.

9. M. Shuman, "The Small-Mart Revolution: How Local Businesses Are Beating the Global Competition," BerrettKoehler Publishers, San Francisco, 2008.

10. Y. Oruwari, M. Jev and P. Owei, "Acquisition of Technological Capability in Africa: A Case Study of Indigenous Building Materials Firms in Nigeria," ATPS Working Paper Series No. 33, African Technology Policy Studies Network, Nairobi, 2002.

11. K. K. Ashraf, "This Is Not a Building! Hand-Making a School in a Bangladeshi Village," Architectural Design, Vol. 77, No. 6, 2007, pp. 114-117. http://dx.doi.org/10.1002/ad.575

12. C. C. Zhou, G. F. Yin and X. B. Hu, "Multi-Objective Optimization of Material Selection for Sustainable Products: Artificial Neural Networks and Genetic Algorithm Approach," Materials & Design, Vol. 30, No. 4, 2009, pp. 1209-1215. http://dx.doi.org/10.1016/j.matdes.2008.06.006

13. P. Zhou, B. W. Ang and D. Q. Zhou, "Weighting and Aggregation in Composite Indicator Construction: A Multiplicative Optimization Approach," Social Indicator Research, Vol. 96, No. 1, 2010, pp. 169-181. http://dx.doi.org/10.1007/s11205-009-9472-3

14. G. Seyfang, "Community Action for Sustainable Housing: Building a Low Carbon Future," Energy Policy, Vol. 38, No. 12, 2010, pp. 7624-7633.

15. M. Malanca, "Green Building Rating Tools in Africa," In: Conference on Promoting Green Building Rating in Africa, Green Building Africa, Nairobi, 4-6 May 2010, pp. 16-25.

16. L. Wastiels, I. Wouters and J. Lindekens, "Material Knowledge for Design: The Architect's Vocabulary, Emerging Trends in Design Research," International Association of Societies of Design Research

(IASDR) Conference, Hong Kong, 16-19 July 2007.

17. M. C. Quinones, "Decision Support System For Building Construction Product Selection Using Life-Cycle Management," A Thesis Presented to the Academic Faculty in Partial Fulfillment of the Requirements for the Degree Master of Science in Building Construction and Facility Management, Georgia Institute of Technology, Atlanta, 2011.

18. W. B. Trusty, "Incorporating LCA in Green Building Rating Systems," Air & Waste Management Association, Ottawa, 2009.

19. W. B. Trusty, "Sustainable Building: A Materials Perspective," Prepared for Canada Mortgage and Housing Corporation Continuing Education Series for Architects, 2003.

20. W. B. Trusty, "Understanding the Green Building Toolkit: Picking the Right Tool for the Job," Proceedings of the USGBC Greenbuild Conference & Expo, Pittsburgh, 2003.

21. W. B. Trusty, J. K. Meril and G. A. Norris, "ATHENA: A LCA Decision Support Tool for the Building Community," Proceedings: Green Building Challenge '98—An International Conference on the Performance Assessment of Buildings, Vancouver, 26-28 October 1998, p. 8.

22. T. Woolley, "Natural Building: A Guide to Materials and Techniques," The Crowood Press Ltd, Ramsbury, Marlborough, Wiltshire, 2006.

23. United States Green Building Council (USGBC), "LEEDLeadership in Energy and Environmental Design: Pilot Credit Library: Pilot Credit 1—Life Cycle Assessment of Building Assemblies and Materials," US Green Building Council, 2010.

24. L. Florez, D. Castro and J. Irizarry, "Impact of Sustainability Perceptions on Optimal Material Selection in Construction Projects," Proceedings of the Second International Conference on Sustainable Construction Materials and Technologies, University Politecnica delle Marche, Ancona, Italy, Coventry University and The University of Wisconsin Milwaukee Centre for By-products Utilization, 28-30 June 2010, pp. 719-727. http://www.claisse.info/Proceedings.htm,

25. L. Florez, D. Castro-Lacouture and J. Irizarry, "Impact of Sustainability Perceptions on the Purchasability of Materials in Construction Projects," Proceedings of the 2009 ASCE Construction Research Congress, Banff, 8-10 May 2010, pp. 226-235

26. D. Castro-Lacouture, J. A. Sefair, L. Florez and A. L. Medaglia, "Optimization Model for the Selection of Materials Using the LEED Green Building Rating System," Proceedings of the 2009 ASCE Construction Research Congress, Seattle, Washington, 5-7 April 2009,

pp. 608- 617.

27. E. Keysar and A. Pearce, "Decision Support Tools for Green Building: Facilitating Selection among New Adopters on Public Sector-Projects," Journal of Green Building, Vol. 2, No. 3, 2007, pp. 153-171. http://dx.doi. org/10.3992/jgb.2.3.153

28. C. Bayer, M. Gamble, R. Gentry and S. Joshi, "AIA Guide to Building Life Cycle Assessment in Practice," The American Institute of Architects, Washington DC, 2010.

29. ATHENA Institute, "The Impact Estimator for Buildings," 2011. http:// athenasmi.org/tools/impactEstimator/

30. ATHENA Institute, "The EcoCalculator for Buildings," 2011. http:// athenasmi.org/tools/ecoCalculator/index.html

31. Z. Kapelan, D. Savic and G. Walters, "Decision-Suppport Tools for Sustainable Urban Development," Proceedings of the Institution of Civil Engineers, Engineering Sustainability, Vol. 158, No. 3, 2005, pp. 135-142.

32. S. Rahman, S. Perera, H. Odeyinka and Y. Bi, "A Knowledge-Based Decision Support System for Roofing Materials selection and Cost Estimating: A Conceptual Framework and Catamodelling," 25th Annual ARCOM Conference, Nottingham, 7-9 September 2009, pp. 1-10.

33. S. Rahman, S. Perera, H. Odeyinka and Y. Bi, "A Conceptual Knowledge-Based Cost Model for Optimising the Selection of Material and Technology for Building Design," In: A. R. J. Dainty, Ed, 24th Annual ARCOM Conference, Association of Researchers in Construction Management, University of Glamorgan, 1-3 September 2008, pp. 217-225.

34. E. Loh, T. Crosbie, N. Dawood and J. Dean, "A Framework and Decision Support System to Increase Building Life Cycle Energy Performance," Journal of Information Technology in Construction, Vol. 15, No. 2, 2010, pp. 337-353.

35. G. K. C. Ding, "Sustainable Construction: The Role of Environmental Assessment Tools," Journal of Environmental Management, Vol. 86, No. 3, 2008, pp. 451-464. http://dx.doi.org/10.1016/j.jenvman.2006.12.025

36. C. Hopfe, C. Struck, et al., "Exploration of Using Building Performance Simulation Tools for Conceptual Building Design," IBPSA-NVL Conference, Delft, 20 October 2005, pp. 1-8.

37. R. S. Perera and U. Fernando, "Cost Modelling for Roofing Material Selection," Built Environment: Srilanka, Vol. 3, No. 1, 2002, pp. 11-24.

38. A. Mohamed and T. Celik, "An Integrated KnowledgeBased System for Alternative Design and Materials Selection and Cost Estimating," Expert Systems with Applications, Vol. 14, No. 3, 1998, pp. 329-339. http://dx.doi.org/10.1016/S0957-4174(97)00086-9

39. M. A. A. Mahmoud, M. Aref and A. Al-Hammad, "An Expert System for Evaluation and Selection of Floor Finishing Materials," Expert Systems with Applications, Vol. 10, No. 2, 1996, pp. 281-303. http://dx.doi.org/10.1016/0957-4174(95)00054-2

40. K. Lam and N. Wong, "A study of the Use of Performance Based Simulation Tools for Building Design and Evaluation in Singapore," IBPSA, Kyoto, 1999.

41. J. L. Chen, S. H. Sun and W. C. Hwang, "An Intelligent Data Base System for Composite Material Selection in Structural Design," Engineering Fracture Mechanics, Vol. 50, No. 5-6, 1995, pp. 935-946. http://dx.doi.org/10.1016/0013-7944(94)E0068-R

42. G. Soronis, "An Approach to the Selection of Roofing Materials for Durability," Construction and Building Materials, Vol. 6, No. 1, 1992, pp. 9-14.

43. I. Giorgetti and A. Lovell, "Sustainable Building Practices for Low Cost Housing: Implications for Climate Change Mitigation and Adaptation in Developing Countries," Giorgetti and Lovell, South Africa, 2010.

44. R. Ellis, "Who Pays for Green Buildings? The Economics of Sustainable Buildings," CB Richard Ellis and EMEA Research, New York, 2009.

45. R. J. Cole, "Building Environmental Assessment Methods: Redefining Intentions and Roles," Building Research and Information, Vol. 35, No. 5, 2005, pp. 455-467.

46. R. J. Cole, G. Lidnsey and J. A. Todd, "Assessing Life Cycles: Shifting from Green to Sustainable Design," Proceedings: International Conference Sustainable Building, Rotterdam, 22-25 October 2000, pp. 22-24.

47. R. K. Yin, "Case Study Research: Design and Methods," 4th Edition, Sage Publications, Los Angeles, 2009.

48. B. Reza, R. Sadiq and K. Hewage, "Sustainability Assessment of Flooring Systems in the City of Tehran: An AHPBased Life Cycle Analysis," Construction and Building Materials, Vol. 25, No. 4, 2011, pp. 2053-2066.

49. D. K. H. Chua, Y. C. Kog and P. K. Loh, "Critical Success Factors for Different Project Objectives," Journal of Construction Engineering

and Management, Vol. 125, No. 3, 1999, pp. 142-150. http://dx.doi. org/10.1061/(ASCE)0733-9364(1999)125:3(142)

50. T. L. Saaty, "Relative Measurement and Its Generalization in Decision Making Why Pairwise Comparisons Are Central in Mathematics for the Measurement of Intangible Factors the Analytic Hierarchy/Network Process," RACSAMRevista de la Real Academia de Ciencias Exactas, Fisicas y Naturales. Serie A. Matematicas, Vol. 102, No. 2, 2008, pp. 251-318.

51. T. L. Saaty, "Time Dependent Decision-Making; Dynamic Priorities in the AHP/ANP: Generalizing From Points to Functions and from Real to Complex Variables," Mathematical and Computer Modelling, Vol. 46, No. 7-8, 2007, pp. 860-891

52. T. L. Saaty, "Decision Making for Leaders: The Analytic Hierarchy Process for Decisions in a Complex World," RWS Publications, Pittsburgh, 2001.

53. T. L. Saaty, "Fundamentals of the Analytic Hierarchy Process," RWS Publications, Pittsburgh, 2000.

54. T. L. Saaty, "Fundamentals of Decision Making and Priority Theory with the Analytic Hierarchy Process," RWS Publishers, Pittsburgh, 1994.

55. T. L. Saaty, "The Analytic Hierarchy Process," McGrawHill, New York, 1980.

56. J. A. Alonso and M. T. Lamata, "Consistency in the Analytic Hierarchy Process: A New Approach," International Journal of Uncertainty, Fuzziness and KnowledgeBased Systems, Vol. 14, No. 4, 2006, pp. 445-459. http://dx.doi.org/10.1142/S0218488506004114

57. P. Gluch and H. Baumann, "The Life Cycle Costing (LCC) Approach: A Conceptual Discussion of its Usefulness for Environmental Decision Making," Building and Environment, Vol. 39, No. 5, 2004, pp. 571-580. http://dx.doi.org/10.1016/j.buildenv.2003.10.008

58. C. J. Kibert, "Sustainable Construction: Green Building Design and Delivery," 2nd Edition, John Wiley and Sons, Inc., Hoboken, 2008.

59. R. Spiegel and D. Meadows, "Green Building Materials: A Guide to Product Selection and Specification," John Wiley & Sons, Inc., New York, 2010, pp. 1-7.

60. M. F. Ashby and K. Johnson, "Materials and Design: The Art and Science of Material Selection in Product Design," Butterworth-Heinemann, Oxford, Boston, 2002.

61. J. F. Hair, R. E. Anderson, R. L. Tatham and W. C. Black, "Multivariate Data Analysis," Prentice Hall, Upper Saddle River, 1998.

Chapter 6

IMPACTS OF AN INNOVATIVE RESIDENTIAL CONSTRUCTION METHOD ON INTERNAL CONDITIONS

Roger Birchmore[1], Andy Pivac[2], and Robert Tait[2]

[1]Department of Construction, Unitec Institute of Technology, Auckland, New Zealand

[2]Department of Building Technology, Unitec Institute of Technology, Auckland, New Zealand

ABSTRACT

New Zealand houses are known for producing sub-optimal internal thermal conditions and unacceptably high internal moisture levels. These contribute to poor levels of health, mould and can coincide with the decay of structural timber frames. A proposed solution is to provide an alternative structure utilising plywood, a vapour check on the internal face of the timber frame and an additional air gap, followed by the internal lining. The internal vapour check is designed to prevent moisture vapour diffusion from inside into the frame and to permit moisture diffusion from outside through the structure to the internal environment. Two full scale houses had temperatures, dew points and humidity levels monitored in passive, unoccupied conditions. The test case house incorporated the innovative construction solution. The control house was of identical design and location, using standard construction practice. The calculated internal moisture content profile appeared to be unrelated to the external moisture content as expected, instead following the profile of the changing internal temperature. Whilst the innovative construction appeared to prevent moisture diffusion into the structure in winter and permit it inside in summer, this resulted in a generally higher internal relative humidity than the control house.

INTRODUCTION

Mackintosh [1] summarises New Zealand's climate as:

"Warm subtropical in the far north to cool temperate climates in the far south, with severe alpine conditions in the mountainous areas. Mean annual temperatures range from 10 °C in the south to 16 °C in the north of New Zealand. Most of New Zealand would have at least 2000 sunshine hours annually."

Auckland, New Zealand's largest centre of population, located in the north island experiences an average of 632 degree days (base 15.5 °C). In the alpine region in the South Island, Queenstown experiences an average of 2137 degree days to the same base [2].

This data does not describe harsh external conditions for much of the population but instances where combinations of low temperatures and high moisture levels lead to poor internal environments, in many parts or the country are documented widely by a number of authors [3,4,5,6]. The World Health Organisation [7] links poor internal conditions to a range of health problems that are also reported in New Zealand research. Reflecting the largely temperate nature of the climate, guidance from the Building Research Association of New Zealand (BRANZ) [8] is that vapour barriers are only a requirement in alpine regions or those with significant internal moisture generation linked with spa pools or other similar sources. Moisture in open roof spaces is also not considered to be a significant problem and New Zealand Building codes currently have no requirements for the ventilation of roof spaces. Since the publication of the guidance obviating the need for vapour barriers other changes to the Building Code have resulted in minimum insulations levels being raised. There are no requirements for the specific airtightness of buildings. In response to the challenges described, there has been research on solutions that tackle the problems of poor conditions and health directly or indirectly through improving the sustainability of homes [9,10,11,12]. This work has tended to focus on the thermal solutions and energy consumption aspects. Su [13,14] extended this and researched the prevention of winter mould growth in occupied New Zealand houses employing primarily passive and active ventilation and thermal insulation prevention measures. Comparing the static and dynamic simulation methods de Groot [3] expanded further to explore in detail the impacts of moisture transfer through the envelope. He cautioned that the increase of thermal insulation without the consideration of interstitial moisture might move the visible mould problem to an invisible one. Simulating alternative retrofit solutions over a three year period in Auckland, he demonstrated that a vapour barrier was effective in preventing interstitial condensation occurring to levels that might encourage mould growth. Leardini and van Raamsdonk [15] added to the concerns of occupant health to include structural degradation. They also outline fears that increasing levels of thermal insulation increases chances of interstitial condensation. Their Wärmer und

Feuchte instationär (WUFI) [16] software simulation of a timber framed house in Auckland, retrofitted with insulation, indicated a clear risk of interstitial conditions when examined over a three year period. This analysis also indicated the tendency for humid summertime conditions to drive moisture inside. The conventional vapour barrier approach risks trapping this moisture vapour into the structure as shown in Figure 1B. They propose that a solution is to provide a vapour check that prevents vapour transfer from moist inside conditions into the wall structure (Figure 1A,C) but also permits externally driven vapour to pass through the structure to the inside (Figure 1D). This vapour check additionally provides all the benefits of an airtight barrier, so whilst reducing the possibility of interstitial condensation, may exacerbate the challenge of increased internal moisture levels and its associated risks. De Groot and Leardini [17] identified a lack of information on the success of retrofit solutions and the general need to improve understanding of the impacts of combining insulation airtightness and humidity control. This paper outlines the early findings of a research project that moves research from desktop simulation to exploring the impact of a construction employing such a vapour check on unoccupied conditions in a real house.

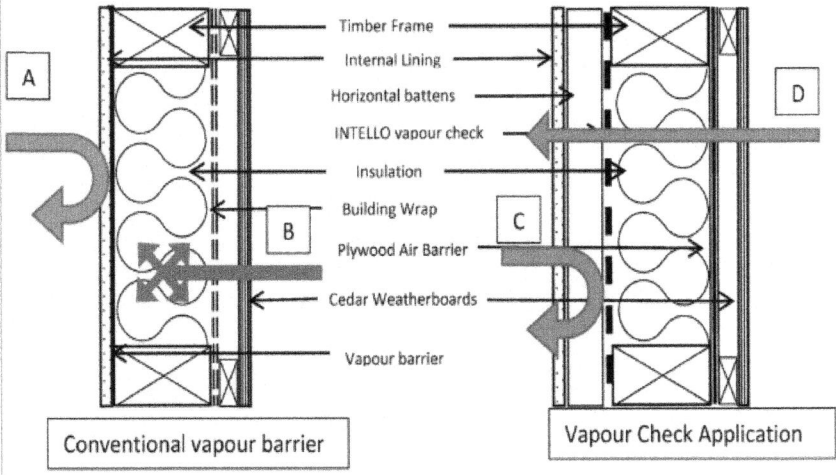

Figure 1: Moisture flows through the structure.

RESEARCH DESIGN AND METHODS

The fundamental aim of this project is to allow comparison testing of individual or combinations of building materials and techniques that have the potential for improving the building performance of this standard New Zealand house type. In particular, this paper compares the use of a vapour check material not widely used in New Zealand, combined with a rigid air barrier, compared to conventional New Zealand construction methods. The particular focus is on impacts on the moisture levels and internal air temperatures. Initially, manual calculations employing the dew point method were conducted to investigate the likelihood of interstitial condensation occurring in the control house. Figure 2 shows the risk of condensation occurring in the structure 100 mm from the inside surface. Figure 3 shows how the vapour check removes this risk. This aligns with the more extensive analysis by Leardini and van Raamsdonk [15]. It was decided to employ the vapour check proposed in their work with a view to initially exploring its impact on the internal spatial conditions of the houses. [18,19] indicate the significant impact that occupants can have on the energy consumption and internal conditions of a space, quoting examples where variations were many times the base consumption figures. The two houses have therefore been tested in passive, unoccupied conditions to remove as many variables as possible. This data would provide the basis for predicting impacts of the construction methods on internal comfort conditions, risks of mould growth and also spatial heating and cooling energy use. This is the first stage of a project that will be followed with analysis of the construction method on the interstitial conditions.

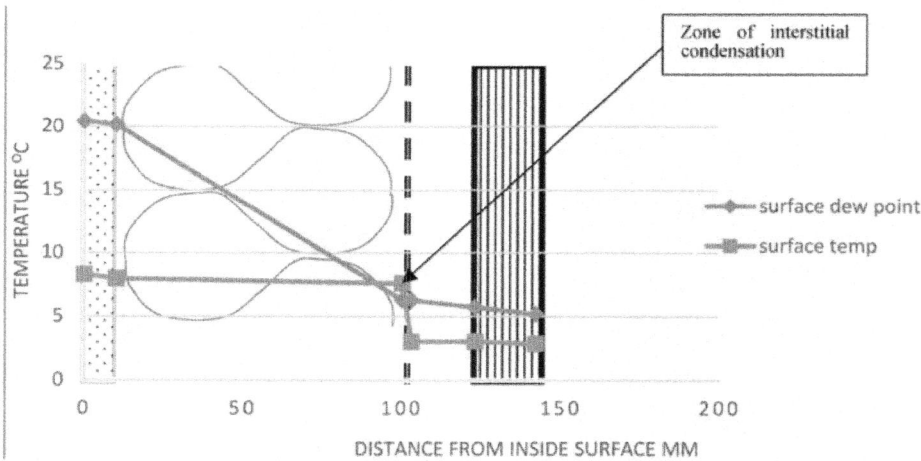

Figure 2: Interstitial condensation calculations in the control house.

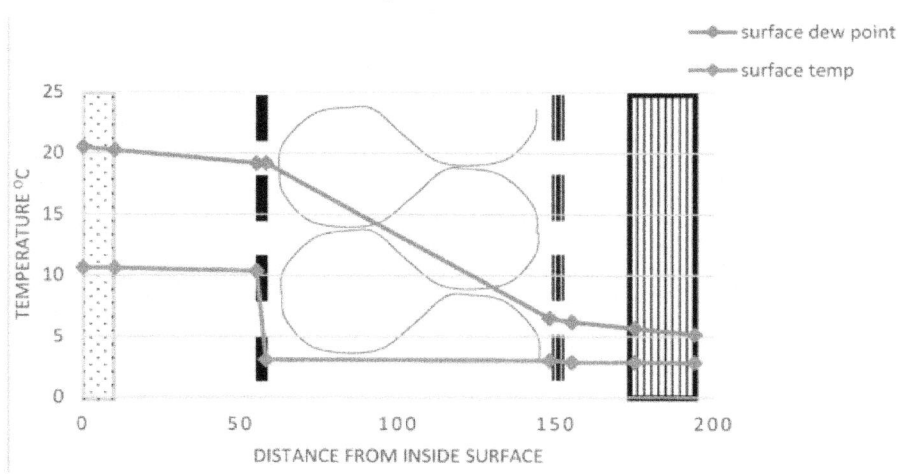

Figure 3: Interstitial condensation calculations in the test house.

Control House

The houses are single storied with three bedrooms and two bathrooms and are constructed as part of the Unitec carpentry programmes. The houses are completed by students to be relocated, and they are undecorated and without floor coverings or wall finishes. Electrical and plumbing fittings are installed but not connected. Figure 4 illustrates their overall nature.

Figure 4: Control house (foreground) and test house.

Table 1 summarises the materials used in the construction of these houses. Overhangs on the north side of the house provide complete shading from direct solar gain through double glazing during the hottest periods of the summer months. These houses are similar in design and construction to thousands of houses found in suburban areas and provide an ideal basis for examining the potential for improvements to a common housing type.

Table 1: Construction details for the control and test houses

Common to Control House and Test House
Timber Frame on pile foundation
Sub floor cladding 150 mm × 25 mm radiata pine boards with 20 mm gaps
Particle board, floor foil insulation draped between 190 mm × 45 mm joists
140 mm polyester ceiling batts, 10 mm plasterboard
Double glazed windows

Control House	Test House
Trussroof (radiata pine treated) Coloursteel roofing on building paper	Trussroof (radiata pine treated) Coloursteel roofing on building paper, INTELLO wrap on bottom chord of trusses
cedar weatherboard cladding, natural finish	cedar weatherboard cladding, natural finish
20 mm cavity battens	20 mm cavity battens
Building wrap	7 mm Plywood
90 mm × 45 mm radiata pine framing	90 mm × 45 mm radiata pine framing
Polyester insulation	Polyester insulation
-	INTELLO Vapour check
-	45 mm × 45 mm battens
10 mm plasterboard	10 mm plasterboard

Table 2 provides dimensional thermal and hygroscopic properties of relevant materials in the external envelopes. These are values associated with readily available material sizes to ensure the overall construction R-values exceed minimums stated in verification method H/VM1 [20], ensuring compliance with the New Zealand Building Code. Minimum R-values for this context are:

- Roof   2.9 m²K/W
- Wall   1.9 m²K/W
- Windows   0.26 m²K/W
- Floor   1.3 m²K/W

Table 2: Thermal and hygroscopic properties of materials

Material	Thickness t	Conductivity k	Resistance R	Vapour Resistivity r_v	Vapour Resistance G
-	mm	W/mK	m²K/W	MNs/gm	MNs/g
External Walls	-	-	-	-	-
Plasterboard	10	0.22	0.05	60	0.6
Air gap (Test House)	45	-	0.18	-	5
Vapour Check (Test House)	3	-	-	-	see Table 3
Insulation	90	0.041	2.20	7	0.63
Building Wrap (Control House)	3	n/a	n/a	2300	6.9
Plywood Air Barrier	7	0.13	0.05	-	2.2
Ventilated Air Gap	20	-	0.09	-	-
Cedar Weatherboard	19	0.11	0.09	-	0.26

Plasterboard	10	0.22	0.05	60	0.6
Air gap (Test House)	45	-	0.18	-	5
Vapour Check (Test House)	3	-	-	-	see Table 3
Air Gap	300	-	0.16	-	5
Building Wrap	3	-	-	2300	6.9
Profiled Sheet Steel Roof	4	-	-	-	2000

Test House

The test house was constructed one year later than the control house and modifications incorporated the use of the INTELLO vapour check proposed by Leardini and van Raamsdonk [15]. In addition modifications were made were to replace the building paper with 7 mm thick plywood barrier treated to H3.2 CCA (Copper Chrome Arsenate) in accordance with AS/NZ 1604.3 [21] to meet AS/NZS 2269.0 [22]. Vertical sheet joints were sealed with flashing tape. This feature was felt to have significant potential as an alternative that provided the functions of bracing and rigid air barrier in a single element that offered enhanced seismic resistance. This had potential for use in the rebuild of houses since the Christchurch earthquakes. The vapour check was placed on the internal surfaces of external walls and ceilings. A 45 mm cavity batten was then added before fixing of the plasterboard. Details are shown in Figure 5 and Figure 6. Ceiling insulation was placed directly on top of the ceiling material in the control house and on top of the INTELLO vapour check in the control house. In both houses the roof space that resulted was an uninsulated cavity. The INTELLO vapour check is intended to have two functions. The first is to control the passage of vapour through the timber to the structure. This minimises the chances of charging the frame with moisture that could then condense during cold spells, eventually resulting in the rotting of the frame. Its high level of resistance shown in Table 3 prevents vapour generated internally from passing into the structure in winter. The low level of resistance in the opposite direction permits the vapour to pass from high humidity conditions present in summer time though the structure to the inside. This again prevents the possible accumulation of moisture in the frame for future condensation. The second function is for the INTELLO vapour check to act as an additional barrier to infiltration through the fabric of the building. The minimisation of infiltration is proposed to be an important contributor to reducing unplanned heat loss and moisture ingress. The additional air gap between the INTELLO and the internal plasterboard lining provides a route for internal services to circulate with minimal penetrations through the INTELLO.

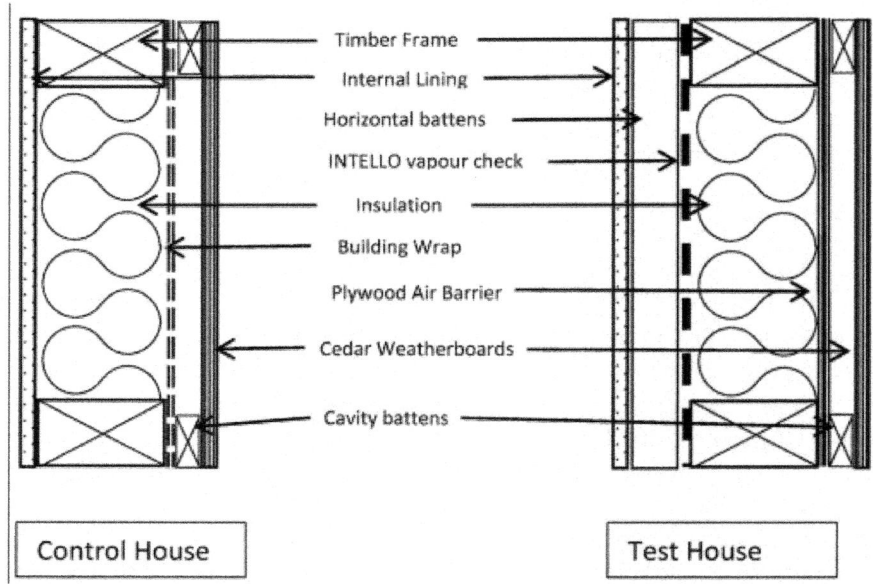

Figure 5: Construction detail—external walls.

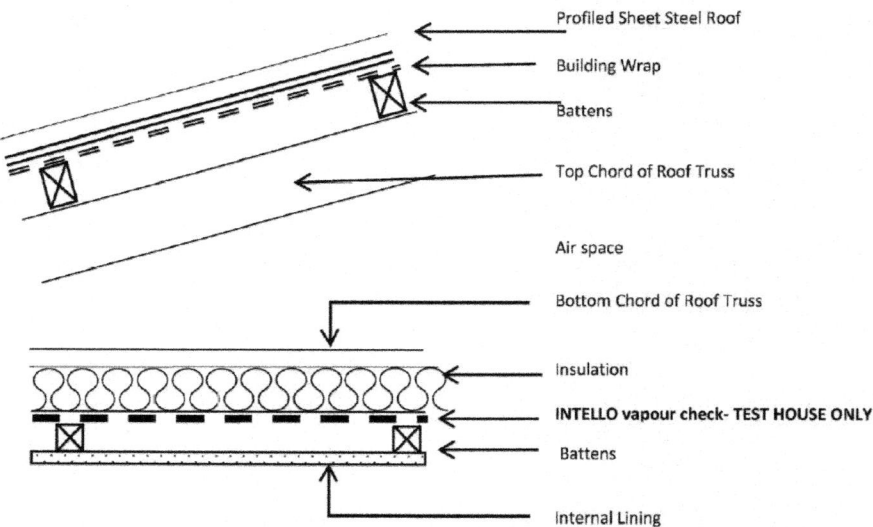

Figure 6: Construction detail—roof.

Table 3: Vapour diffusion resistance of the INTELLO membrane [23]

Testing Conditions	Vapour Diffusion Resistance (MNs/g)	
Average Ambient Humidity	20%	85%
Direction of Diffusion Flow	Out towards the air barrier	Inwards towards the air barrier
INTELLO	125	1.25

Site

The site is on the Unitec Institute of Technology campus in Mt Albert Auckland. The site is relatively exposed with an open grassed area to the northwest. Surrounding buildings are reasonably distant to the south, north and east. Behind the houses to the southeast is a hilly incline and the student building yard. The houses are located with identical orientations but separated to avoid mutual shading.

Monitoring Process

Temperature sensors have been set up to sample the internal air temperature at hourly intervals. Sensors used are Lascar EL-USB-2 Humidity and Temperature USB data loggers. These measure and store relative humidity (RH), dew point (DP) and dry bulb air temperature readings (DB) over 0% RH to 100% RH and −35 °C to +80 °C measurement ranges. Sensors were located identically in the two houses to align with practice outlined by Barley *et al.* [24] at a height of 1500 mm above ground level suspended from the ceiling by builders twine. Sensor layout is given in Figure 7. In order to check the appropriate test location for the sensor, a second sensor was located at the edge of the room to check initial operation and determine the degree of variability experienced across each space. It was found that the average variation between measurements from the centre of the room and from the edge of the room vary by an average of 0.2 °C over the 168 hourly measurements, with the maximum variation less than 0.5 °C. This is well within the accuracy stated for the sensors which is stated to be ±0.5 °C [25] and indicates that a single measurement in the chosen position is representative of the overall room conditions. Dew point measurements have been used as this single figure provides an indicator of absolute moisture content. Localised weather data is measured at a weather station indicated in Figure 8. They are monitored in a passive, unoccupied condition.

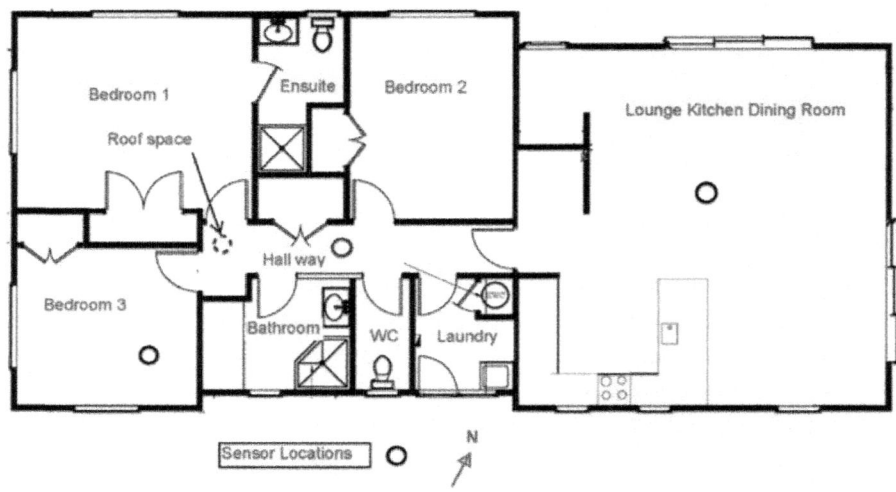

Figure 7: House plan with sensor locations.

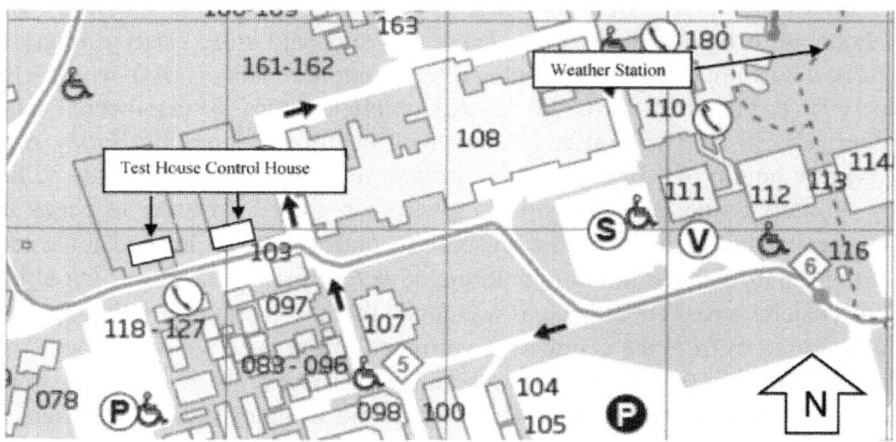

Figure 8: House and weather station location details.

Air Tightness

Both houses were tested for air tightness using the standard blower door test following European standard EN 13829:2000 [26]. Openings associated with extract ventilation and unconnected waste pipes were sealed for air pressure testing and for subsequent monitoring of air temperature, dew point and relative humidity. Table 4 shows the air changes per hour of the whole house volume under the standard test conditions of 50 Pa pressurisation and 50 Pa depressurisation. It indicates that the test house has an air leakage rate of less than a third of the control house. The control house sits just outside the airtight classification for New Zealand houses which peaks at an airtightness of 5 ac/h [27]. The test house is comfortably in the airtight classification but is still well above the requirements of the Passive House Institute [28].

Table 4: Results of airtightness testing

Testing Conditions	Control House (ac/h)	Test House (ac/h)
Depressurisation	6.58	1.92
Pressurisation	6.93	2.10
Average	6.75	2.01

Rooms Being Analysed

The rooms chosen for analysis in this paper were the lounge kitchen dining room, bedroom 3, hall and the roofspace. Floor and wall areas are shown in Table 5. The lounge kitchen dining room is the largest space in the house and is has external walls on the South, East and North Face with glazing in each with a window to wall ratio of 29%. Its inclusion of the kitchen also examines a space where occupancy may generate significant additional internal moisture. Bedroom 3 is on the opposite, coldest corner of the house receiving minimal solar gain. With a window to wall ratio of 41% the hall is internal with the exception of its floor and ceiling. The roofspace provides data on the external side of the vapour check in the test house and the external side of the insulation in both houses.

Table 5: Details of rooms being analysed

Room	Floor Area (m^2)	Net External Wall Area (m^2)
Lounge Kitchen Dining room	46.4	33.6
Bedroom 3	9.8	11.07
Hall	5.6	Nil
Roofspace	120.5	Nil

RESULTS AND DISCUSSION

Seasonal Data

The means of dry bulb (DB) and dew point temperatures, (DP) measured at hourly intervals for the winter and summer seasons are summarised for each building in Figure 9. The summer season is defined as 1 December–28 February, spring as 1 March–31 May, winter as 1 June–31 August and spring as 1 September–30 November. The data indicates that as the mean temperatures drop towards winter and as expected rise up again in spring. Also as expected, the range of temperatures in the roofspace exceed those in occupied spaces indicating the smoothing effect of in wall and under floor insulation. The trend and interrelationship holds for all three internal spaces suggesting that the airspace in each room within each house is reasonably homogenous and that orientation and wall to window ratios have a small effect on mean temperatures. Absolute moisture content and dry bulb temperature are not parameters that are inherently connected. As the buildings have no internal moisture generation it was expected that absolute moisture contents, represented by the dew point temperature levels would follow external moisture readings. Figure 9 indicates that in fact they seem to follow the internal dry bulb readings. The detail of a few sample days in Figure 10 and Figure 11 re-enforces this and indicates little or no connection between internal and external dew point temperatures.

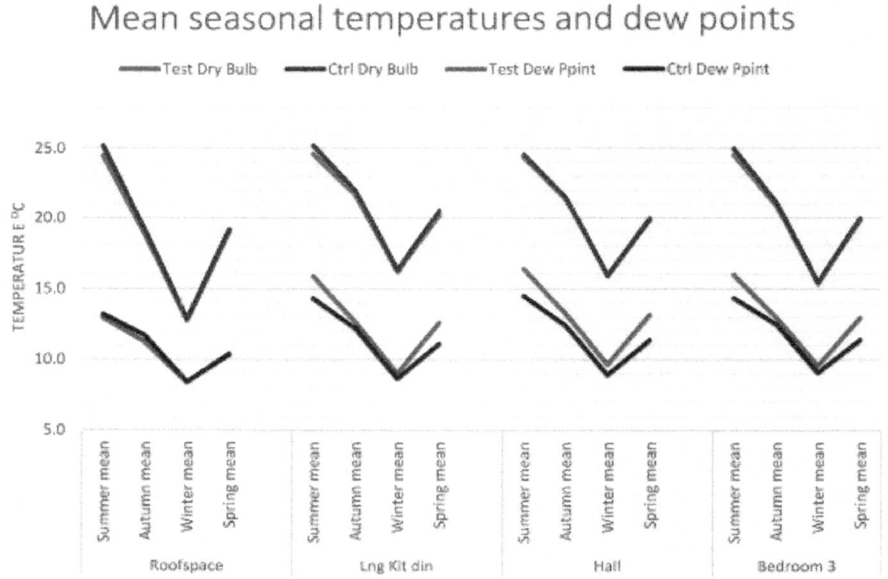

Mean seasonal temperatures and dew points

Figure 9: Comparison of space temperature and dew point seasonal means.

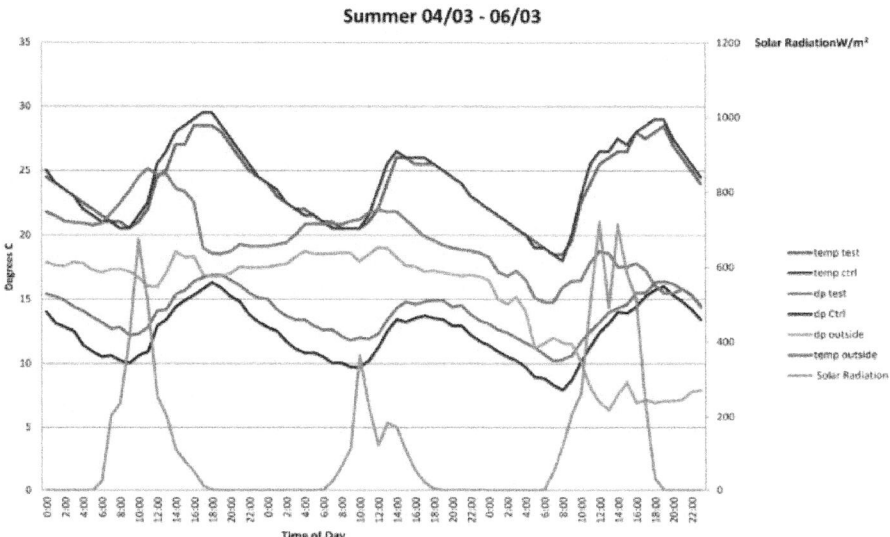

Figure 10: Dry bulb and dew point comparisons for the lounge kitchen dining room with external summer conditions.

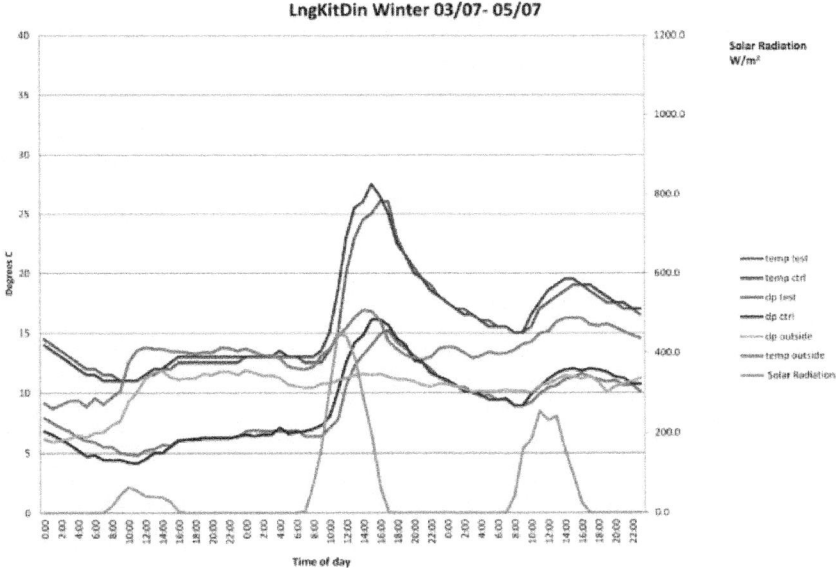

Figure 11: Temperatures and dew point comparisons for the lounge kitchen dining room with external winter conditions.

The significance of the differences between seasonal measures of control and test dry bulb temperatures and between control and test dew point temperatures were calculated using a two tailed *T*-test. Calculations were completed on two spaces, the roofspace and the lounge kitchen dining room being the most different spaces. Calculations were completed on seasonal means using an alpha of 0.05 with a hypothesis that the difference between two population means would be zero. *P*scores are quoted to four significant figures.

Scores where $p \geq 0.05$ indicates that the hypothesis is correct and that there is no significant difference between the populations. Scores where $p \leq 0.05$ indicates the hypothesis is incorrect and that there is a significant difference between the populations. Table 6 shows that in all spaces with the exception of the roofspace during winter and for dry bulb temperatures in spring the difference between the test and control house was statistically significant with a confidence of 95%.

Table 6: Summary of statistical significance of differences between control and test houses.

Season	Space	Dry Bulb	Dew Point
-	-	*p*	*p*
Summer	LoungKitDin	0	0
	Roofspace	0	0
Autumn	LoungKitDin	0	0
	Roofspace	0	0
Winter	LoungKitDin	0	0
	Roofspace	0.1007	0.3745
Spring	LoungKitDin	0	0
	Roofspace	0	0.0263

Detailed Results of Selected Days

Summer

The detailed results of a few days in each season are shown in Figure 10, Figure 11, Figure 12 and Figure 13. They have been chosen to illustrate the apparent strong influence of solar gain on internal conditions. This indicates that the internal dry bulb temperatures follow the cyclical pattern of the external solar gain. The periods of the two cycles appear identical but with the dry bulb lagging between four and 6 h. The internal dew point also follows with the same period but with slightly reduced amplitudes. There is a very weak connection if any, with the dew point of the external air. As there were no occupant generated sources of internal moisture, the changing dew points over a daily cycle could be resulting from residual construction moisture. As the temperature rises, moisture still present in the construction evaporates into the air. As it cools it is re-absorbed into the structure. Whilst the test house vapour check is designed to reduce this, the floors of both houses are exposed particle board and the unpainted gypsum board will have some absorption capacity.

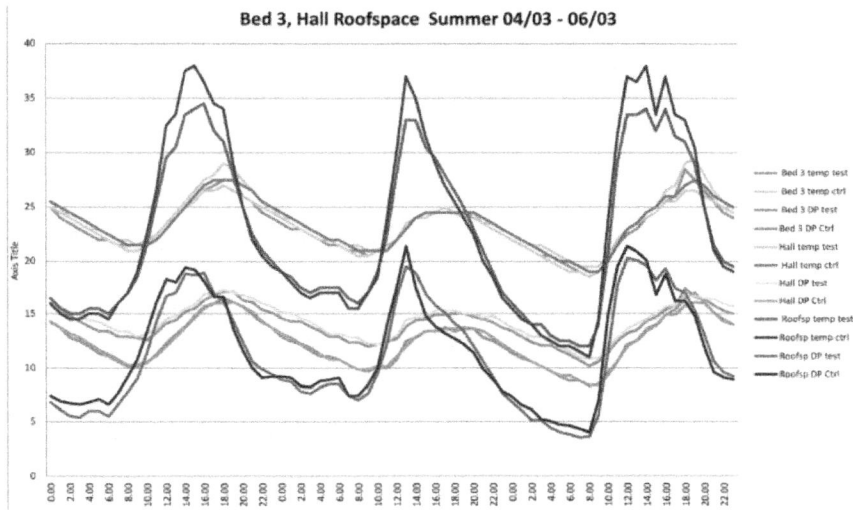

Figure 12: Dry bulb temperatures and dew point comparisons for the bedroom 3 hall and roofspace with external summer conditions.

Figure 12 shows the performance for three other spaces during the same period. The external conditions are the same as inFigure 10 but have been removed for clarity. Reflecting the seasonal means in Figure 9, the roofspace exhibits a wider range of variations than other internal spaces. The dry bulb and dew point temperatures in the bedroom 3 and hall cluster close together within similar ranges and cycle to those of the lounge kitchen dining room. This demonstrates that the seasonal homogeneity is reflected on a daily cycle. The reduction of extreme internal temperatures by thermal insulation in the roofspace is also very clear on this daily cycle. The four to 6 h lag between peak external temperature and internal temperature observed in the lounge kitchen dining room remains for other internal spaces. This is reduced to two hours for the roofspace. The reduction in lag is expected as the temperature is being measured on the external side of the insulation. Differences between dry bulb temperatures in the control house and the test house are small, being largest at extremes of conditions and very small between. Differences between the dew point readings in the houses appear more significant and more evenly spread over the daily cycle.

Winter

The data in Figure 11, Illustrates the strong link between solar gain, internal dry bulb and internal dewpoint temperatures where the solar gain on the second day leads and influences the plots of other parameters. Notably the thermal lag in both houses has reduced to approximately 2 h.

Figure 13 reflects parallels with Figure 12 but with lower temperature levels associated with the winter season. The clustering of the graphs continues to support the homogeneity of the internal spaces. The roofspace extremes are reduced. Time lags have also reduced to a maximum of two h and minimums of zero. Differences between control and test house dry bulb temperatures and dew points are very small especially on the first day with low solar radiation.

Moisture Contents, in kg of moisture per kg of dry air, can be read from a Psychrometric chart for given dew points. Using the mean density of the air and the space volume, the volume of moisture being evaporated and re-absorbed on a daily cycle can be estimated. 1 December was chosen as the day with the maximum daily change in dew point temperature and Table 7 indicates the result as 0.52 L.

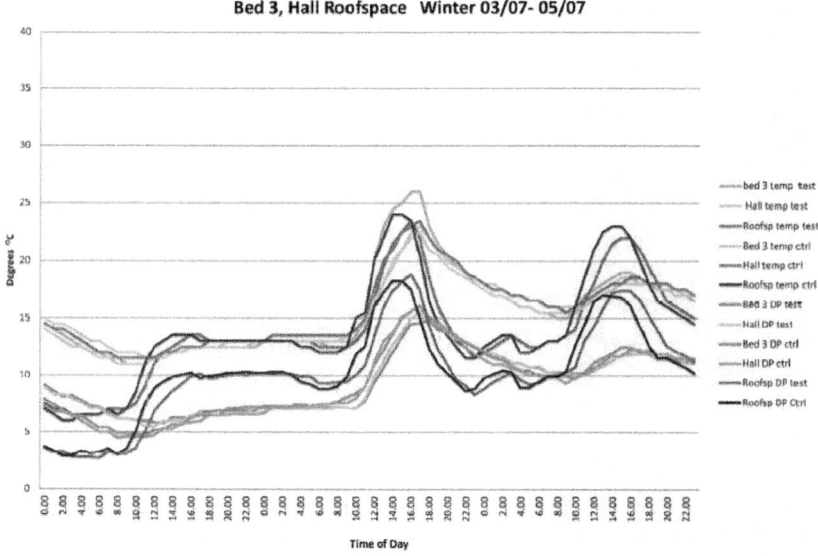

Figure 13: Dry bulb temperatures and dew point comparisons for the bedroom 3 hall and roofspace with external winter conditions.

Table 7: Data for estimation of the moisture volume being evaporated and re-absorbed

	Summer 1st Dec		
Dry Bulb (°C)	**Dew Point (°C)**	**Density (kg/m³)**	**Moisture Content (kg/kg)**
Max 25.5	15.4	1.163	0.0109
Min 15.5	8.6	1.211	0.0069
Difference	6.8	-	0.004
Mean density (kg/m³)	-	1.187	-
Space volume (m³)	108.54	-	-
Space mass (kg)	128.809	-	-
Moisture mass (kg)	0.52	-	-
Moisture volume (L)	0.52	-	-

CONCLUSIONS

The internal seasonal spatial data in Figure 9 along with the statistical analysis supports the expected performance of the test house structure regarding internal moisture levels. The summer season trends are reinforced by the daily analysis inFigure 10. In winter, higher internal dew points in the test house compared to the control house suggest that internal vapour is held within the occupied space and is not being permitted to enter the structure. The risk of interstitial condensation should therefore be reduced. In summer, higher internal dew points in the test house compared to the control house suggest that vapour is being allowed to pass through the structure to the inside. This prevents the moisture getting trapped within the timber frame, reducing the potential to cause interstitial condensation when the climate permits. The roofspace data supports this further. As the readings are on the external side of the vapour check, vapour passing through the check in summer should result in lower dew point temperatures within the control house roofspace compared to the test house.Figure 10 indicates this. They should also be lower in winter but this is not as evident when looking at seasonal means or the test of significance, but the daily analysis in Figure 12 does support this. Overall this differentiates the performance of the vapour check from a conventional vapour barrier which would trap the moisture within the timber structure, preventing it from entering the occupied space.

Compared to the control house, the test house offered improved thermal resistance from the additional sealed air gap and significantly reduced infiltration due to the vapour check properties. It was felt that this would result in warmer test house conditions in winter and cooler conditions in summer. Overall differences are not large however. Small differences between the average dry bulb temperatures over the winter and spring seasons between the control and test Houses suggest that the different properties are having a small effect on unoccupied thermal conditions. Figure 11 shows that on cloudy winter days the difference between internal and external temperatures was minimal and so differences between control and test houses could also be very small. However on sunny winter days, internal temperatures rose nearly ten degrees higher than outside. Under these winter conditions the test house was actually slightly cooler than the control house. As the test house infiltration rate was one third of the control house rate, this suggests that the heat losses due to infiltration through the closed structure in an unoccupied condition might be much smaller than anticipated or are not being prevented by increased airtightness of the envelope.

The calculated daily variation of 0.52 L of water is a small proportion of the possible moisture generation in a house. TenWolde and Pilon [29] suggest that a family of four would produce up to 15 L per day. However, 0.52 L represents the amount of moisture left to be absorbed by the air. The moisture absorptive properties on internal linings, quantified by Mitamura, Rode and Schultz [30] and termed moisture buffering, could account for a larger volume of daily moisture change in the unoccupied condition.

The tendency of the vapour check to produce slightly higher internal dew points emphasises the need to combine this type of structure with minimal ventilation rates to ensure the moisture is ventilated to outside. Internal surface temperatures are likely to be even lower than space temperatures, especially in winter. The combination of this fact and that moisture generated by occupation may aggravate conditions, further underlines the importance of ventilating moisture sources to outside.

Whilst the positive differences in conditions provided by the innovative structure are statistically significant they are still small and might be unlikely to be financially beneficial in the short term. Work is underway with the detailed monitoring of temperature and humidity in each layer of the wall construction of each building. This will enable tracking in detail of the passage of vapour throughthe envelope and the identification of interstitial condensation risk and help allocate the differences to the performance of the internal vapour check and the plywood rigid air barrier. Further work will include monitoring conditions under controlled, active heating, cooling and moisture generation which should provide a stiffer test and opportunity for the structure to show a financially beneficial difference.

ACKNOWLEDGMENTS

This research has been carried out with the support of Thomas van Raamsdonk of ProClima and the Unitec Internal Research Fund.

AUTHOR CONTRIBUTIONS

Roger Birchmore contributed to the research design and carried out the collection and analysis of the data. Robert Tait and Andy Pivac contributed to the research design experimentation set up and conducting measurements.

REFERENCES

1. Mackintosh, L. NIWA Overview of New Zealand Climate. 2001. Available online: https://www.niwa.co.nz/education-and-training/schools/resources/climate/overview (accessed on 11 June 2014).

2. BizEE Degree Days. Weather Data for Energy Professionals. Available online: http://www.degreedays.net/# (accessed on 22 December 2014).

3. De Groot, H. Indoor Air Quality and Health. An Analysis of the Indoor Air Quality and Health in New Zealand's Homes. 2009. Available online: http://www.archigraphic.de/_originals/pdfs/indoor_air_quality_and_health.pdf(accessed on 11 June 2014).

4. Howden-Chapman, P.; Saville-Smith, K.; Crane, J.; Wilson, N. Risk factors for mould in housing: A national survey.*Indoor Air* 2005, *15*, 469–476.

5. Howden-Chapman, P.; Crane, J.; Chapman, R.; Fougere, G. Improving health and energy efficiency through community-based housing interventions. *Int. J. Public Health* 2011, *56*, 583–588.

6. New Zealand Business Council for Sustainable Development (NZBCSD). Media Release: $20 Billion Cost of Fixing Country's Homes Less than 4% of Their Value. 2008. Available online: http://www.scoop.co.nz/stories/BU0812/S00020.htm (accessed on 11 June 2014).

7. World Health Organisation. Guidelines for Indoor Air Quality: Dampness and Mould. 2009. Available online: http://site.ebrary.com/lib/unitech/Doc?id=10367463&ppg=55 (accessed on 11 June 2014).

8. Sargent, S. Getting Clear on Vapour Barriers and Underlays. 2007; Build 99. Available online: http://www.buildmagazine.org.nz/articles/show/getting-clear-on-vapour-barriers-and-underlays/ (accessed on 4 February 2015).

9. Howden-Chapman, P.; Matheson, A.; Crane, J.; Viggers, H.; Cunningham, M.; Blakely, T.; Cunningham, C.; Woodward, A.; Saville-Smith, K.; O'Dea, D.; *et al.* Effects of insulating existing houses on health inequality: Cluster randomised study in the community. *Br. Med. J.* 2007, *334*, 460.

10. Easton, L.; Saville Smith, K. Homesmart Renovations—Testing Tools to Promote Sustainable Renovation. In Proceedings of the New Zealand

Sustainable Building Conference, Wellington, New Zealand, 26–28 May 2010.

11. Callau, M. Upgrading Housing in NZ for Thermal Efficiency. In Proceedings of the New Zealand Sustainable Building Conference, Wellington, New Zealand, 26–28 May 2010.

12. Burgess, J.C.; Buckett, N.R.; Camilleri, M.J.T.; Burrough, L.J.; Pollard, A.R. Papakowhai Retrofit Project—Improving the Thermal Envelope and Space Heating. In Proceedings of the New Zealand Sustainable Building Conference, Wellington, New Zealand, 26–28 May 2010.

13. Su, B. Prevention of inter mould growth in housing. *Archit. Sci. Rev.* 2006, *49*, 385–390.

14. Su, B. Indoor moisture control of Auckland houses with different ventilation systems. *Int. J. Civ. Archit. Sci. Eng.* 2013, 7, 411–415.

15. Leardini, P.M.; van Raamsdonk, T. Design for Airtightness and Moisture Control in New Zealand Housing. In Proceedings of the New Zealand Sustainable Building Conference, Wellington, New Zealand, 26–28 May 2010.

16. WUFI Pro, 2D, Plus Software 2013. Fraunhofer-Institute fur Bauphysik (IBP): Holzkirchen. Available online: http://www.wufi.de/index_e.html (accessed on 21 January 2015).

17. De Groot, H.; Leardini, P.M. Indoor Air Quality and Health in New Zealand's Traditional Homes. In Proceedings of the 44th Annual Conference of the Architectural Science Association, Unitec Institute of Technology, Auckland, New Zealand, 24–26 November 2010.

18. Love, J. Mapping the Impact of Changes in Occupant Heating Behaviour on Space Heating Energy Use as a Result of UK Domestic Retrofit. Retrofit Conference, University of Salford, UK; 2012. Available online: http://www.salford.ac.uk/__data./pdf_file/0005/142385/023-Love.pdf (accessed on 21 January 2015).

19. Clevenger, C.M.; Haymaker, J. The Impact of Building Occupation on Energy Modelling Simulations. Joint International Conference on Computing and Decision Making in Civil and Building Engineering, Montreal, Canada; 2006. Available online: http://web.stanford.edu/group/peec/cgi-bin/docs/people/profiles/The%20Impact%20of%20the%20Building%20Occupant%20On%20Energy%20Modeling%20Simulations.pdf (accessed on 21 January 2015).

20. Department of Building and Housing Compliance document for New Zealand Building Code Clause H1 Energy Efficiency—Third Edition. Available online: http://www.dbh.govt.nz/UserFiles/File/Publications/

Building/Compliance-documents/H1-energy-efficiency-3rd-edition-amendment-2.pdf (accessed on 21 January 2015).

21. *Specification for Preservative Treatment Part 3: Plywood*; SNZ 2012 AS/NZ 1604.3; Standards New Zealand: Wellington, New Zealand, 2012.

22. *Plywood—Structural Part 0: Specifications*; SNZ, 2012b, AS/NZS 2269.0:2012; Standards New Zealand: Wellington, New Zealand, 2012.

23. Moll, L.; van Raamsdonk, T. *A New Zealand Based Study on Airtightness and Moisture Management*; ProClima, NZ Ltd: Wellington, New Zealand, 2009.

24. Barley, D.; Deru, M.; Pless, S.; Orcellini, P. *Procedure for Measuring and Reporting Commercial Building Energy Performance*; Technical Report NREL/TP-550–38601; National Renewable Energy Laboratory, Battelle: Golden, CO, USA, 2005.

25. LASCAR Certificate of Calibration. Available online: http://www.lascarelectronics.com/pdf-usb-datalogging/datalogger0237687001331303304.pdf (accessed on 21 January 2015).

26. *Thermal Performance of Buildings—Determination of Air Permeability of Buildings—Fan Pressurization Method*; European Committee for Standardization CEN/TC 89 2000. EN 13829:2000; British Standards Institution: London, UK, 2001.

27. Stoecklein, A.; Bassett, M. *ALF3—A Design Tool for Energy Efficient Houses*; Building Research Association of New Zealand: Judgeford, New Zealand, 1999.

28. Passive House Institute Passive House Requirements. Available online: http://www.passiv.de/en/02_informations/02_passive-house-requirements/02_passive-house-requirements.htm (accessed on 11 June 2014).

29. TenWolde, A.; Pilon, C. The Effect of Indoor Humidity on Water Vapor Release in Homes. Available online: http://web.ornl.gov/sci/buildings/2012/2007%20B10%20papers/071_TenWolde.pdf (accessed on 11 June 2014).

30. Mitamura, T.; Rode, C.; Schultz, J. Full-Scale Testing of Indoor Humidity and Moisture Buffering in Building Materials. In Proceedings of the Indoor Air Quality 2001, Moisture, Microbes, and Health Effects: Indoor Air Quality and Moisture in Buildings; ASHRAE: Atlanta, GA, USA, 2001.

Chapter 7

ENERGY-EFFICIENT TECHNOLOGIES AND THE BUILDING'S SALEABLE FLOOR AREA: BUST OR BOOST FOR HIGHLY-EFFICIENT GREEN CONSTRUCTION?

Agnieszka Zalejska-Jonsson[1], Hans Lind[1], and Staffan Hintze[2]

[1]Real Estate and Construction Management, School of Architecture and the Built Environment, KTH Royal Institute of Technology, Brinellvägen 1, Stockholm 10044, Sweden

[2]Highway and Railway Engineering, School of Architecture and the Built Environment, KTH Royal Institute of Technology, Brinellvägen 23, Stockholm 10044, Sweden

ABSTRACT

When the external measurements of a building are fixed, an increase in external wall thickness caused by additional insulation, for example, will lead to loss of saleable floor area. This issue has to be taken into account in the evaluation of investment profitability. This paper examines how technologies used in energy-efficient residential building construction affect the available saleable floor area and how this impacts profitability of investment. Using a modeled building and an analysis of the average construction cost, we assessed losses and gains of saleable floor area in energy-efficient buildings. The analysis shows that the impact of potential losses or gains of saleable floor area should be taken into account when comparing investment alternatives: building energy-efficient green dwellings or building conventional ones. The results indicate that constructing energy-efficient buildings and introducing very energy-efficient technologies may be energy- and cost-effective even compared with conventional buildings. Employing new products in energy-efficient construction allows benefit to be drawn from lower energy consumption during the life cycle of the building, but also from the increase in saleable floor area.

INTRODUCTION

There are ambitious goals in the EU to reduce energy consumption in the building stock and a crucial question is to what extent the investment in energy-efficient technologies is profitable and whether further political measures are necessary.

The process of decision-making in simple terms is based on valuing benefits against costs and against alternative solutions. In the case of investment in new property projects, initial and future costs are weighed against expected income. If we consider a scenario where a developer has the choice of constructing the same building as a conventional or as a high-performance green building, we can expect the decision to be dependent on investment viability. Research shows that initial construction costs for energy-efficient green building are generally higher than for conventional building. The difference can vary from 0% to as much as 20% [1,2,3,4,5,6,7,8,9]. The variation in investment cost depends on climate conditions, the developer's experience, environmental goals and the designed energy efficiency and is often related to higher material, labor and/or design costs.

On the other hand, the operation costs for a high-performance building are expected to be up to 40%–50% lower than for conventional buildings [9,10], where the predicted cost reduction depends mainly on the energy-efficiency of the building.

Finally, literature brings forward evidence that green buildings transact at 3%–12% higher prices than conventional buildings on the commercial [10,11,12] and the residential market [13,14,15,16,17].

This type of data can be used to calculate the profitability of the investment as is done in [9]. However, the reliability of an analysis depends on the accuracy of its assumptions. The literature has indicated a gap between the recorded and calculated maintenance and operation costs of green buildings (e.g., [18,19]). Moreover, the outcomes of the analysis have also proved to be highly sensitive to the expected rate of return [20,21,22] and presumed energy prices [23,24].

In this paper, we *first* show that these calculations can be misleading if they do not take into account that the choice of building with higher energy efficiency, e.g., a passive house, can reduce the saleable floor area. Thus, if the external measurements of the building are fixed, a building with thicker walls will entail less saleable floor area. *Secondly*, we show that technological development in recent years has reduced this loss and that this has contributed to the profitability of energy efficient buildings. This is shown by comparing technologies and prices from 2002 with those from 2012.

Specifications that facilitate energy-efficiency gains include compact construction, minimum thermal bridge value, a very well thermally insulated building envelope, energy-efficient windows and adequate choice of heating and ventilation systems [25,26]. For highly energy-efficient buildings, it is essential that the building envelope is airtight and very well insulated. The latter may have significant impact on the width of the external walls, roof and foundation. Consequently, external walls in energy-efficient buildings may require more floor area than those in conventional buildings. In the case of the scenario described above, where the external measurements of a building are fixed or limited, construction of an energy-efficient building has a direct effect on the amount of saleable floor area and consequently on the developer's potential income from rent or sale. Therefore, analyses that compare energy-efficient and conventional building projects should factor in the total floor area that is available for sale or rent. A significant difference in floor area availability may have an effect on potential income, but also on the potential customer segment. As far as the authors know, no earlier studies have quantified gains and loss related to saleable floor spaces in these buildings.

This paper contributes to the discussion on the profitability of energy-efficient solutions in green buildings [9,10,27,28] by investigating the possible impact of introducing more energy-efficient products on the economic attractiveness and profitability of constructing highly energy-efficient buildings.

ASSUMPTIONS AND ANALYSIS

Modeled Buildings

The investigation started by modeling a building that was a typical terraced house in Northern Europe. The building consists of six dwellings, each of them two-level apartments, with total external measurements of approximately 12 m × 35 m (for details, see drawings in Figure 1 below). Initially, this building is based on drawings and information about the first passive house built in Sweden.

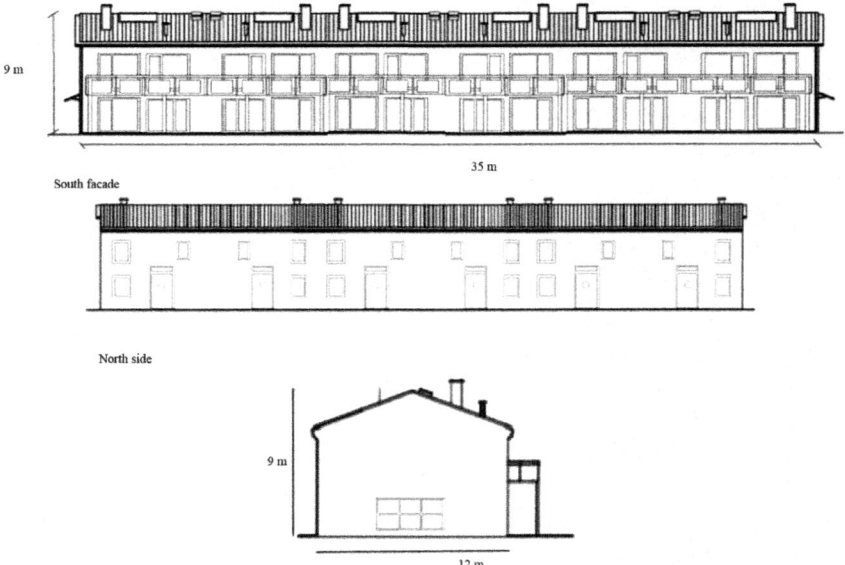

Figure 1: The model house.

Two building construction types are analyzed:

(A)   A timber wall construction;

(B)   A lightweight concrete brick wall construction.

For each of these two types of building, conventional and energy-efficient cases are modeled.

The conventional building follows the specified energy requirement in the current Swedish Building Regulations [29] for residential buildings with electrical heating, climate zone south, and therefore it is assumed that the maximum energy requirement (space heating) is 55 kWh/m². The basic notation for these is CVN-A (timber) and CVN-B (brick) (see Table 1).

The second building is an energy-efficient building for which the calculated annual space heating is 26 kWh/m². A building that fulfils this requirement is considered by current Swedish Building Regulations [29] to be a very low energy building. Additionally, it is assumed that the primary energy requirement inclusive of household electricity for the energy-efficient building is not expected to exceed 110 kW/m². During modeling, passive house principles were used [2,30]. The basic notation for these is EE-A (timber) and EE-B (brick) (see Table 1).

Table 1: Cases analyzed in this paper

Notation	Explanation
CNV-A	conventional building, timber construction, insulation mineral wool, lambda 0.036 W/(mK)
CNV B	conventional building, brick construction, insulation mineral wool, lambda 0.036 W/(mK)
EE-A1	energy-efficient building, timber construction, insulation mineral wool, lambda 0.036 W/(mK)
EE-A2N	new technology, energy-efficient building, timber construction, insulation mineral wool, lambda 0.033 W/(mK)
EE-A3N	new technology, energy-efficient building, timber construction, insulation pir (polyisocyanurate), lambda 0.024 W/(mK)
EE-B1	energy-efficient building, brick construction, insulation mineral wool, lambda 0.036 W/(mK)
EE-B2N	new technology, energy-efficient building, brick construction, insulation mineral wool, lambda 0.033 W/(mK)
EE-B3N	new technology, energy-efficient building, brick construction, insulation pir (polyisocyanurate), lambda 0.024 W/(mK)

In the course of the analysis, the energy-efficient building envelope is adjusted so that these values stay constant. No changes are made in the construction of roof and foundation, for which U-values are U(foundation) = 0.10 W/(m²K) and U(roof) = 0.08 W/(m²K). The airtightness of the building envelope is assumed to be the following for the conventional and the energy efficient building: 0.6 and 0.4 h^{-1}, at +/−50 Pa. It is further assumed possible to use air heating and heat recovery ventilation with an efficiency of 75% in both buildings. It is also assumed that, if necessary, the supplementary electric heating may be used in the buildings. The main assumptions are summarized in Table 2.

The energy-efficient building is modified step by step by applying new technology and using products with low thermal conductivity (described in the paper as lambda). It was essential that all the products used in the modeling were available on the market. Prototypes and early innovations were not considered. The reason for selecting innovative products that had already entered the market was to examine cost and potential benefits of using these products.

The important rule for this exercise was that, regardless of construction type and the novelty of the products, the building had to fulfill specified energy requirements. This premise allows for changes in the envelope (external wall), and consequently the benefits of using more energy-efficient products can be quantified. In the exercise, we have used the PHPP program (The Passive House Planning Package).

Table 2: Basic assumptions

Assumed requirements	Conventional building	Energy-efficient building
Building dimensions (external)	12 m × 35 m	12 m × 35 m
Height (to the roof top)	9 m	9 m
Number of apartments	6	6
Number of levels	2	2
Basement	no	no
Annual space heating (kW/m²) *	55	26
Annual primary energy including household electricity (kW/m²)	not specified	110
Airtightness (at +/−50 Pa)	0.6 (h⁻¹)	0.4 (h⁻¹)
U-value (fundament) [W/(m²K)]	0.14	0.10
U-value (roof) [W/(m²K)]	0.14	0.08

Note: * calculated according to guidelines in Swedish Building Regulations.

Construction Cost

In this stage of the analysis, we calculated the average cost for producing our modeled buildings. All the prices used in the calculations are based on average market prices, which means that no special offers or discounts were considered. A price discount is possible to negotiate, but it is safe to assume that the same discount can be negotiated on all the products and therefore not relevant in the present analysis. The analysis excludes taxes and labor costs. The costs of constructing our model buildings were calculated using construction material prices from 2002/2003 and 2012/2013, as available in Sweden (sources: Sektionsfakta NYB 02/03 and NYB 12/13 [31,32]). All prices from 2002/2003 were adjusted for inflation. The cost assessment of new energy-efficient products was based on prices received from suppliers or sales representatives in 2013.

Difference in Floor Area

The next step of the analysis aimed at identifying losses and gains of saleable floor area caused by the difference in external wall measurements. The different technological improvements described in this paper may have an impact on energy requirement or on available living space. Considering that the energy requirement in our modeled buildings must be the same, regardless of the technology that has been applied, the building envelope was adjusted and this determines the effect that particular innovations may have on available living area. First, only the impact of different insulations was analyzed—see Table 3—and, secondly, the impact of better windows on possible adjustments to the building envelope was analyzed. In order to simplify the presentation, the second case is reported in Appendix 1–3 only. It is possible that some solutions may involve higher risks regarding such aspects as airtightness guarantee, mold issues or fire safety, and these problems are commented on in the discussion section below.

Table 3: (a) Loss in floor area between conventional and energy-efficient building, year 2002. (b) Loss in floor area between conventional and energy-efficient building, year 2013

Year	Type	Loss in floor area building as a whole, m^2	Compared to
(a) 2002	EE-A1	−12.8	CVN-A
	EE-B1	−18.4	CVN-B
(b) 2013	EE-A1	−12.8	CVN-A
	EE-B1	−13.8	CVN-B

Appraising Economic Losses and Gains Based on Saleable Floor Area

In order to assess whether the living floor area gains can defray the costs of construction, it was assumed that the developer can sell or rent one square meter of floor area at a given price. Two different price levels are used: p_s1 represents the average price that the developer can sell a dwelling for in midsized cities or in the suburbs of large cities (p_r1—assumed rental fee); p2 represents the average price at which the developer can expect to sell a dwelling located in the city centre in major cities like Stockholm, Goteborg or Malmö (p_r2—assumed rental fee); see Table 4. The prices are based on the current situation but they are applied both for 2002 and 2012 in order not to introduce more aspects than necessary. The role of price changes is commented upon in the discussion. The assessed income losses or gains in relation to difference in saleable living area are presented as a total value, *i.e.*, as a result of multiplying the difference in saleable area and price per square-meter.

Table 4: Assumed selling and renting prices for new residential construction in the city centre and in the suburbs

Location	Assumed selling and renting prices
Sale price of m^2 in the suburbs (p_s1)	2500 (Euro/m^2)
Sale price of m^2 in the city centre (p_s2)	6000 (Euro/m^2)
Rent price per year of m^2 in the suburbs (p_r1)	100 (Euro/m^2)
Rent price per year of m^2 in the city centre (p_r2)	150 (Euro/m^2)

There are reasons to believe that the square-meter price of an energy-efficient building may be higher than that of a conventional building [16,17,33]; however, for better comparability, the price of one square meter is the same regardless of building type or energy-efficiency level. It is assumed that there is no extra willingness to pay for the energy-efficient building.

STUDY RESULTS AND DISCUSSION

It is possible to make calculations for an almost infinite number of cases based on the assumptions above, therefore only the cases that seem most interesting are reported below. Results for cases where we also take into account the effect of window quality are reported in Appendix 1–3 (Table A1, Table A2, Table A3, Table A4, Table A5), but they are also included in the discussion.

Conventional versus Energy-Efficient Building with Standard Products—Difference in Floor Area

The floor area benefits or losses are presented in the form of difference in total living floor area (m²) calculated for the whole building. The comparison is made between a conventional building (CNV-A or CVN-B1) and an energy-efficient building with old techniques (EE-A1 and EE-B1). The results are reported in Table 3 below and in Table A1 for different assumptions about windows.

In the case of timber construction, floor area lost to external walls in energy-efficient buildings in 2002 was 12.8 m², but for brick construction, the difference was 18.4 m² (Table 3a). In the ten-year period, new dimensions of lightweight concrete bricks became available on the market. The greater range of products affected prices and allowed adjustments in brick wall construction. With products available on the market in 2012/2013, we were able to reduce the latter gap to 13.8 m² (Table 3b).

The Situation in 2002

Timber Houses

Table 5 below reports the cost difference between conventional and energy-efficient timber houses in 2002, indicating that cost difference in construction was approximately 14,000 Euro (as calculated for the whole building, with prices adjusted for inflation to year 2013). The assessed income losses in relation to difference in saleable living area indicate that for the house constructed in the suburbs, where m² prices are relatively lower, the anticipated income loss is approximately 32,000 Euro, but the income decrement is even higher in the city centre 76,800 Euro.

Table 5: Cost difference and assessed living area lost between conventional and energy-efficient building constructed in 2002, timber house

Difference in cost, living floor area and income	CVN A-EE-A1 2002
Construction cost difference *	
Cost difference (Euro/m^2 wall section)	−6.56
Cost difference (Euro, total wall construction)	−3,256
Cost difference windows (Euro)	−10,997
Total cost difference (windows + wall) (Euro)	−14,253
Gains/losses in living floor area (m^2)	−12.8
Assessed income losses/gains due to area difference *	
$p_s1 = 2,500$ Euro/m^2	−32,000
$p_s2 = 6,000$ Euro/m^2	−76,800
$p_r1 = 150$ Euro/m^2	−1,920
$p_r2 = 100$ Euro/m^2	−1,280

Note: * Cost difference and assessed income loss/gains are presented as a total value for a modeled building.

Brick Houses

Table 6 below summarizes the result from conventional and energy-efficient brick houses in 2002 and the result shows that, taking into account loss of saleable area, the cost for the energy-efficient building was 110,400 Euro higher in the central location and nearly 46,000 Euro higher in the suburban location.

Table 6: Cost difference and assessed living area lost between conventional and energy-efficient building constructed in 2002, brick house

Difference in cost, living floor area and income	CVN B-EE-B1 2002
Construction cost difference	
Cost difference (Euro/m^2 wall section)	−26.25
Cost difference (Euro, total wall construction)	−13,025
Cost difference windows (Euro)	−10,997
Total cost difference (windows + wall) (Euro)	−24,022
Gains/losses in living floor area (m^2)	−18.4
Assessed income losses/gains due to area difference	
$p_s1 = 2,500$ Euro/m^2	−46,000
$p_s2 = 6,000$ Euro/m^2	−110,400
$p_r1 = 150$ Euro/m^2	−2,760
$p_r2 = 100$ Euro/m^2	−1,840

The Situation in 2012

Timber Houses

Table 7 below reports a cost difference between conventional and energy-efficient timber houses in 2012 of 5500 Euro, indicating that the construction cost difference are lower than that in 2002. The relative price of the more energy-efficient products had fallen. The optimal envelope for the modeled building in 2002 and in 2012 was the same; therefore, the difference in living floor area between conventional and energy-efficient building was the same (12.8 m^2). Consequently, the result shows that when taking into account loss of saleable area, the cost for the energy-efficient building was 76,800 Euro higher in the central location and 32,000 Euro higher in the suburban location. Results for different assumptions about windows are reported in Table A2, Table A3.

Table 7: Cost difference and assessed living area lost between conventional and energy-efficient building constructed in 2013, timber house

Difference in cost, living floor area and income	CVN A-EE-A1 2013
Construction cost difference	
Cost difference (Euro/m^2 wall section)	−7.66
Cost difference (Euro, total wall construction)	−3,801
Cost difference windows (Euro)	−1,746
Total cost difference (windows + wall) (Euro)	−5,547
Gains/losses in living floor area (m^2)	−12.8
Assessed income losses/gains due to area difference	
$p_s 1 = 2,500$ Euro/m^2	−32,000
$p_s 2 = 6,000$ Euro/m^2	−76,800
$p_r 1 = 150$ Euro/m^2	−1,920
$p_r 2 = 100$ Euro/m^2	−1,280

Brick Houses

The relative costs for the energy-efficient building in 2012 are lower than in 2002 due to greater product availability. Considering cost efficiency and product range, we were able to reduce the gap in saleable living floor area between conventional and energy-efficient building from 18.4 m^2 in 2002 to 13.8 m^2 in 2012. Table 8 below reports the result from conventional and energy-efficient brick houses in 2012 and the result shows that, taking into account loss of saleable area, the cost for the energy-efficient building was 82,800 Euro higher in the central location and 34,500 Euro higher in the suburban location.

At this point, it is important to discuss how the initial assumptions could have affected building envelope construction in 2002. First, it was assumed quite strictly that buildings must be airtight, delivering 0.4 h^{-1} at +/−50 Pa. A decade of learning and sharing the experience of energy-efficient building construction resulted in a significant improvement in the airtightness of new buildings in the Nordic countries. Ten years of experience translate into a reduction of labor hours to perform highly accurate work. Secondly, for convenience of analysis, it was assumed that products like tapes, foil or thermal-free bridge connections are available and commonly used. It is possible that the construction cost of an energy-efficient airtight building could have been much higher in 2002 due to the higher cost and lower availability of those products on the market.

Table 8: Cost difference and assessed living area lost between conventional and energy-efficient building constructed in 2013, brick house

Difference in cost, living floor area and income	CVN B-EE-B1 2012
Construction cost difference	
Cost difference (Euro/m^2 wall section)	−28.98
Cost difference (Euro, total wall construction)	−14,380
Cost difference windows (Euro)	−1,746
Total cost difference (windows + wall) (Euro)	−16,127
Gains/losses in living floor area (m^2)	−13.8
Assessed income losses/gains due to area difference	
$p_s1 = 2,500$ Euro/m^2	−34,500
$p_s2 = 6,000$ Euro/m^2	−82,800
$p_r1 = 150$ Euro/m^2	−2,070
$p_r2 = 100$ Euro/m^2	−1,380

Energy-Efficient Building with New Products—Difference in Floor Area

The floor area benefits or losses are presented in the form of difference in total living floor area (m^2) calculated for the whole building. The comparison is made between a conventional building (CNV-A or CVN-B) and an energy-efficient building with old techniques (EE-A1 and EE-B1), as well as an energy-efficient building with newly developed products (EE-A2N, EE-A3N, EE-B2N, EE-B3N). The results are reported in Table 9 below and in Appendix 2 and 3 for different assumptions about windows.

Table 9: Loss in floor area for different energy-efficient technologies, only changes in wall construction, 2013

Type	Loss in floor area building as a whole, m^2	Compared to
EE-A1 *	−12.8	CVN-A
EE-A2N	−9.1	CVN-A
EE-A3N	3.9	CVN-A
EE-B1	−13.8	CVN-B
EE-B2N	−9.2	CVN-B
EE-B3N	20.3	CVN-B

Note: * Area loss/gains calculated as a difference in total living area between conventional and energy-efficient building.

Timber Houses

The analysis shows that applying new energy-efficient solutions in the construction helps achieve energy goals and may also be more profitable for the developer. By applying more energy-efficient components in constructing the building envelope, it was possible to adjust external wall width so that the very low space heating level was maintained and the gap in living floor area between conventional and energy-efficient building decreased to approximately 9 m² in the case of EE-A2N (insulation at lambda = 0.033). In the case of EE-A3N (insulation at lambda = 0.024), the saleable floor area increase was almost 4 m² more than in a conventional building (Table 9).

If account is taken of the gain in saleable area, the energy-efficient building with new products (EE-A3N) can generate 23,700 Euro more income in the central location than the conventional building (Table 10), which is enough to defray the extra cost. The income generated from the gain of saleable area (EE-A3N) in the suburban location was calculated at 9700 Euro (Table 10).

When highly energy-efficient windows were also applied in the buildings, the increase in living floor area was 3.9 m² for EE-A2N and 7.4 for EE-A3N more than that in conventional building (CVN-A) (Table A1) and was sufficient to defray the higher cost of highly energy-efficient windows and new insulation in the case of EE-A3N (Table A2, Appendix 2). Taking into account loss of saleable area, the income for the energy-efficient building (EE-A3N) was approximately 44,000 Euro higher in the central location and 18,500 Euro higher in the suburban location (Table A2). The results imply that constructing energy-efficient buildings with highly energy-efficient components may be more attractive than producing conventional buildings.

Table 10: Cost difference and assessed living area lost between conventional and energy-efficient building constructed with new product, timber house

Difference in cost, living floor area and income	CNV A-EE-A2N	CNV A-EE-A3N
Construction cost difference		
Cost difference (Euro/m² wall section)	−7.3	−38.5
Cost difference (Euro, total wall construction)	−3,635	−19,084
Cost difference windows (Euro)	−1,746	−1,746
Total cost difference (windows + wall) (Euro)	−5,381	−20,831
Gains/losses in living floor area (m²)	−9.1	3.9
Assessed income losses/gains due to area difference		
$p_s1 = 2,500$ Euro/m²	−22,750	9,750
$p_s2 = 6,000$ Euro/m²	−54,600	23,400
$p_r1 = 150$ Euro/m²	−1,365	585
$p_r2 = 100$ Euro/m²	−910	390

The analysis shows that using highly energy-efficient new components in the construction of energy-efficient timber houses results in an increase in saleable floor area and is often more profitable (Table 11). Furthermore, according to the results (Table A2, Table A3, Appendix 2), by applying both highly-energy efficient windows and new insulation, a developer can build an energy-efficient instead of a conventional building, which allows more living space to be sold and consequently increases income. This is even before considering potential energy and environmental savings.

Table 11: Cost difference and assessed living area lost between energy-efficient and energy-efficient building constructed with new product, timber house

Difference in cost, living floor area and income	EE-A1-EE-A2N	EE-A1-EE-A3N
Construction cost difference		
Cost difference (Euro/m^2 wall section)	0.3	−34.8
Cost difference (Euro, total wall construction)	166	−17,268
Cost difference windows (Euro)	−1,746	−1,7464
Total cost difference (windows + wall) (Euro)	−1,580	−19,014
Gains/losses in living floor area (m^2)	3.7	16.7
Assessed income losses/gains due to area difference		
$p_s1 = 2,500$ Euro/m^2	9,250	41,750
$p_s2 = 6,000$ Euro/m^2	22,200	100,200
$p_r1 = 150$ Euro/m^2	555	2,505
$p_r2 = 100$ Euro/m^2	370	1,670

Brick Houses

In the case of a brick wall construction, potential saleable floor area increases when highly energy-efficient products are employed in the building envelope construction. Applying the new technological solutions enables the developer to increase income by as much as 50,000 Euro in the suburbs and approx. 121,000 Euro in the city centre (Table 12). Using the new products in energy-efficient building construction increases saleable floor area, which in the case of EE-B2N was 4.6 m^2 and in EE-B3N 34 m^2, compared with energy-efficient building using old technologies (Table 13).

Adopting more energy-efficient windows and new insulation also encouraged favorable changes in light-concrete wall construction. In the case of EE-B3N (light-concrete brick construction with PIR insulation at lambda 0.024), by adopting windows with average U = 0.7 W/(m^{2}K), it was possible to re-design the external wall so that gains in living floor area could defray the additional cost of the new component. The benefit is 30 m^2 greater living floor area compared with a conventional building (Table A3).

Table 12: Cost difference and assessed living area lost between conventional and energy-efficient building constructed with new product, brick house

Difference in cost, living floor area and income	CNV B-EE-B2N	CNV B-EE-B3N
Construction cost difference		
Cost difference (Euro/m² wall section)	−30.4	−62.1
Cost difference (Euro, total wall construction)	−15,100	−30,802
Cost difference windows (Euro)	−1,746	−1,746
Total cost difference (windows + wall) (Euro)	−16,846	−32,548
Gains/losses in living floor area (m²)	−9.2	20.3
Assessed income losses/gains due to area difference (Euro)		
p_s1 = 2,500 Euro/m²	−23,000	50,750
p_s2 = 6,000 Euro/m²	−55,200	121,800
p_r1 = 150 Euro/m²	−1,380	3,045
p_r2 = 100 Euro/m²	−920	2,030

Table 13: Cost difference and assessed living area lost between energy-efficient and energy-efficient building constructed with new product, brick house

Difference in cost, living floor area and income	EE-B1-EE-A2N	EE-A1-EE-B3N
Construction cost difference		
Cost difference (Euro/m² wall section)	−1.5	−33.1
Cost difference (Euro, total wall construction)	−720	−16,422
Cost difference windows (Euro)	−1,746	−1,746
Total cost difference (windows + wall) (Euro)	−720	−16,422
Gains/losses in living floor area (m²)	4.6	34.0
Assessed income losses/gains due to area difference (Euro)		
p_s1 = 2,500 Euro/m²	11,500	85,000
p_s2 = 6,000 Euro/m²	27,760	204,000
p_r1 = 150 Euro/m²	690	5,100
p_r2 = 100 Euro/m²	460	3,400

The advantage of applying new highly energy-efficient components might also be qualitative: for example, more advanced window solutions help to minimize the thermal bridges, which reduces heat loss and the risk of draughts, and consequently delivers better indoor comfort for occupants. However, there are certain risks which should be discussed, for example, risks related to density of insulation material, airtightness of the building envelope and the moisture level of other components used in the construction, particularly organic material like timber. Checking for moisture level is as important as ensuring that the building envelope is airtight. One of the consequences of failure to produce an airtight building is heat loss and therefore an increase in energy consumption; however, sealing a building envelope with a high moisture level may also lead to problems with moisture and even mould. Ensuring that the moisture level in a building construction does not exceed safe parameters is essential for occupants' well-being and a healthy indoor environment.

Limitations

This paper has shown the effect of employing new technologies on the profitability of producing energy-efficient buildings; however, the analysis has certain limitations. During the investigation, it became clear that the innovative products are still in the prototype phase. Their use and availability on the market is relatively low. The standard energy-efficient products available on the market and used in this exercise as new technology were launched as few as 8–10 years ago. Unfortunately, solutions presented at building fairs or in manufacturers' catalogues were so new that detailed descriptions of the product or prices were sometimes not available. Detailed technical information was obtainable only on request, often directed or re-directed to the manufacturer.

It is unclear what total impact new energy-efficient technologies may have on the environment and peoples' health, as life cycle analysis and toxicity analysis of the presented solutions are outside the scope of this paper, but we hope that future studies will address those issues. Furthermore, the presented results are based on a simulation exercise, where certain assumptions had to be made, for example, regarding building positioning or installation system. It should be pointed out that there are virtually endless design alternatives among which we have presented only a few. The differences in saleable floor gains or losses depend on comparable design alternatives. Finally, prices used in the cost assessment are only based on purchasing material prices; costs of logistics, labor and external works were not considered.

CONCLUDING COMMENTS

The intention of this paper was to investigate how new energy-efficient products affect construction cost and profit. As noted in the literature (see extensive literature on economics of energy efficiency, innovation and technological development for example [34,35,36,37]), one of the greatest barriers to diffusion and commercialization of new environmental technologies is that benefits are spread out over time (e.g., energy savings) or not observable directly (e.g., environmental impact). It is thus important to demonstrate that implementing new energy-efficient technologies in the construction of buildings can have a more direct effect, which may positively impact on the profitability of highly energy-efficient buildings in the form of saleable floor area.

The impact of potential losses or gains of saleable floor area should be taken into account when comparing investment alternatives: building energy-efficient green dwellings or building conventional ones. The paper shows that constructing energy-efficient buildings and introducing very energy-efficient technologies may be both energy- and cost-effective when compared with conventional buildings. Employing new products in energy-efficient construction allows not only for benefits to be drawn from lower energy consumption during the life cycle of the building, but also from the increase in saleable floor area. This may have a significant effect on investment appraisal, particularly for projects in the city centre and other areas with high prices.

ACKNOWLEDGMENTS

The presented study is part of a research project which is funded by SBUF, The Development Fund of the Swedish Construction Industry.

Appendix 1. Loss in Floor Area in Relation to Windows of Different Quality

The calculations are made in the same way as in the main text. The table below presenting results for EE-A2N, EE-A3N, EE-B2N and EE-B3N include changes in insulation and highly energy-efficient windows. No changes were made to CNV A, CVN B and EE-A1 and EE-A2.

An average energy-efficiency (U) value for windows used in conventional buildings CVN-A and CVN-B was approximately 1.1 $W/(m^2K)$; an average energy-efficiency (U) value for windows used in energy-efficient houses was 0.9 $W/(m^2K)$; In this stage windows of 0.9 $W/(m^2K)$ in cases EE-A2N, EE-A3N, EE-B2N and EE-B3N were replaced with more energy-efficient windows where average U value was 0.7 $W/(m^2K)$. The average energy-efficiency value (U) for EE-A1 and EE-B1 were kept the same, *i.e.*, 0.9 $W/(m^2K)$. The simulation is done only for 2013 construction.

Table A1: Loss in floor area between conventional and energy-efficient building and between energy-efficient building with different technologies, year 2013

Type	Loss in floor area building as a whole, m^2	Compared to
EE-A1	−12.8	CVN-A
EE-A2N, $U_{windows} = 0.7\ W/(m^2K)$	3.9	CVN-A
EE-A3N, $U_{windows} = 0.7\ W/(m^2K)$	7.4	CVN-A
EE-B1	−13.8	CVN-B
EE-B2N, $U_{windows} = 0.7\ W/(m^2K)$	25.8	CVN-B
EE-B3N, $U_{windows} = 0.7\ W/(m^2K)$	29.5	CVN-B
EE-A2N, $U_{windows} = 0.7\ W/(m^2K)$	16.7	EE-A1
EE-A3N, $U_{windows} = 0.7\ W/(m^2K)$	24.1	EE-A1
EE-B2N, $U_{windows} = 0.7\ W/(m^2K)$	39.6	EE-B1
EE-B3N, $U_{windows} = 0.7\ W/(m^2K)$	43.3	EE-B1

Appendix 2. Results When Taking Window Quality into Account

The Situation in 2002

Highly energy-efficient windows are considered as new products; therefore, simulation could not be performed.

The Situation in 2013

Timber Houses

Table A2: Cost difference and assessed living area lost between conventional and energy-efficient building with new products constructed in 2013, timber house

Difference in cost, living floor area and income	CNV A-EE-A2N	CNV A-EE-A3N
Construction cost difference		
Cost difference (Euro/m² wall section)	−0.3	−38.5
Cost difference (Euro, total wall construction)	−146	−19,084
Cost difference windows (Euro)	−7,233	−7,233
Total cost difference (windows + wall) (Euro)	−7,380	−26,317
Gains/losses in living floor area (m²)	3.9	7.4
Assessed income losses/gains due to area difference (Euro)		
p_s1 = 2,500 Euro/m²	9,750	18,500
p_s2 = 6,000 Euro/m²	23,400	44,400
p_r1 = 150 Euro/m²	585	1,110
p_r2 = 100 Euro/m²	390	740

Brick Houses

Table A3: Cost difference and assessed living area lost between conventional and energy-efficient building constructed in 2013, brick house

Difference in cost, living floor area and income	CNV B-EE-B2N	CNV B-EE-B3N
Construction cost difference		
Cost difference (Euro/m² wall section)	−4.6	−27.3
Cost difference (Euro, total wall construction)	−2,267	−13,558
Cost difference windows (Euro)	−7,233	−7,233
Total cost difference (windows + wall) (Euro)	−9,500	−20,792
Gains/losses in living floor area (m²)	25.8	29.5
Assessed income losses/gains due to area difference (Euro)		
p_s1 = 2,500 Euro/m²	64,500	73,750
p_s2 = 6,000 Euro/m²	154,800	177,000
p_r1 = 150 Euro/m²	3,870	4,425
p_r2 = 100 Euro/m²	2,580	2,950

Appendix 3. Comparison of Energy-Efficient Buildings with Different Technology

Table A4: Cost difference and assessed living area lost between energy-efficient and energy-efficient building constructed with new product, timber house

Difference in cost, living floor area and income	EE-A1-EE-A2N	EE-A1-EE-A3N
Construction cost difference		
Cost difference (Euro./m² wall section)	7.4	−30.8
Cost difference (Euro, total wall construction)	3,654	−15,283
Cost difference windows (Euro)	−7,233	−7,233
Total cost difference (windows + wall) (Euro)	−3,578	−22,516
Gains/losses in living floor area (m²)	16.7	24.1
Assessed income losses/gains due to area difference (Euro)		
p_s1 = 2,500 Euro/m²	41,750	60,250
p_s2 = 6,000 Euro/m²	100,200	144,600
p_r1 = 150 Euro/m²	2,505	3,615
p_r2 = 100 Euro/m²	1,670	2,410

Table A5: Cost difference and assessed living area lost between energy-efficient and energy-efficient building constructed with new product, brick house

Difference in cost, living floor area and income	EE-A1-EE-B2N	EE-A1-EE-B3N
Construction cost difference		
Cost difference (Euro/m² wall section)	24.4	1.7
Cost difference (Euro, total wall construction)	12,113	821
Cost difference windows (Euro)	−7,233	−7,233
Total cost difference (windows + wall) (Euro)	4,879	−6,411
Gains/losses in living floor area (m²)	39.6	43.3
Assessed income losses/gains due to area difference		
p_s1 = 2,500 Euro/m²	99,000	108,250
p_s2 = 6,000 Euro/m²	237,600	259,800
p_r1 = 150 Euro/m²	5,949	6,495
p_r2 = 100 Euro/m²	3,960	4,330

REFERENCES

1. Stegall, N.; Dzombak, D. *Cost Implications of LEED Silver Certification for New House Residence Hall at Carnegie Mellon University*; Senior Honors Research Project. Carnegie Mellon University: Pittsburgh, PA, USA, 2004. Available online:http://www.cmu.edu/environment/ campus-green-design/green-buildings/images/newhouse_report.pdf (accessed on 2 April 2013).

2. Schnieders, J.; Hermelink, A. CEPHEUS results: Measurements and occupants' satisfaction provide evidence for passive houses being an option for sustainable building. *Energy Policy* 2006, *34*, 151–171.

3. Karlsson, J.F.; Moshfegh, B. A comprehensive investigation of a low-energy building in Sweden. *Renew. Energy* 2007,*32*, 1830–1841.

4. Matthiessen, L.; Morris, P. *Cost of Green Revisited: Re-Examining the Feasibility and Cost Impact of Sustainable Design in the Light of Increased Market Adoption*; Davis Langdon: New York, NY, USA, 2007.

5. Audenaert, A.; de Cleyn, S.H.; Vankerckhove, B. Economic analysis of passive houses and low-energy houses compared with standard houses. *Energy Policy* 2008, *36*, 47–55.

6. Mahdavi, A.; Doppelbauer, E.-M. A performance comparison of passive and low-energy buildings. *Energy Build.* 2010,*42*, 1314–1319.

7. Zhang, X.; Platten, A.; Shen, L. Green property development practice in China: Costs and barriers. *Build. Environ.*2011, *46*, 2153–2160.

8. Miller, N.; Spivey, J.; Florance, A. Does green pay off? Available online: http://www.costar.com/ josre/doesGreenPayOff.htm (accessed on 2 January 2010).

9. Zalejska-Jonsson, A.; Lind, H.; Hintze, S. Low-energy *versus* conventional residential buildings: Cost and profit. *J. Eur. Real Estate Res.* 2012, *5*, 211–228.

10. Ries, R.; Bilec, M.M.; Gokhan, N.M.; Needy, K.L. The economic benefits of green buildings: A comprehensive case study. *Eng. Econ.* 2006, *51*, 259–295.

11. Eichholtz, P.; Kok, N.; Quigley, J. Doing well by doing good? Green office buildings. *Am. Econ. Rev.* 2010, *100*, 2494–2511.

12. Fuerst, F.; McAllister, P. Green noise or green value? Measuring the effects of environmental certification on office values. *Real Estate Econ.* 2011, *39*, 45–69.

13. Ott, W.; Baur, M.; Jakob, M. *Direct and Indirect Additional Benefits of Energy Efficiency in Residential Buildings*; publication 260001; Study by Econcept and CEPE ETH Zurich on Behalf of the Research Programme EWG (www.ewg-bfe.ch) of the Swiss Federal Office of Energy: Bern, Switzerland, 2006.

14. Salvi, M.; Horehajova, A.; Syz, J. Are Households Willing to Pay for Green Buildings? An Empirical Study of the Swiss Minergie Label. In Proceedings of the European Real Estate Society (ERES) Conference, Milan, Italy, 23–26 June 2010.

15. Bloom, B.; Nobe, M.; Nobe, M. Valuing green home designs: A study of ENERGY STAR Homes. *J. Sustain. Real Estate* 2011, *3*, 109–126.

16. Brounen, D.; Kok, N. On the economics of energy labels in the housing market. *J. Environ. Econ. Manag.* 2011, *62*, 166–179.

17. Banfi, S.; Farsi, M.; Filippini, M.; Jakob, M. Willingness to pay for energy-saving measures in residential buildings.*Energy Econ.* 2008, *30*, 503–516.

18. Bordass, B.; Cohen, R.; Standeven, M.; Leaman, A. Assessing building performance in use 2: Technical performance of the probe buildings. *Build. Res. Inf.* 2001, *29*, 103–113.

19. Leaman, A.; Bordass, B. Assessing building performance in use 4: The probe occupant surveys and their implications.*Build. Res. Inf.* 2001, *29*, 129–143.

20. Corum, K.R.; O'Neal, D.L. Investment in energy-efficient houses: An estimate of discount rates implicit in new home construction practices. *Energy* 1982, *7*, 389–400.

21. Stocks, K.J. Discount rate for technology assessment: An application to the energy sector. *Energy Econ.* 1984, *6*, 177–185.

22. Howarth, R.B. Discounting and uncertainty in climate change policy analysis. *Land Econ.* 2003, *79*, 369–381.

23. Boonekamp, P.G.M. Price elasticities, policy measures and actual developments in household energy consumption—A bottom up analysis for the Netherlands. *Energy Econ.* 2007, *29*, 133–157.

24. Nässén, J.; Sprei, F.; Holmberg, J. Stagnating energy efficiency in the Swedish building sector—Economic and organisational explanations. *Energy Policy* 2008, *36*, 3814–3822.

25. Smeds, J.; Wall, M. Enhanced energy conservation in houses through high performance design. *Energy Build.* 2007, *39*, 273–278.

26. Krope, J.; Goricanec, D. Energy Efficiency and Thermal Envelope. In *A Handbook of Sustainable Building Design and Engineering: An Integrated Approach to Energy, Health, and Operational Performance*; Mumovic, D., Santamouris, M., Eds.; Earthscan: London, UK, 2009; pp. 23–34.

27. Kneifel, J. Life-cycle carbon and cost analysis of energy efficiency measures in new commercial buildings. *Energy Build.* 2010, *42*, 333–340.

28. Mohammadpourkarbasi, H.; Sharples, S. The eco-refurbishment of a 19th century terraced house: Energy and cost performance for current

and future UK climates. *Buildings* 2013, *3*, 220–244.

29. Boverket, *Regelsamling för byggande*; (in Swedish). Boverket: Karlskrona, Sweden, 2011.

30. Kildsgaard, I.; Jarnehammar, A.; Widheden, A.; Wall, M. Energy and environmental performance of multi-story apartment buildings built in timber construction using passive house principles. *Buildings* 2013, *3*, 258–277.

31. *Sektionsfakta–NYB 02/03 Wikells Byggberäkningar*; (in Swedish). Elanders Svenskt Tryck: Växjö, Sweden, 2003.

32. *Sektionsfakta–NYB 12/13 Wikells Byggberäkningar*; (in Swedish). Elanders Svenskt Tryck: Växjö, Sweden, 2013.

33. Zalejska-Jonsson, A. *Value of Environmental Factors and Willingness to Pay for Green Apartments in Sweden*; Working Paper Series, No. 13/1; Department of Real Estate and Construction Management & Centre for Banking and Finance (cefin), Royal Institute of Technology: Stockholm, Sweden, 2013.

34. Jaffe, A.; Newell, R.; Stavins, R. The Economics of Energy Efficiency. In *Encyclopedia of Energy*; Cleveland, C., Ed.; Elsevier: Amsterdam, the Netherlands, 2004; pp. 79–90.

35. Howarth, R.B.; Andersson, B. Market barriers to energy efficiency. *Energy Econ.* 1993, *15*, 262–272.

36. Jaffe, A.B.; Newell, R.G.; Stavins, R.N. A tale of two market failures: Technology and environmental policy. *Ecol. Econ.*2005, *54*, 164–174.

37. Popp, D.; Newell, R.G.; Jaffe, A.B. Energy, the Environment, and Technological Change. In *Handbook of the Economics of Innovation 2*; Elsevier: Amsterdam., the Netherlands, 2010; pp. 873–937.

Chapter 8

WIRELESS LASER RANGE FINDER SYSTEM FOR VERTICAL DISPLACEMENT MONITORING OF MEGA-TRUSSES DURING CONSTRUCTION

Hyo Seon Park[1, 2], Sewook Son[2], Se Woon Choi[2], and Yousok Kim[2]

[1]Department of Architectural Engineering, Yonsei University, 134 Shinchon-dong, Seoul 110-732, Korea

[2]Center for Structural Health Care Technology in Buildings, Yonsei University, 134 Shinchon-dong, Seoul 110-732, Korea

ABSTRACT

As buildings become increasingly complex, construction monitoring using various sensors is urgently needed for both more systematic and accurate safety management and high-quality productivity in construction. In this study, a monitoring system that is composed of a laser displacement sensor (LDS) and a wireless sensor node was proposed and applied to an irregular building under construction. The subject building consists of large cross-sectional members, such as mega-columns, mega-trusses, and edge truss, which secured the large spaces. The mega-trusses and edge truss that support this large space are of the cantilever type. The vertical displacement occurring at the free end of these members was directly measured using an LDS. To validate the accuracy and reliability of the deflection data measured from the LDS, a total station was also employed as a sensor for comparison with the LDS. In addition, the numerical simulation result was compared with the deflection obtained from the LDS and total station. Based on these investigations, the proposed wireless displacement monitoring system was able to improve the construction quality by monitoring the real-time behavior of the structure, and the applicability of the proposed system to buildings under construction for the evaluation of structural safety was confirmed.

INTRODUCTION

Recently, the need for the aesthetics and commercial intent of buildings to coincide with the development of their construction technique has led to an increasing trend for high-rise and irregular buildings. This trend increases the difficulty of construction due to the design complexities inherent in increasingly detailed construction [1–3]. Furthermore, the management of the construction process must ensure high-quality construction to prevent safety accidents caused by substandard construction [4–7]. Therefore, construction monitoring using various sensors is urgently needed for both more systematic and accurate safety management and high-quality productivity in construction. In addition, there is growing interest in structural health monitoring (SHM) techniques and the establishment of real-time monitoring systems based on sensor technology [8–12].

In terms of the type of sensors employed in SHM, accelerometers are mainly employed in a traditional SHM system to identify changes in dynamic characteristics, such as natural frequency, mode shape, and modal damping, induced from damage in the structure [13]. However, it is difficult to perform a quantitative evaluation on the health condition and safety of structures based on the change of dynamic characteristics because these characteristics are easily affected by non-structural elements and environmental conditions. Furthermore, another confounding factor is that damage is typically a local phenomenon, which is captured by higher frequency modes. However, the vibration-based damage identification method, which generally relies on lower frequency modes, tends to capture the global response of the structure and is less sensitive to local structural changes [14,15]. For these reasons, strain-type sensors are adopted to directly measure the strain of a structural element [16–20]. The stress distribution estimated from a strain measurement can be utilized in the safety assessment of an element by comparing the stress distribution with the yield stress of the materials or the design strength of the structural members. Several types of strain gauges are used to monitor structural responses, including electrical strain gauges (ESGs), fiber optic sensors (FOSs), and vibrating wire strain gauges (VWSGs). The major issue with the data obtained from strain-type sensors is that the observed strain value represents a damage condition restricted to a small area of the member. Therefore, a large number of strain gauges are needed to improve the accuracy of a structural damage evaluation.

To overcome the aforementioned limitations involving acceleration and strain values, there have been efforts to directly measure the displacement of a structure using a global positioning system (GPS), a vision-based system, and laser Doppler vibrometers. The measured displacement data can be utilized as a damage index, from which the structural stability and quality of construction can be estimated. GPS represents a good alternative to a displacement measuring system [21–24]. However, its applicability is limited to flexible and high-rise building structures because its accuracy is limited to 1 cm in the horizontal direction and 2 cm in the vertical direction [25]. The other main limitation in the application of GPS is that this system is restricted to outdoor or open spaces because signals are received from satellites. Vision-based systems are also proposed to measure the displacement of a structure and demonstrated an acceptable accuracy in SHM [26,27]. However, the light requirement in a vision-based monitoring system makes it difficult to use such systems at night; although night vision technology exists, its use is not yet feasible in SHM. Laser Doppler vibrometers also perform very well, but, these instruments are not suitable for long-term monitoring systems [28]. In addition, laser displacement sensors have been employed for measuring the displacement of structures subjected various loads such as wind and earthquake load in previous researches [29–31]. However, these measurement systems were used in laboratory experiments where environmental conditions can be easily controlled, and therefore their applicability to the real structures has not been validated. Because the ultimate goal of SHM involves the continuous and automatic monitoring the structural behavior induced from various loadings, the stable performance of monitoring system applied in actual structures is important and urgently needed. In particular, long-term monitoring of deflection in elements resulting from sudden change of loading condition, which frequently occurs in construction process and is a crucial factor in the safety of structure, is very challenging task, and there have been limited studies conducted using the existing displacement monitoring system.

In this study, a wireless laser range finder system is employed to directly measure the deflection of structural members in an irregular building that is currently under construction. The monitored irregular building is composed of large cross-sectional members, such as mega-columns, mega-trusses, and edge truss, which secure the large spaces. The mega-trusses and edge truss that support this large space are of the cantilever type. Thus, any vertical displacement occurring at the free end of these members was set as a major monitoring target and directly measured using a laser displacement sensor (LDS). In addition, the salient feature of this monitoring system is that a manager can perform real-time monitoring automatically, and the data are sent to a sensor node and then transmitted to a remote host server via wireless communications with good applicability to the structure under construction. The wireless network employed in this measurement system also overcomes the weakness of existing wired monitoring systems, which require large budgets to build the cable network and are inconvenient to manage [32].

To validate the accuracy and reliability of the deflection data measured from the LDS, the vertical deflections of the mega-trusses and edge truss were also measured using a total station as a sensor for comparison with the LDS during the bent removal, which was thought to cause the largest deflection during this construction process. In addition, the deflection of the mega-truss upon bent removal was analyzed using a structural analysis program and was compared with the deflection measurements obtained from the LDS and total station. Finally, the solutions for several issues discovered through the application of such wireless monitoring systems using an LDS were proposed for this practical application while continuing to perform structural safety monitoring using the data measured using this sensor technology.

DISPLACEMENT MONITORING SYSTEM

A wireless measurement system was built to monitor the vertical deflections that occur in cantilever-type mega-trusses in an irregular building. The automatic wireless displacement measurement system consists of a sensor (LDS) and a sensor node. The sensor node is equipped with a battery, which supplies the power, and a code-division multiple-access (CDMA) driving circuit that delivers the data. The process for measuring data from the LDS can be divided into three steps, as illustrated in Figure 1.

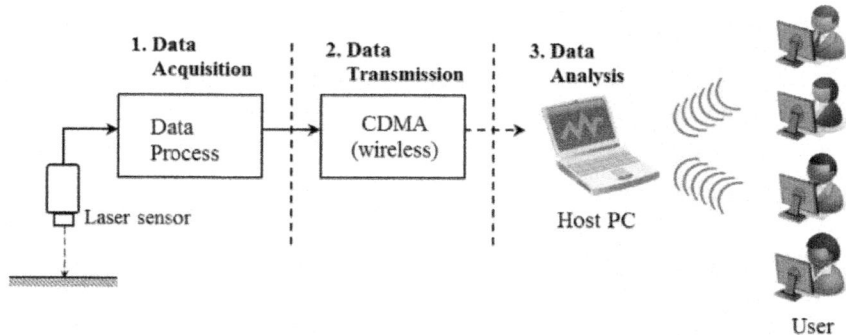

Figure 1: Data transmission process for the LDS.

Laser Displacement Sensor

The principle of non-contact laser sensors is largely divided into: (1) eddy-current, (2) optical, and (3) ultrasonic waves [33]. The eddy-current technique has a short measurement range and is therefore not suitable for long-distance measurements. The ultrasonic wave style has a long measurement range and can measure all objects; however, it is greatly affected by environmental factors, such as wind and temperature. Thus, when outside use is intended, the optical LDS technique is the most appropriate option.

In this research, an LDS was used to measure displacement in an economical and stable manner. An optical laser sensor (LLD-0100 model, JENOPTIK AG, Jena, Germany, Table 1) with a maximum sampling rate of 50 Hz was used in this study. This model has a built-in processor function such that the displacement data are outputted while power is supplied according to the preset data acquisition period. Thus, the distance between the LDS and object can be determined without a separate reader.

Table 1: Specifications of the LDS

Model	Measurement Range (m)	Accuracy (mm)	Resolution (mm)	Temperature Range (°C)	Max. Sampling Rate
LLD-0100	0.2–35	±2	0.1	−10–60	50 Hz

As depicted in Figure 2, the measurement principle of the optical LDS technique is that a laser beam, often with a diameter on the order of millimeters, is scattered when the target is reached, and this scattered beam creates an image on a one-dimensional position-sensing device that is then converted into an electrical signal. The distance between the LDS and target can be triangulated from the positional information of the imaged laser beam.

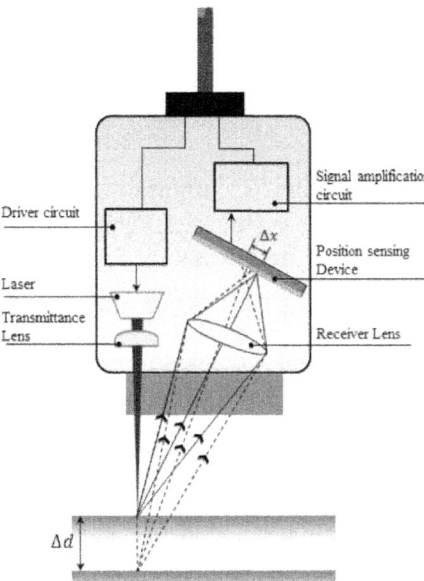

Figure 2: Principle of the LDS.

In a laboratory experiment, the environmental conditions can be easily controlled for the LDS application. However, when an LDS is used at a construction site, the measurement range, material of the object, and surrounding environment should be considered. First, a long reference distance and measurement range between the monitoring target and LDS are needed, considering the various obstacles that hinder stable measurement at a construction site. The measurement range of the LDS used in this research is 0.2–35 m, as indicated in Table 1, and can reach up to 150 m when reflecting plates are used. The accuracy of the LDS is 2 mm, which is relatively coarse. However, the measurement range is the primary requisite in this application, where the monitored structure is a large-scale irregular building structure during construction. High-accuracy LDSs with an accuracy on the order of micrometer are also available [34,35]; however, their short reference distance and measurement range which are less than 1 m are not applicable in this research. Therefore, a deflection accuracy on the order of millimeters is regarded as acceptable considering the trade-off between accuracy and the measurement range.

Wireless Monitoring System

The sensor node with the four-channel sensor interface can simultaneously receive data from four LDSs, which are connected through a cable for RS-422 communication. The remote wireless communication method (CDMA) [36] was used to send the data measured by the LDS from the construction site to the host personal computer (PC) in a remote location. The data sent to the host PC could be verified online in real-time via the integrated management software.

A continual power supply is important for the stable operation of automatic wireless monitoring systems. A cable layout plan is necessary for a direct power supply at a construction site, and the cables for such a power supply can reduce the applicability to a construction site from an operational and management perspective. Therefore, the wireless monitoring system developed in this research uses both a technique with low power consumption for the processor and a timer control to switch between the operating mode (data acquisition and transmission) and the sleep mode (power consumption minimized), which are synchronized with the data measurement periods to minimize power consumption. The processor is connected to a power source through the two circuits Regulator 1 and Regulator 2. Regulator 1 constantly supplies power to the processor so that it can operate with a minimum amount of power in sleep mode. Regulator 2 operates during the data measurement periods and activates the sensor node by supplying power to the sensor and

CDMA drive circuit [37]. In other words, the lifespan of the battery was extended by minimizing the power consumed during times of inactivity rather than uniformly consuming power at all times. The power supply consisted of a rechargeable lithium-ion battery and was replaced approximately once every three months. The specifications for the sensor node are provided in Table 2.

Table 2: Sensor node specification

Number of Channels	Power Consumption	Data Output	Software
4	DC 7.4–24 V 100 mA (operation)	CDMA	LDSMS Client

APPLICATION TO A CONSTRUCTION SITE

Target Structure (D Building)

This research applied the proposed wireless displacement monitoring system to and conducted real-time monitoring of a building under construction (D building). The building is located in Seoul, the capital of South Korea; its construction began in April 2009 and is scheduled to be completed in 2013. As observed in Figure 3, the D building is an irregularly shaped building that will be used for various exhibitions as a museum, experience facility, convention hall, designed corporate offices and a sky lounge and consists of three underground floors and four stories.

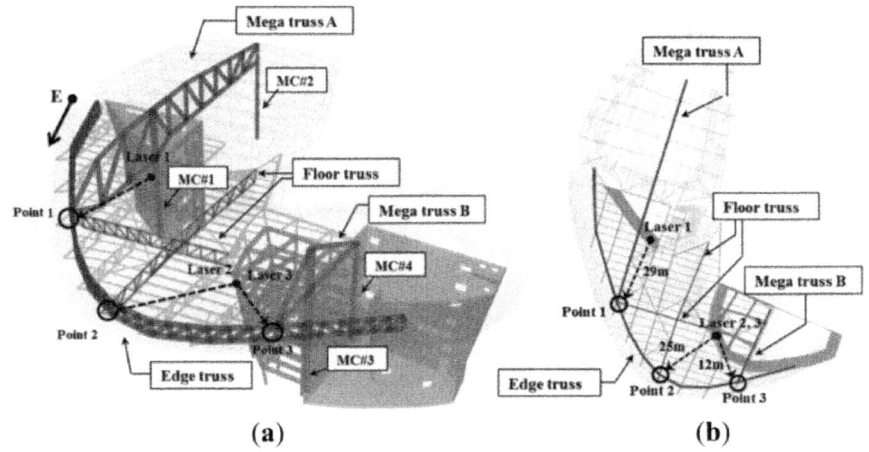

(a)   (b)

Figure 3: Target structure, D building. (a) Three-dimensional view of the main elements. (b) Cross-sectional plan of the main elements.

The exhibition zone uses steel frames to present a large free space, and steel trusses with particularly large cross-sections were used to secure this space. These members can be grouped into mega-trusses A and B, an edge truss, and a floor truss according to their location. The composition of the complex space consists of four mega-columns (#1, 2, 3, and 4) that support mega-trusses A and B with additional smaller supports consisting of diverse steel numbers and cross-sectional sizes (Figure 3). As observed in Figure 4, mega-truss A was fabricated in several parts and constructed on-site through sequential erection and welding. The span length of the edge truss in Figure 5 is approximately 142 m; thus, it was divided into 10 parts for fabrication and similarly constructed via a sequential erection and welding process. To ensure structural safety while welding each part, a temporary bent was installed below the welding spots to support the weight of the structural member such as a mega-truss or edge truss.

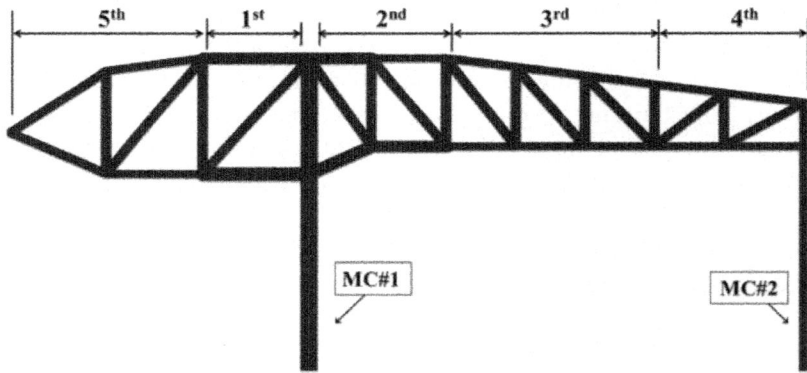

Figure 4: Mega-truss A.

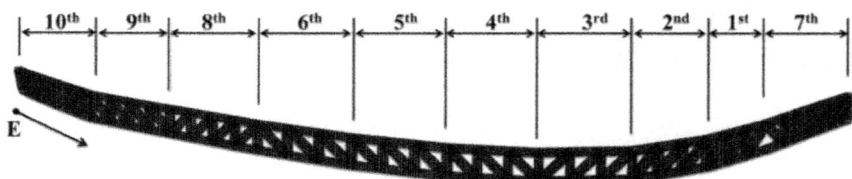

Figure 5: Edge truss.

In addition, each of the cores, which consist of mega-column #1 for one and mega-columns #3 and 4 for the other, is a reinforced-concrete (RC) structure, and these two cores are connected through the fourth floor slab used by the exhibition hall. As there is no column under the fourth floor slab, the dead load is sustained by the floor truss, which supports the center of this slab, and the edge truss supports its border and flows into the connected mega-truss, which is supported by the mega-column and RC core.

The distance from mega-columns #1 and #2 to the edge truss is 35 m and 12.3 m, respectively, and the mega-truss connected to the edge truss is of the cantilever type. The gravity load of the space frame, which is installed on the exterior panel, combines with that of the fourth floor slab, particularly during construction; thus, the deflection of the mega truss is a priority control and monitoring target during construction. In particular, the variations in the deflection of the edge truss were continually monitored as the 10 temporary bents that had been installed to support the construction of both the mega-truss and edge truss were individually removed following the order shown in Figure 6.

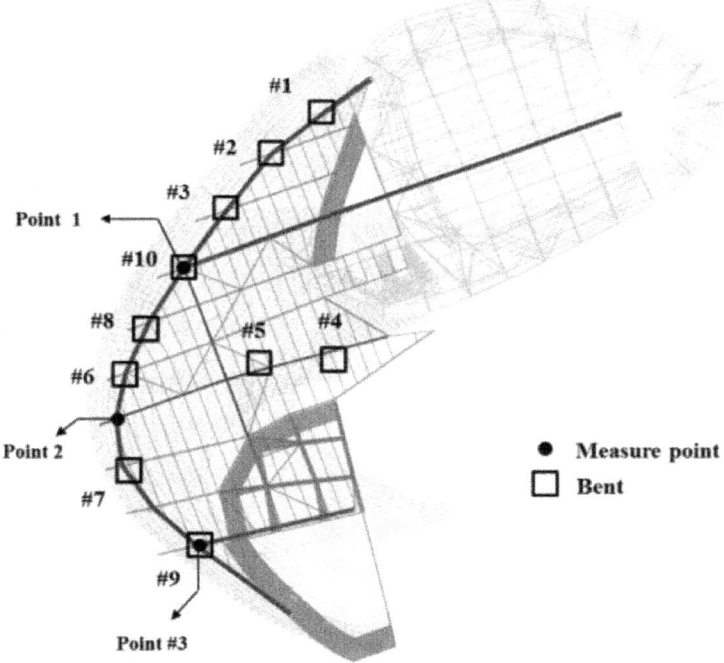

Figure 6: Removal schedule for temporary bents.

Measurement Setup

The vertical deflections of the edge truss were measured from three points (points 1, 2, and 3 in Figure 3). Point 1 is the connection point between mega-truss A and the edge truss, point 2 is the connection point between the edge truss and floor truss, and point 3 is the connection point between mega-truss B and the edge truss. The deflections that can occur at these three points due to the dead and live loads only were calculated using a commercial structural analysis program (MIDAS/GEN" ver.800, [38]), and the predicted camber of each point is listed in Table 3.

Table 3: Camber of each measurement point

Location	Point 1	Point 2	Point 3
Camber (DL + LL) (mm)	155.1	164.4	44.3

When measuring vertical deflection, it is best to measure the displacement by installing an LDS on the ground and reflecting it off of the target member. However, due to the conditions of the construction site, such as the various construction equipment and movements, and because the measurement range of the LDS is limited (maximum of 35 m), it was impossible to install these sensors on the ground. Therefore, in this research, the LDSs were installed horizontally within the building in the areas least affected by the construction to measure the data in a stable manner (Figure 7).

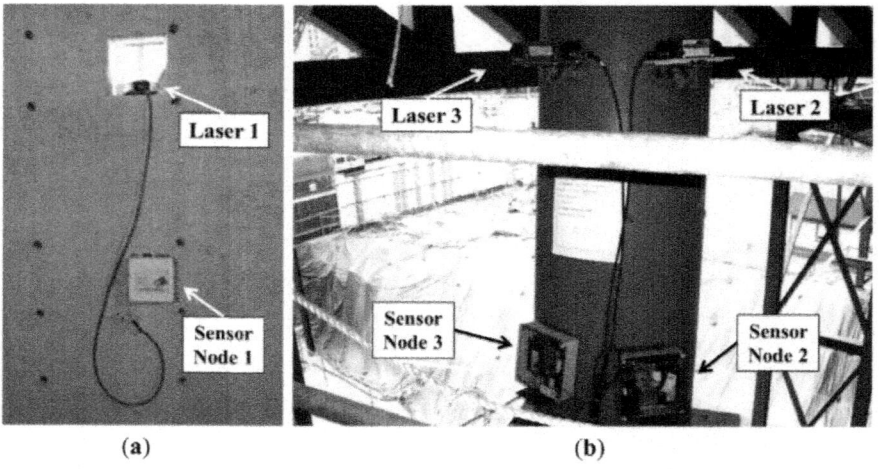

(a)   (b)

Figure 7: Installation location. (a) Laser sensor 1. (b) Laser sensors 2 and 3.

To determine the vertical displacement via the horizontal displacement measurements, a triangle module was welded to the target points, as illustrated in Figure 8. A module (300 × 300 × 300 mm) with a height (h) of 300 mm that forms a 45° angle to its base was welded onto the measurement point.

The module was planned such that the horizontal displacement of the LDS and the measurement point would create a vertical deflection. In other words, because the triangle module was at 45°, determining its vertical displacement by measuring the horizontal displacement was possible using the principle that the distance (Δx) and vertical variation (Δz) were the same when the vertical deflection occurred in the z direction.

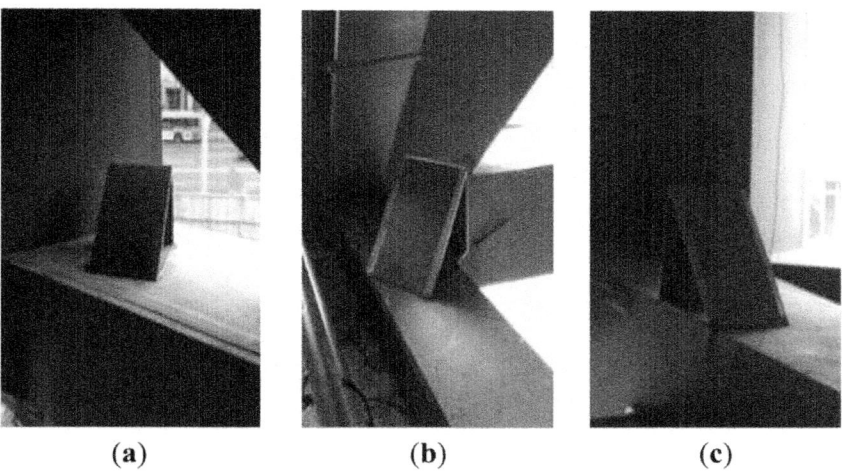

(a)   (b)   (c)

Figure 8: Triangle module installed at the measured point: (**a**) point 1, (**b**) point 2, and (**c**) point 3.

Total Station

To verify the accuracy and applicability of the deflections measured by the proposed wireless monitoring systems, the vertical deflections were also measured using a total station as a sensor for comparison with the LDS during the bent removal, which was thought to cause the largest deflection during this construction process.

The total station combines an electro-optical instrument (electrical discharge machine (EDM)) to measure the angle with a theodolite to simultaneously measure distance. This device can determine the three-dimensional coordinates of the target point (Figure 9). The total station consists of four parts: a detector measuring the vertical angle of the up and down motions, a detector measuring the horizontal angle of the right and left motions, a distance gauge measuring the distance from the body to the target point, and a tilting sensor that measures and revises the level of the body. Prisms are installed at the target points, and a light wave with a known wavelength and frequency is transmitted (Figures

10 and 11). The distance can be determined from the time it takes the light wave to return, and the three-dimensional coordinates can be determined based on the azimuth and altitude. Therefore, the data measured using the total station are more accurate than those measured with the LDS, which only measures the one-dimensional distance. Thus, this research used the total station measurement as the real deflection when comparing and analyzing the results obtained using the LDS.

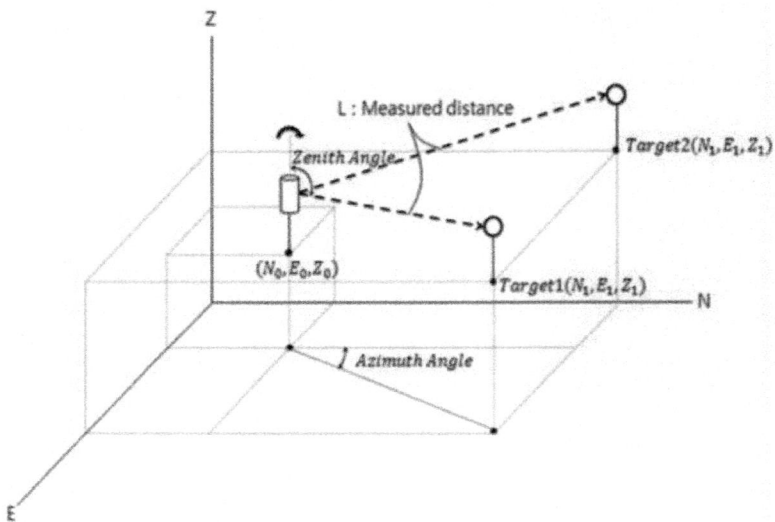

Figure 9: Measurement method of the total station.

(a)   (b)

Figure 10: Total station measurement: (a) total station and (b) prism.

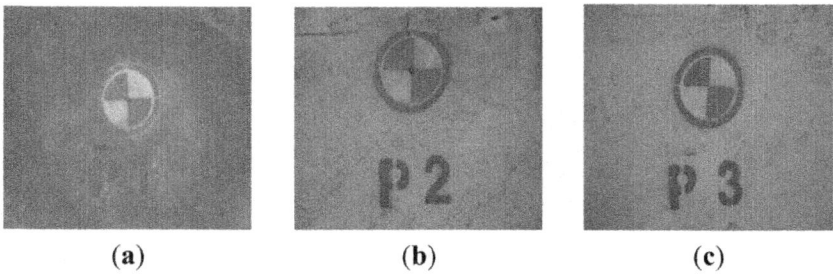

(a)   (b)   (c)

Figure 11: Prism: (**a**) point 1 mark, (**b**) point 2 mark, and (**c**) point 3 mark.

In this research, the prism was installed above the point where the LDS measured the deflection and only used the total station for measurements during the bent removal period. Figures 10 and 11 provide images of the installation of the total station at the construction site and the prism at the target point, respectively.

EVOLUTION OF DISPLACEMENT DURING CONSTRUCTION

Evolution of Deflection at Each Point during Construction

After the welding connections of the fabricated mega-truss and edge truss were completed, the temporary bents were removed. The dead load of the structure that had been supported by the bents was then directly supported by these members, and the edge truss began to deflect. A total of 10 bends had supported the edge truss, and these bents were sequentially removed according to the construction schedule, as observed in Figure 6.

The deflection of the edge truss upon bent removal was analyzed using a structural analysis program (MIDAS/GEN" ver.800, [38]) and compared with the deflection measurements obtained from the LDS and total station. When modeling the structure, the space frame, or external panel, was excluded to reflect the construction stage at the time of measurement, and the mega-trusses, edge truss, and floor trusses were modeled as rigid connections considering their large sectional size and welding construction. The analysis was performed by removing each temporary bent in accordance with the construction process, and the deflections at points 1, 2, and 3 for each step were analyzed. Figure 12 presents plots of the calculated vertical displacement of the structure both before and after the removal of the temporary bents. The maximum displacement occurs at point 2 on the edge truss after bent removal.

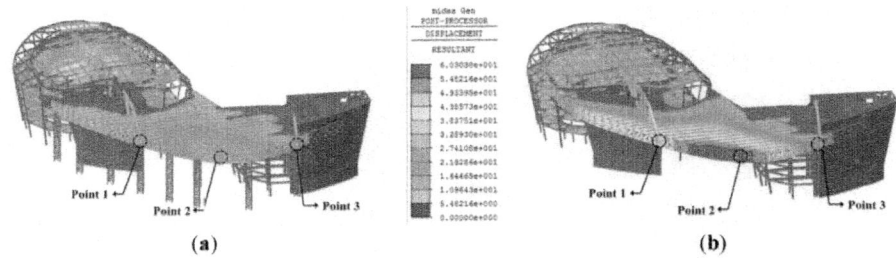

Figure 12: Calculated displacement of building: (**a**) before and (**b**) after bent removal.

The plots in Figure 13 compare the values obtained from both the structural analysis and vertical deflection as measured by the LDS and total station for each measuring point. The LDS continuously measured each point every 30 min, and the total station was measured while each bent was being removed.

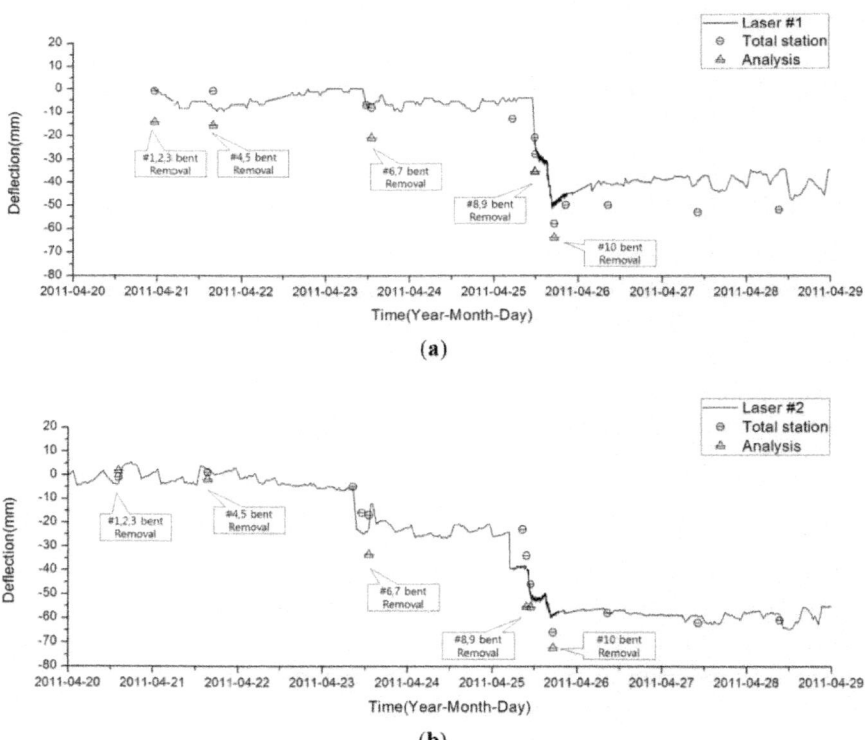

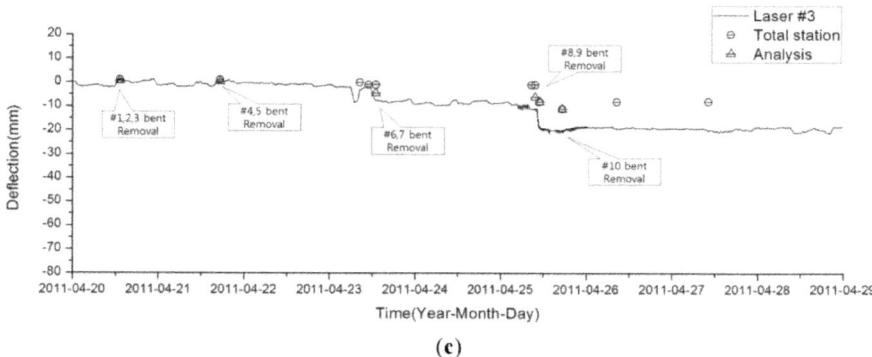

(c)

Figure 13: Variations of deflection during bent removal: (**a**) point 1, (**b**) point 2 and (**c**) point 3.

The difference in the deflection between the LDS and total station at point 2 was less than 5 mm, and changes occurred as the deflection increased with time (sequential removal of the bent). The vertical displacement determined for each point from the analysis overestimated the LDS and total station measurements. Nevertheless, the increase in sequential deflection due to bent removal was reproduced in the structural analysis results.

Comparison of the Deflection Values obtained from the LDS and Total Station

As demonstrated in Figure 13, the total station measurements were larger than those of the LDS at point 1; however, the results obtained at point 3 exhibited an opposite trend. The total station can measure the three-dimensional displacement of the target; however, the LDS can only provide data regarding the distance between the sensor and target point. This type of error generated in a pointer measuring system was previously investigated in pioneering research [39], and in this study, the discrepancy between two measuring systems is also investigated in the following manner. When structural deformation occurs, the triangular module on the target object also moves, and the error from the horizontal displacement and rotational components may be included in the measurement data with the absolute vertical displacement. In other words, the difference in distance caused by the rotation of the marked axis shown in Figure 14 can be obtained using Equations (1), (2), and (3).

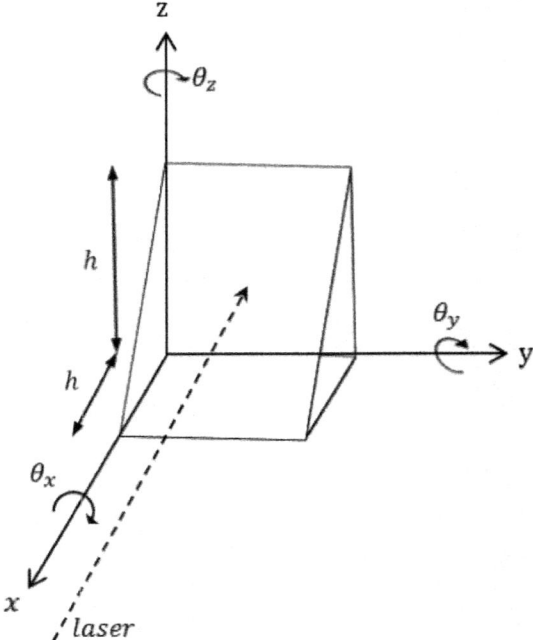

Figure 14: Variations in the measured value according to the rotation of the triangle module.

When the member and triangular module are attached to the x and y planes, Equation (1) provides the difference in distance between the LDS and the slope face of the module when the module rotates θ in the θx direction. Equations (2) and(3) provide the difference in distance between the LDS and module for the y and z axes, respectively:

About θx, $\Delta x = h/2(\sin\theta+\cos\theta-1)$ (1)

About θy, $\Delta x = h\{(1-\cos\theta)/(\sin\theta-\cos\theta)\}$ (2)

About θz, $\Delta x = h/2\{(1+\sin\theta-\cos\theta)/\cos\theta$ (3)

As demonstrated in Figure 15, the maximum distance between the LDS and target occurred when the module was rotated around the z axis. The minimum distance between the LDS and target occurred when the module was rotated around the y axis; however, this variation induced from the rotation around the y axis was small compared to the variations of the other axes.

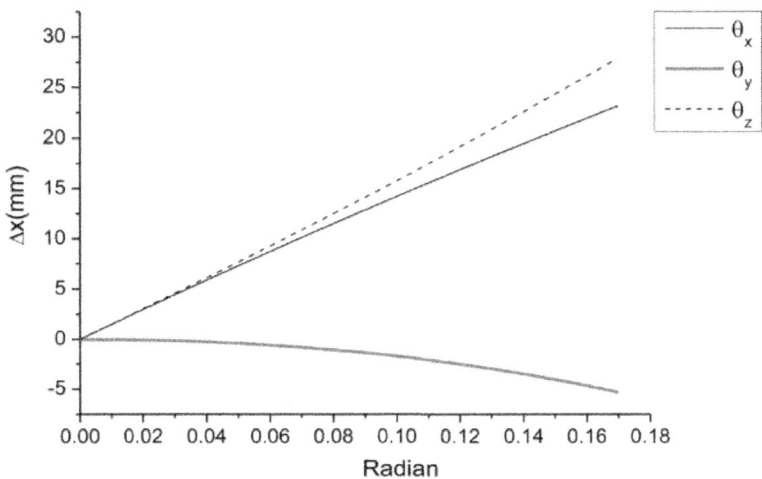

Figure 15: Change in horizontal distance due to rotation around each axis.

As these results and the measurement data demonstrate, the largest vertical displacement occurred at point 2. Therefore, because the module at point 1 can be observed to rotate in the $-\theta x$ direction, the horizontal distance between the LDS and target module decreases, and a value smaller than the real deflection observed from the total station is measured. In contrast, at point 3, the module rotates in the $+\theta x$ direction, and the LDS measures a higher value than the real deflection. Therefore, the LDS measurement in Figure 13(a) is larger than that of the total station, whereas the LDS measurement inFigure 13(c) is smaller than that of the total station. That is, the effects of rotation are included in the results measured by the LDS in addition to the displacement by pure vertical deflection.

Detection of the Real-Time Behavior of Structures

April 25, 2011 was the last day of temporary bent removal, and bents 8, 9, and 10 were removed. The location of bent 9 was the connection point between mega-truss B and the edge truss (point 3). Bent 8 was the connection between mega-truss A and the edge truss (point 1). Thus, the locations of these bents have the largest effect on the LDS measurements. Furthermore, the deflection that occurs after removing all of the bents is caused by the dead load and can be used to estimate the safety of the construction.

Figure 16 presents the measurement results at each point during the day of bent removal. At 10 AM, when the removal of bent 8 began, the LDS measurement interval was reduced from 30 min to 1 min to continuously acquire data from and to verify the deflection of each point in real time. However, a null-data period can be observed from 7 AM to 11 AM and from 2 AM to 3:30 PM at points 1 (Figure 16(a)) and 3 (Figure 16(c)), respectively. These null-data periods result from lost measurements due to the presence of obstacles, such as construction equipment, between the LDS and target during bent removal. After 5 PM, when all of the bents had been removed, stable measurements were obtained without large variations in the deflection. From the above results, it can be observed that the safety monitoring of the structures is possible, even when there are sudden changes to the load (e.g., bent removal in this study), via the flexible control of measurement intervals. Furthermore, the maximum deflection at each point after bent removal was smaller than the initial camber value (Table 3) determined from the predicted vertical deflections of 52, 67, and 13 mm at points 1, 2, and 3, respectively. Thus, the safety of the edge truss and mega-truss was confirmed.

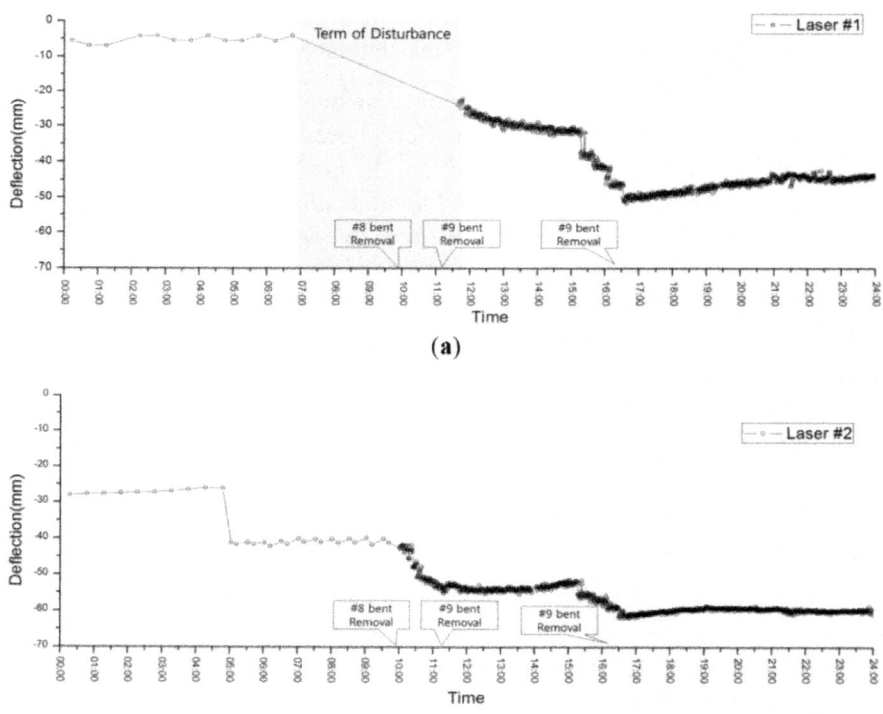

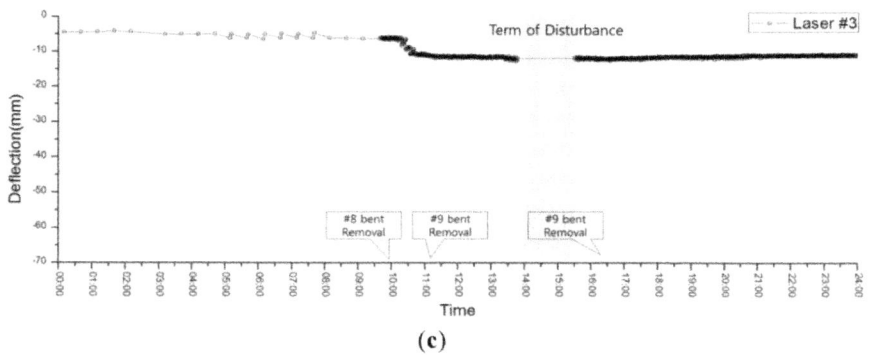

(c)

Figure 16: Measured deflection on April 25: (**a**) point 1, (**b**) point 2, and (**c**) point 3.

CONCLUSIONS

This research proposed a displacement monitoring system using an LDS to perform real-time monitoring of an irregular large-scale building under construction, which consisted of structural members (namely, mega-trusses, mega-columns, and an edge truss) for large spaces. The boundary deflection of the broad slab, which was supported by a cantilever-type mega-truss and the edge truss, was targeted for the primary monitoring. The data measured by the LDS were sent to the sensor node, processed by a built-in processor, and finally transmitted via a CDMA method to a remote host PC server. This entire process was successfully automated. The following results were obtained from the deflection data measured from the proposed monitoring system

The deflection data measured from the LDSs accurately captured the cantilever behavior (*i.e.*, deflection at the free end) caused by the removal of bents, which were temporarily installed at the free end of the mega-truss and edge truss during construction to prevent the deflection of the free end of the slab. Furthermore, the maximum deflection at each point after bent removal was smaller than the initial camber value determined from the predicted vertical deflections. Thus, the structural safety of the edge truss and mega-truss were confirmed.

From the deflection data collected from the total stations, which allowed for more accurate distance measurements, slight discrepancies between the two measurements were observed. The source of these permissible differences can be attributed to the movement of the triangular module on the target object, and the error from the horizontal displacement and rotational components might be included in the results measured by the LDS in addition to the pure vertical deflection. In the numerical simulation results, the increase in sequential deflection due to bent removal was also reproduced, although these results overestimated the LDS and total station measurements.

According to the above results, the proposed wireless displacement monitoring system based on the LDS was able to improve construction quality by monitoring the real-time behavior of the structure, and the applicability of the proposed system to buildings under construction for the evaluation of structural safety was also confirmed.

ACKNOWLEDGMENTS

This work was supported by the National Research Foundation of Korea (NRF) grant funded by the Korea government (Ministry of Education, Science and Technology, MEST) (No. 2011-0018360).

REFERENCES

1. Hegger, J. High-strength concrete for a 186 m high office building in Frankfurt, Germany. *Eng. Struct.* 1996, *11*, 850–854.

2. Tatsuya, W.; Noriyuki, F.; Yasuo, I.; Takashi, S. Automated construction system for high-rise reinforced concrete buildings. *Autom. Constr.* 2000, *9*, 229–250.

3. Abid, A.; Ribakov, Y. Recent trends in steel fibered high-strength concrete. *Mater. Des.* 2011, *32*, 4122–4151.

4. Ikeda, Y.; Harada, T. Application of the Automated Building Construction System Using the Conventional Construction Method Together. Proceedings of the 23rd International Symposium on Automation and Robotics in Construction, Tokyo, Japan, 3–5 October 2006.

5. Hu, Z.; Zhang, J. BIM- and 4D-based integrated solution of analysis and management for conflicts and structural safety problems during construction: 2. Development and site trials. *Autom. Constr.* 2011, *20*, 167–180.

6. Zhou, W.; Whyte, J.; Sacks, R. Construction safety and digital design: A review. *Autom. Constr.* 2012, *22*, 102–111.

7. Zhang, S.; Teizer, J.; Lee, J.; Charles, M.E.; Manu, V. Building Information Modeling (BIM) and safety. Automatic safety checking of construction models and schedules. *Autom. Constr.* 2012, *29*, 183–195.

8. Park, H.S.; Lee, H.M.; Adeli, H.; Lee, I. A new approach for health monitoring of structures: Terrestrial laser scanning. *Comput. Aided Civil Infrastr. Eng.* 2007, *22*, 19–30.

9. Lee, H.M.; Park, H.S. Estimation of deformed shape of beam structure using 3D coordinate information from terrestrial laser scanning. *Comput. Model. Eng. Sci.* 2008, *29*, 29–44.

10. Ni, Y.Q.; Li, B.; Lam, K.H.; Zhu, D.; Wang, Y.; Lynch, J.P.; Law, K.H. In-construction vibration monitoring of a super-tall structure using a long-range wireless sensing system. *Smart Struct. Syst.* 2010, *7*, 83–102.

11. Wu, Z.F.; Gao, F. Application and research of steel structure construction monitoring of costa rica state stadium canopy with measurement robot. *Energy Procedia* 2011, *13*, 2794–2801.

12. Xia, Y.; Ni, Y.Q.; Zhang, P.; Liao, W.Y.; Ko, J.M. Stress development of a supertall structure during construction: Field monitoring and numerical analysis. *Comput. Aided Civil Infrastr. Eng.* 2011, *26*, 1–8.

13. Salawu, O.S. Detection of structural damage through changes in frequency: A review. *Eng. Struct.* 1997, *19*, 718–723.

14. Doebling, S.W.; Farrar, C.R.; Prime, M.B.; Shevitz, D.W. *Damage Identification and Health Monitoring of Structural and Mechanical Systems from Change in Their Vibration Characteristics: A Literature Review*; Technical Report No. LA-13070-MS; Los Alamos National Laboratory: Los Alamos, NM, USA, 1996.

15. Carden, E.P.; Fanning, P. Vibration based condition monitoring. A review. *Struct. Health Monit.* 2004, *3*, 355–377.

16. Maaskant, R.; Alavie, T.; Measures, R.M.; Tadros, G.; Rizkalla, S.H.; Guha-Thakurta, A. Fiber-optic bragg grating sensors for bridge monitoring. *Cem. Concr. Composites* 1997, *19*, 21–33.

17. Coutts, D.R.; Wang, J.; Cai, J.G. Monitoring and analysis of results for two strutted deep excavations using vibrating wire strain gauges. *Tunn. Underground Space Technol.* 2001, *16*, 87–92.

18. Park, H.S.; Jung, H.S.; Kwon, Y.H.; Seo, J.H. Mathematical models for assessment of the safety of steel beams based on average strains from long gage optic sensors. *Sens. Actuators A Phys.* 2005, *125*, 109–113.

19. Park, H.S.; Jung, S.M.; Lee, H.M.; Kwon, Y.H.; Seo, J.H. Analytical models for assessment of the safety of multi-span steel beams based on average strains from long gage optic sensors. *Sens. Actuators A Phys.* 2007, *137*, 6–12.

20. Lee, H.M.; Park, H.S. Measurement of maximum strain of steel beam structures based on average strains from vibrating wire strain gages. *Exp. Technol.* 2013, *37*, 23–29.

21. Nakamura, S. GPS measurement of wind-induced suspension bridge girder displacements. *J. Struct. Eng.* 2000, *126*, 1413–1419.

22. Celebi, M.; Eeri, M.; Sanli, A. GPS in pioneering dynamic monitoring of long-period structures. *Earthq. Spcetra* 2002,*18*, 47–61.

23. Tamurra, Y.; Matsui, M.; Pagnini, L.C.; Ishibashi, R.; Yoshida, A. Measurement of wind-induced response of building using RTK-GPS. *J. Wind Eng. Ind. Aerodyn.* 2002, *90*, 1783–1793.

24. Park, H.S.; Sohn, H.G.; Kim, I.S.; Park, J.H. Application of GPS to monitoring of wind-induced responses of high-rise buildings. *Struct. Des. Tall Spec. Build.* 2007, *17*, 117–132.

25. Breuer, P.; Chmielewski, T.; Gorski, P.; Konopka, E. Application of GPS technology to measurement of displacement of high-rise structures due to weak winds. *J. Wind Eng. Ind. Aerodyn.* 2002, *90*, 223–230.

26. Fraser, C.S.; Riedel, B. Monitoring the thermal deformation of steel beams via vision metrology. *J. Photogramm. Remote Sens.* 2000, *55*, 268–276.

27. Lee, J.-J.; Shinozuka, M. A vision-based system for remote sensing of bridge displacement. *NDT E Int.* 2006, *39*, 425–431.

28. Nassif, H.H.; Gindy, M.; Davis, J. Comparison of laser doppler vibrometer with contact sensors for monitoring bridge deflection and vibration. *NDT E Int.* 2005, *38*, 213–218.

29. Balendra, T.; Anwar, M.P.; Tey, K.L. Direct measurement of wind-induced displacement in tall building models using laser positioning technique. *J. Wind Eng. Ind. Aerodyn.* 2005, *93*, 399–412.

30. Xu, Y.L.; Zhang, J.; Li, J.C.; Xia, Y. Experimental investigation on statistical moment-based structural damage detection method. *Struct. Health Monit.* 2009, *8*, 555–571.

31. Xu, B.; Song, G.; Masri, S.F. Damage detection for a frame structure model using vibration displacement measurement. *Struct. Health Monit.* 2012, *11*, 281–192.

32. Lynch, J.P.; Loh, K.J. A summary review of wireless sensors and sensor networks for structural health monitoring.*Shock Vib. Dig.* 2006, *38*, 91–128.

33. Shieh, J.; Huber, J.E.; Fleck, N.A.; Ashby, M.F. The selection of sensors. *Prog. Mater. Sci.* 2001, *46*, 461–504.

34. Zhang, F.; Qu, X.; Ouyang, J. An automated inner dimensional measurement system based on a laser displacement sensor for long-stepped pipes. *Sensors* 2012, *12*, 5824–5834.

35. Keyence Global Home. Available online: http://www.keyence.com/ (accessed on 2 May 2013).

36. Viterbi, A.J. *CDMA: Principles of Spread Spectrum Communication*; Addison-Wesley: Reading, MA, USA, 1995.

37. Lee, H.M.; Kim, J.M.; Sho, K.; Park, H.S. A wireless vibrating wire sensor node for continuous structural health monitoring. *Smart Mater. Struct.* 2010.

38. Midas User Support System. Available online: http://en.midasuser.com/ (accessed on 2 May 2013).

39. Casciati, F.; Domaneschi, M.; Faravelli, L. Design and implementation of a pointer system controller. *Nonlinear Dyn.* 2004, *36*, 203–215.

Chapter 9

INFLUENCE ANALYSIS OF A NEW BUILDING TO THE BRIDGE PILE FOUNDATION CONSTRUCTION

Jing Ma

School of Civil Engineering & Architecture, Chongqing Jiaotong University, Chongqing, China

ABSTRACT

This paper is based on the analysis of an industrial factory building to the bridge pile foundation construction stability, and it researches the influence of a new building to the bridge pile foundation internal force by the finite element analysis software ANSYS. By calculating the changes of displacement and internal force of the bridge pile foundation, the deformation can be better controlled. Furthermore, comparing the data of numerical analysis with one of monitor measurements, we conclude that a new building has a small influence on the deformation under load action and the stress variation of a bridge pile foundation. That is to say, the bridge pile foundation is safe and stable under load action.

INTRODUCTION

Since the capital construction increasingly develops and improves in China, more and more new buildings are built on their neighboring existing buildings [1] [2], which have a certain influence on existing buildings. All these situations, including a foreign-style house on the shallow tunnel, a tunnel under high-rise construction, or a deep foundation ditch around the bridge [3], require a strict computational analysis to provide reliable data for the influence extent of new buildings to existing ones and estimate the force change of building structure.

ENGINEERING SITUATION

The new industrial factory is located on a high slope, part of the tectonic denudation hilly topography. According to the original relief map, the terrain is flat in the lows, with a gradient of 35. And the slope is a little steep, with a gradient of 15 or 20. Due to a consequent bedding rock landslide, a 25-meter-high and 30-meter-long fill slope is formed on the section of 10'-10' - 15'-15', whose interface obliquity is about 20 degree, consistent with the dip angle of rock stratum.

Currently, a support reinforcement has been applied to the slope by a pile sheet wall. The length of the slope retaining wall is 587.54 meters. Fifty piles are arranged in the middle of the slope, including the bridge pile foundation support and bolt structure beam protection. Specific plans are shown in Figure 1.

ANALYSIS OF THE FINITE ELEMENT MODEL

Computation Module

To reduce the boundary effect and guarantee the accuracy in computation, the model size is that: length along slope to the factory building (X-direction) is 120 m. Width along slope to Y-direction is 50 m. Height from the lower boundary to the surface (Y-direction) is 58 m.

The whole computation module is simulated with a total of 56,326 planar units and 10,659 nodes in the finite element grid. And the finite element grid is divided as Figure 2.

Design Conditions

The model is calculated and analyzed by using Drucker-Prager Yield Criterion in ANSYS [4] , and material parameters are determined based on data from geological survey report. The results are shown in Table 1.

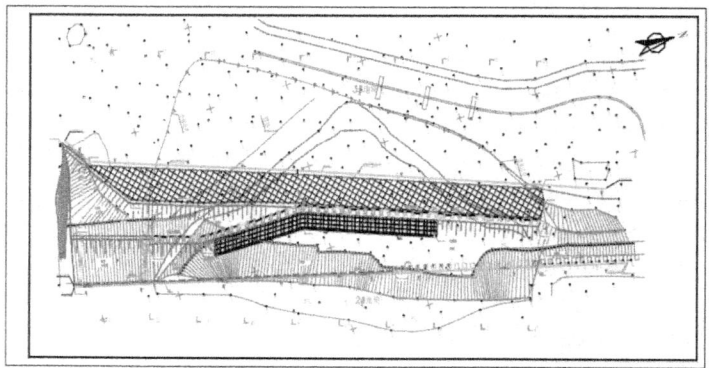

Figure 1: Master plan.

The finite element simulation is computed under the load of self-weight stress and additional stress respectively. We divide the jump into two phases:

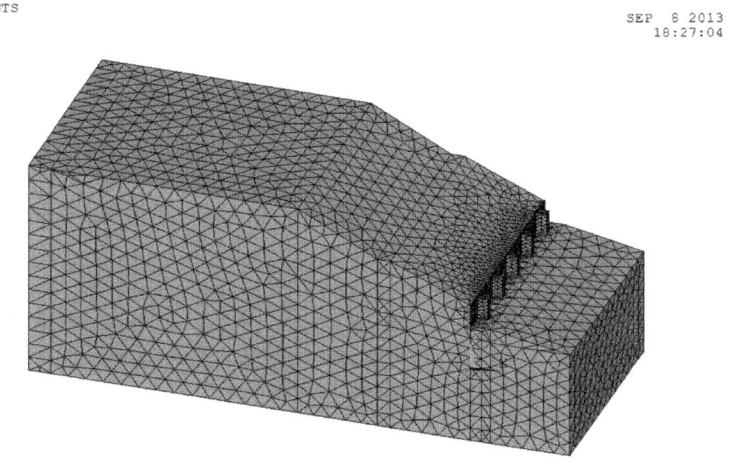

ELEMENTS

SEP 8 2013
18:27:04

Figure 2: The finite element computation and analysis module.

Step 1: self-weight stress loading;

Step 2: factory loading.

Because the factory loading in process is subject to banded model, these loads are equivalently applied to the whole area of industrial buildings in the worst situation, and the force is 250 KN/M^3. Results are shown in Figure 3.

Results

Simulation of the Results of Self-Weight Stress to the Bridge Pile Foundation

Maximum displacement and stress values in all directions of the bridge pile foundation under self-weight stress are shown in Table 2.

The nephogram of maximum displacement and stress values in all directions of the bridge pile foundation under self-weight stress are shown in Figures 4-9.

Under self-weight stress, the displacement and stress values of the bridge pile foundation are both small. The maximum values of displacement and stress of the bridge pile foundation are both in Y-direction, while the ones are small in Z-direction.

Simulation of the Results of Load Action to the Bridge Pile Foundation

Maximum displacement and stress values in all directions of the bridge pile foundation under load action are shown in Table 3.

The nephogram of maximum displacement and stress values in all directions of the bridge pile foundation under load action are shown in Figures 10-15.

Under load action, the values of displacement and stress in each direction increase, especially in Y-direction, which is consistent with reality. The results verify the correctness of the simulation.

Then, the stresses of bridge pile foundation are less than concrete compression strength in each direction, and the displacements under load action all meet the load bearing requirements, which makes it reasonable and feasible to an build an industrial factory near this bridge pile foundation.

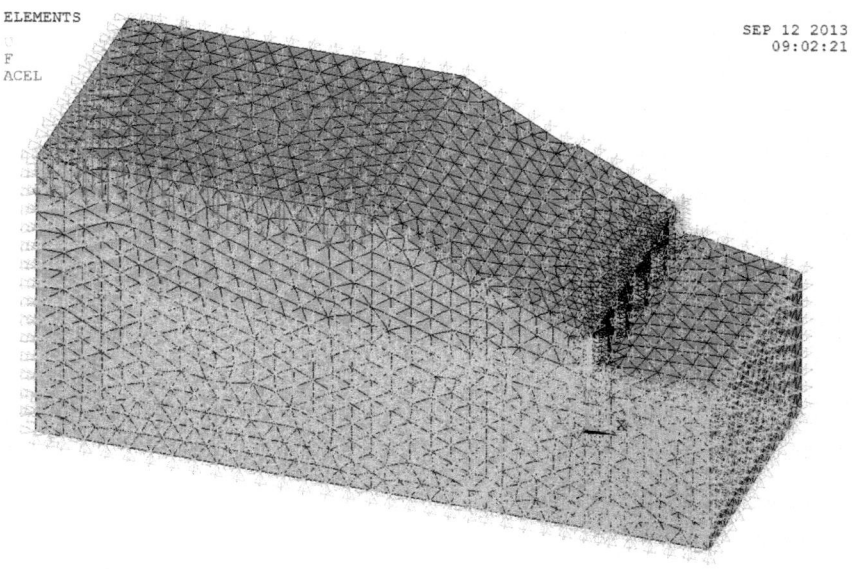

Figure 3. The loading model.

Table 1: Physical property parameter of material

Name	Multiplicity γ kg/m³	Internal frictional angle $\Phi/^\circ$	Elasticity Modulus E/GPa	Poisson ratio μ
Backfill	2000	28	1.7e−3	0.35
Mudstone	2200	35	0.2	0.23
Concrete	2400	—	30	0.20

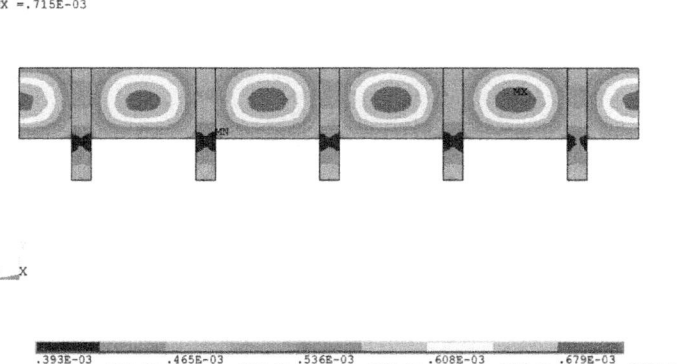

Figure 4: Displacement diagram under self-weight stress in X-direction.

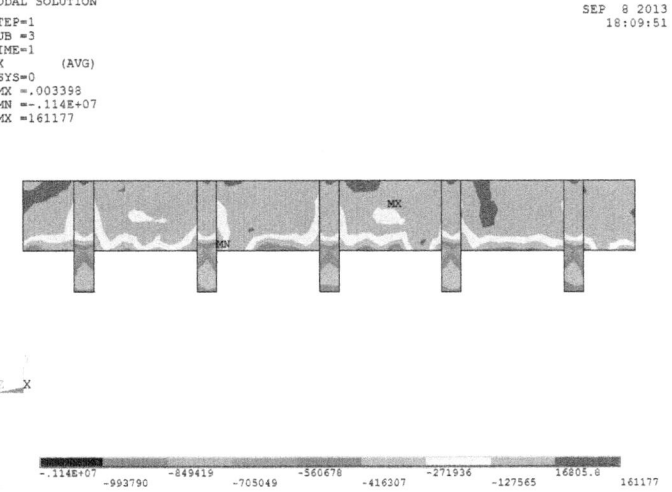

Figure 5: Stress diagram under self-weight stress in X-direction.

Table 2: Bridge pile foundation displacement and stress values under self-weight stress.

		X-direction	Y-direction	Z-direction
	Displacement (mm)	0.715	−3.347	−0.345
Stress (MPa)	Maximum	0.161	0.070	0.568
	Minimum	−1.140	−1.730	−0.730

RELATIVE ANALYSIS OF THE NUMERICAL COMPUTATION AND MONITOR MEASUREMENT

Since the only data we can measure is the bridge pile foundation deformation, we set up three stations along the bridge pile foundation and conducted a long-term monitoring.

Table 3: Bridge pile foundation displacement and stress values under load action.

		X-direction	Y-direction	Z-direction
	Displacement (mm)	0.748	−3.354	−0.360
Stress (MPa)	Maximum	0.265	0.720	0.596
	Minimum	−1.160	−1.750	−1.190

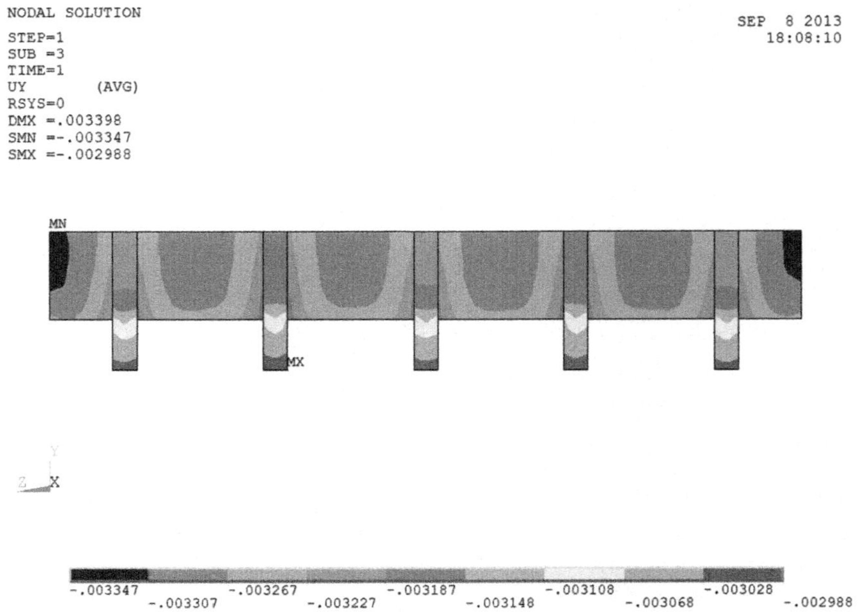

```
NODAL SOLUTION                                          SEP  8 2013
STEP=1                                                    18:08:10
SUB =3
TIME=1
UY        (AVG)
RSYS=0
DMX =.003398
SMN =-.003347
SMX =-.002988
```

```
-.003347      -.003267      -.003187      -.003108      -.003028
      -.003307      -.003227      -.003148      -.003068      -.002988
```

Figure 6: Displacement diagram under self-weight stress in Y-direction.

We got primary data and current data of the bridge pile foundation, before and after setting up the industrial factory respectively. And the average value of three stations is chose as computed displacement value increment [5] . We compare the data of numerical simulation with that of monitor measurement to verify the reliability of the numerical simulation. Computed and measured values are shown in Table 4.

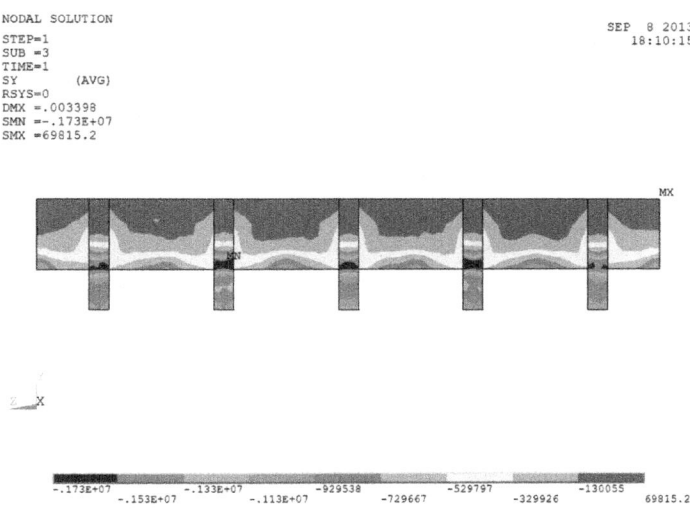

Figure 7: stress diagram under self-weight stress in Y-direction.

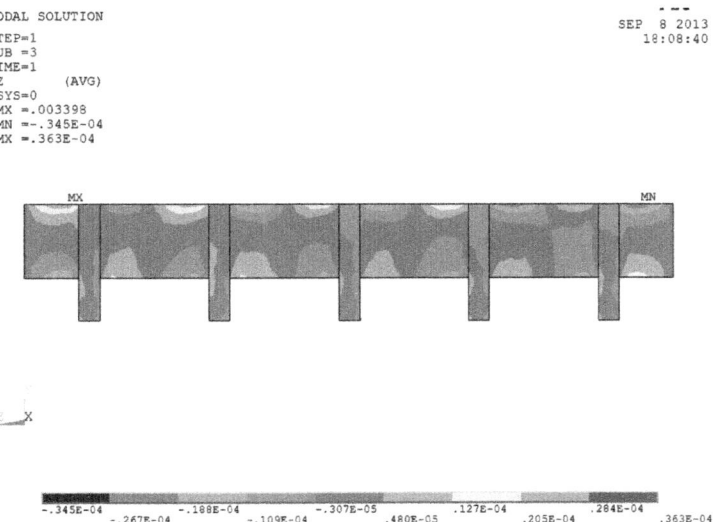

Figure 8: Displacement diagram under self-weight stress in Z-direction.

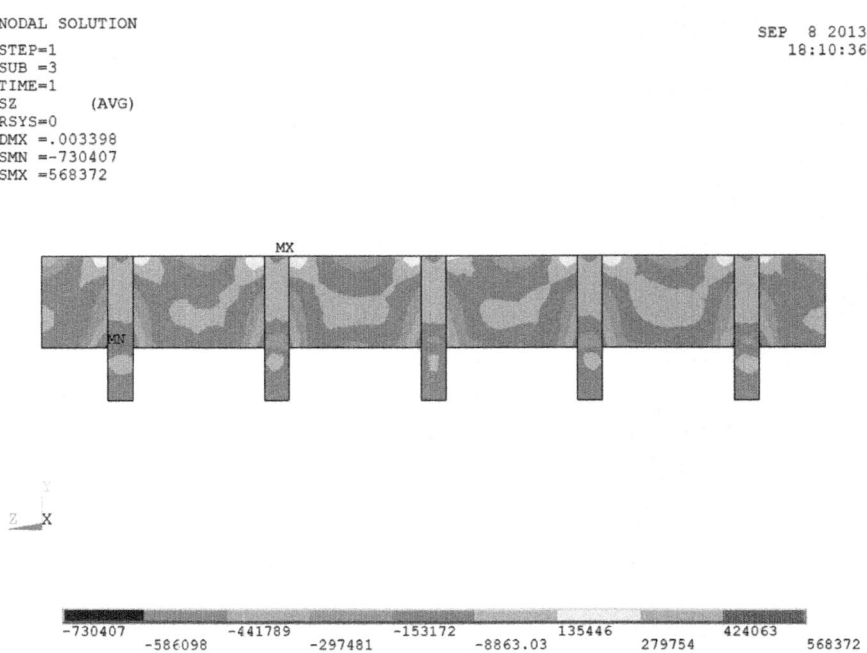

Figure 9: Stress diagram under self-weight stress in Z-direction.

Table 4: Bridge pile foundation displacement and stress values under load action.

		X-direction	Y-direction	Z-direction
Computed stress (MPa)	Maximum increment	0.104	0.650	0.028
	Minimum increment	−0.020	−0.020	−0.460
Computed displacement value increment (m)		0.033	−0.007	−0.091
Measured displacement value (m)		0.030	−0.005	−0.082

The main displacement is that in X-direction (horizontal direction) under load action, and it differs by 0.003 m. The maximum stress change of the bridge pile foundation is 0.65 MPa in Y-direction (vertical direction), which differs by 0.002 m between computed and measured values. The value in Z-direction is perpendicular to the horizontal direction, which differs by 0.009 m between computed and measured values.

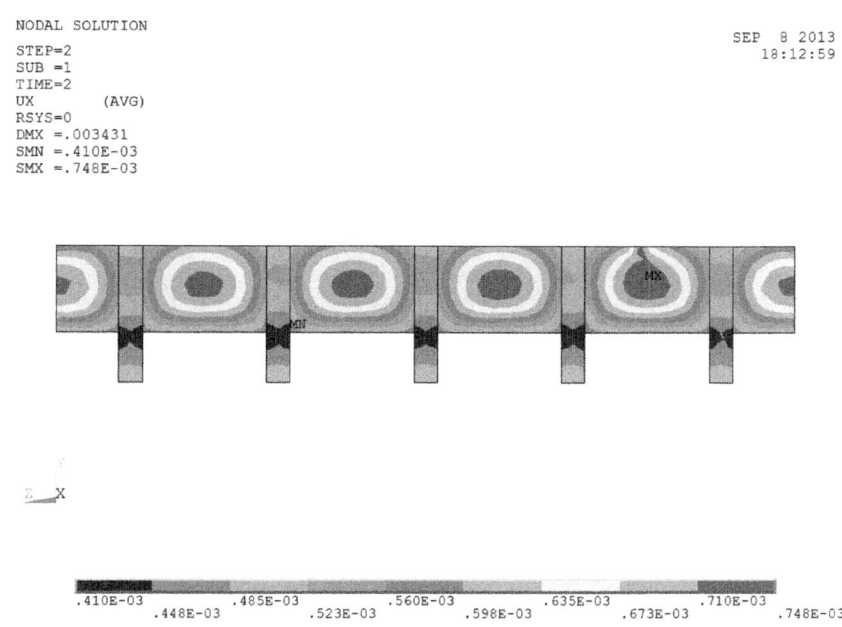

```
NODAL SOLUTION
STEP=2                                                    SEP  8 2013
SUB =1                                                       18:12:59
TIME=2
UX        (AVG)
RSYS=0
DMX =.003431
SMN =.410E-03
SMX =.748E-03
```

```
.410E-03        .485E-03        .560E-03        .635E-03        .710E-03
        .448E-03        .523E-03        .598E-03        .673E-03        .748E-03
```

Figure 10: Displacement diagram under load action in X-direction.

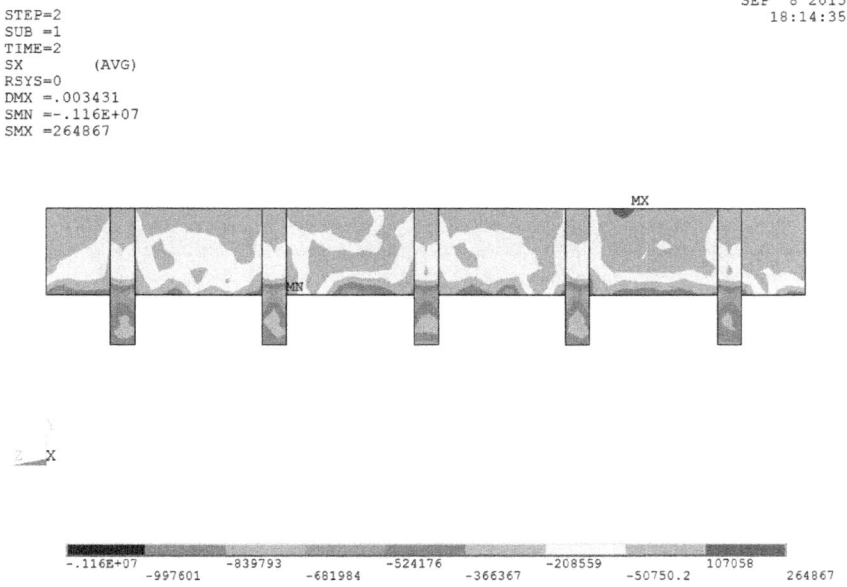

```
NODAL SOLUTION
STEP=2                                                    SEP  8 2013
SUB =1                                                       18:14:35
TIME=2
SX        (AVG)
RSYS=0
DMX =.003431
SMN =-.116E+07
SMX =264867
```

```
-.116E+07        -839793        -524176        -208559        107058
        -997601        -681984        -366367        -50750.2        264867
```

Figure 11. Stress diagram under load action in X-direction.

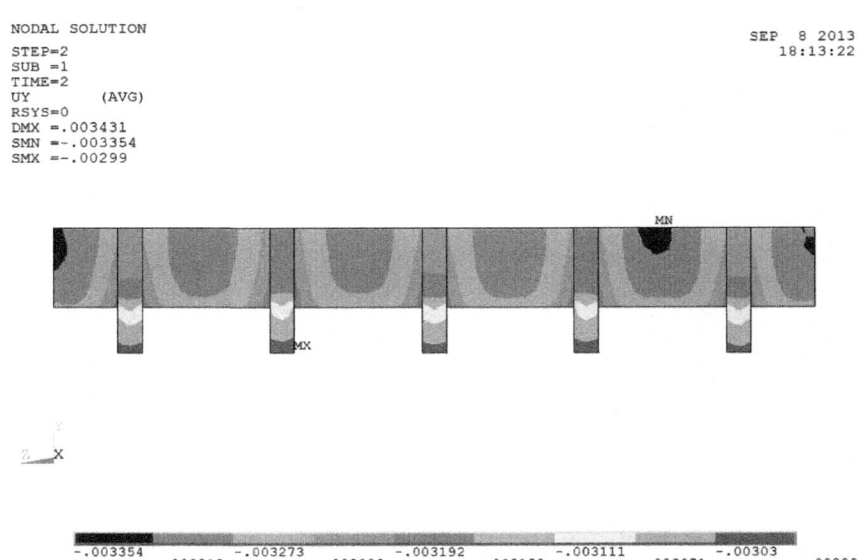

Figure 12: Displacement diagram under load action in Y-direction.

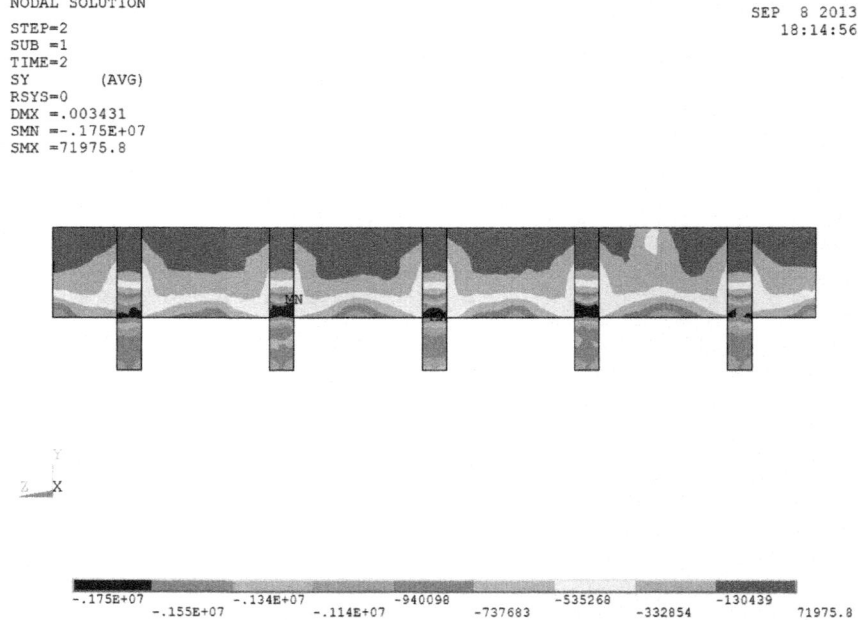

Figure 13. Stress diagram under load action in Y-direction.

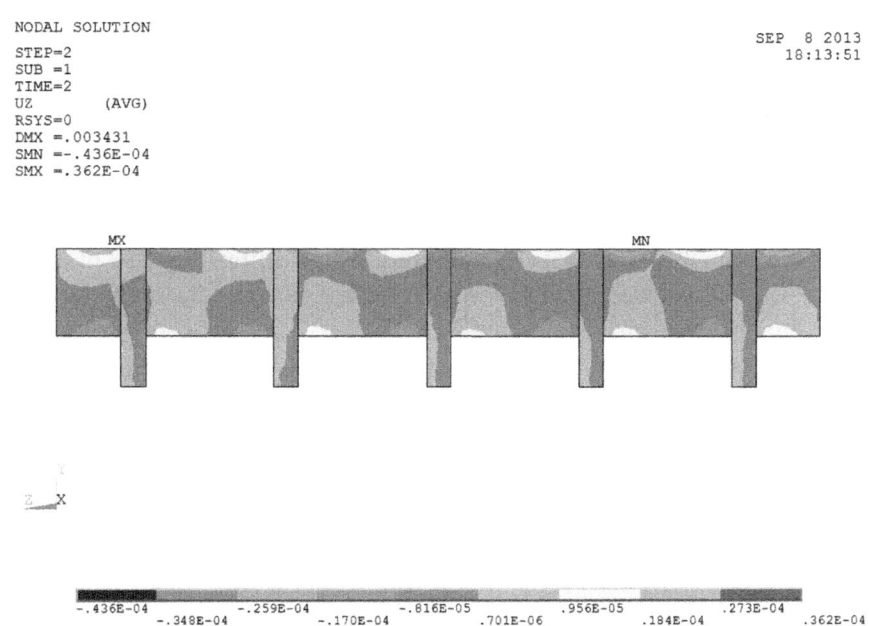

NODAL SOLUTION

STEP=2
SUB =1
TIME=2
UZ   (AVG)
RSYS=0
DMX =.003431
SMN =-.436E-04
SMX =.362E-04

SEP 8 2013
18:13:51

MX   MN

Y
Z___X

-.436E-04   -.259E-04   -.816E-05   .956E-05   .273E-04
      -.348E-04   -.170E-04   .701E-06   .184E-04   .362E-04

Figure 14. Displacement diagram under load action in Z-direction.

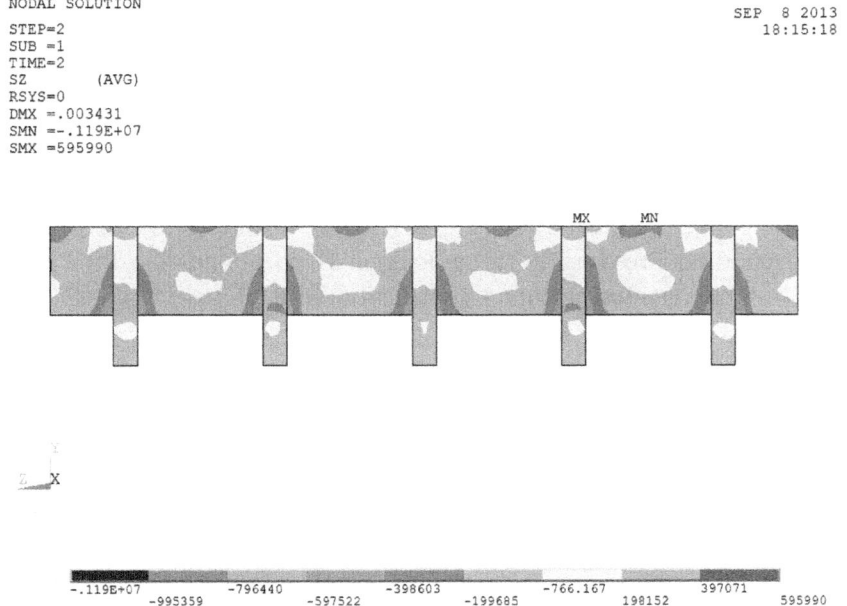

NODAL SOLUTION

STEP=2
SUB =1
TIME=2
SZ   (AVG)
RSYS=0
DMX =.003431
SMN =-.119E+07
SMX =595990

SEP 8 2013
18:15:18

MX   MN

Y
Z___X

-.119E+07   -796440   -398603   -766.167   397071
      -995359   -597522   -199685   198152   595990

Figure 15. Stress diagram under load action in Z-direction.

CONCLUSIONS

Stress and displacement values of supporting structure are computed and analyzed by the finite element software ANSYS. The comparisons between computed and measured values are illustrated as below.

1. There is little difference between computed and measured values in all directions of the bridge pile foundation under load action, particularly in the main displacement (X-direction), which differs by 0.003 m. It shows that the load we applied about 250 c is rational and the computed values are reliable.

2. The maximum stress change of bridge pile foundation under load action is 0.46 MPa in Z-direction, less than concrete compression strength. Hence the bridge pile foundation meets the load bearing requirements.

3. The maximum tensile stress change of bridge pile foundation under load action is 0.65 MPa in Y-direction (vertical direction), less than tensile strength of concrete. Hence the bridge pile foundation meets the load bearing requirements.

Analysis above illustrates that the impact on the displacement and stress change of bridge pile foundation under load action is small. The bridge pile foundation structure under load action is safe and stable.

REFERENCES

1. Liu, J.H. (2002) Several Theories and Calculated Methods of Underground Engineer Construction Mechanics. Railway Standard Design.

2. Hu, B. (2013) Application of Reinforced Concrete Filling Pile Technology in Deep Foundation Pit Supporting. Shandong Metallurgy, 35, 74-75.

3. Wang, Y.G., Li, D.J. and Ye, K.P. (2005) Application of Steel-Tube Supporting Prestressed Technology in Deep Foundation Pit Project in Shanghai Xianlesi Square. Construction Technology, 35, 45-48.

4. Mo, H.O., Crèpe, H.X. and Lai, A.P. (2001) Optimization Design in Foundation Pit Supporting Pile Structure. Rock Mechanics and Engineering, 23, 23-25.

5. Paw-paw, H.Y. and Huang, J.Z. (2001) 3D Finite Element Analysis and Simulation of Deep Foundation Pit Supporting Structure. Shanghai Jiaotong University, 35, 610-613.

Chapter 10

MATERIAL EFFICIENCY OF BUILDING CONSTRUCTION

Antti Ruuska and Tarja Häkkinen

VTT Technical Research Centre of Finland, Tekniikantie 4, 02044 VTT Finland

ABSTRACT

Better construction and use of buildings in the European Union would influence 42% of final energy consumption, about 35% of our greenhouse gas emissions and more than 50% of all extracted materials. It could also help to save up to 30% of water consumption. This paper outlines and draws conclusions about different aspects of the material efficiency of buildings and assesses the significance of different building materials on the material efficiency. The research uses an extensive literature study and a case-study in order to assess: should the depletion of materials be ignored in the environmental or sustainability assessment of buildings, are the related effects on land use, energy use and/or harmful emissions significant, should related indicators (such as GHGs) be used to indicate the material efficiency of buildings, and what is the significance of scarce materials, compared to the use of other building materials. This research suggests that the material efficiency should focus on the significant global impacts of material efficiency; not on the individual factors of it. At present global warming and greenhouse gas emissions are among the biggest global problems on which material efficiency has a direct impact on. Therefore, this paper suggests that greenhouse gas emissions could be used as an indicator for material efficiency in building.

INTRODUCTION

Resource efficiency means efficient use of energy, natural resources, and materials, in order to create products and services with lesser resources and environmental impacts. It is based on life-cycle thinking and comprises of energy efficiency and material efficiency. Whereas the energy efficiency considers sparing use of energy, and ratio of energy use and production, material efficiency is about sparing use of natural material resources, effective management of side-streams, reduction of waste, and recycling [1].

Natural resources underpin the functioning of the European and global economies and the quality of life. These resources include raw materials, such as fuels, minerals and metals, as well as food, soil, water, air, biomass, and ecosystems [2]. A roadmap to a resource-efficient Europe [3] highlights the buildings sector as one of the three key sectors for improvements. Better construction and use of buildings in the European Union would influence 42% of final energy consumption, about 35% of our greenhouse gas emissions and more than 50% of all extracted materials. It could also help to save up to 30% of water consumption.

The importance of material efficiency and the need to improve it can be studied from several perspectives. Limited availability or scarcity of materials may lead to threats to the economy, and the production processes of materials can have significant environmental impacts. The extraction of raw materials and the production of materials may also be energy and/or labor intensive and very costly, and the extraction of materials may lead to land use changes and related impacts.

This article presents an overview of the different aspects of resource and material efficiency in building construction. The paper also presents the results of a case study and analyses the significance of building materials in terms of material scarcity.

Classification of Resources (and Aspects of Scarcity)

Natural resources can be divided into renewable and non-renewable resources. Non-renewable resources are those that can only be harvested once. These are often referred to as stocks (e.g., iron ore) or resources that form extremely slowly (e.g., crude oil) [4]. Azapagic [5] divides the minerals industry into energy minerals (e.g., coal, oil), metallic minerals (e.g., iron, copper and zinc), construction minerals (e.g., natural stone, aggregates, sand, gravel, gypsum), and industrial minerals (e.g., borates, calcium carbonates, kaolin, plastic clays, talc).

A reserve is defined as that part of the reserve base that could be economically extracted or produced at the time of determination (in accordance with the terminology used by the European Commission [6]). The reserves of the most common building materials (aggregates, clay, lime and stone, gypsum, and quartz) are either large or very large [4]. However, buildings also consume materials whose reserves are more limited, for example, coal, oil, and metallic minerals.

The usability of resources depends specifically on the economy and the available technology. Resources that have previously been uneconomical to extract may become usable because of rising values and improved extraction technologies. Political situations and the effects of extraction on the landscape and environment may also affect the usability of resources. Scarcity always has a time dimension: it can be interpreted as a change in availability over time [7]. Steen [8] claims that many life cycle impact assessment (LCIA) approaches mix scarcity with issues such as difficulty of extraction. This can be viewed as double counting, as the effects thereof, such as high energy demand, are accounted for in other categories. Metals in use can also be seen as a global inventory of available metals. Virgin metal is added when necessary to this inventory [9]. Future backup technologies will probably require significantly less energy and other resources than the extraction of virgin metal.

Meadows *et al.* [10] identify that the increasing cost of resources is becoming a major problem for societies. As resources become scarcer, this may influence the quality of life in some parts of society. This, in turn, may have negative impacts on human health as a specific area of protection [11]. It may therefore be important not to separate the environmental and economic aspects. Yellyshetty *et al.* [12] argue that resource depletion needs to be considered in LCAs from the perspective of time, environmental and economic aspects of mineral extraction, and future consequences of decreased availability of mineral resources for a region. Steen [8] highlights three issues that should be considered when drawing conclusions about the inclusion of resource depletion in LCAs: (1) the time perspective when evaluating impacts on abiotic resources; (2) the separation of environmental and economic aspects; and (3) whether the consequences of decreased availability should form part of the LCI or the LCIA. The socio-economic value of mineral extraction can be significant in some regions, and changes in the extraction industry can have important social consequences [13]. Söderholm and Tilton argue that economic depletion will occur long before physical depletion [14].

Another way of looking at the issue of mineral resource scarcity is the surplus cost method, which assumes that future increases in mining volume will lead to increasing production costs per metal or mineral extracted. This is defined as the marginal cost increase (MCI). When the MCI is multiplied by future resource demand, the future costs to society can be determined [15,16].

Indicators for Resource Efficiency and Material Efficiency in the Building Industry

Resource efficiency can be defined with a number of indicators. Each indicator has a specific definition, which contains only certain aspects of the issue.

Resource efficiency may be defined, for example, in terms of land area that an economy requires [17], human impacts on natural processes [18], impacts on land use [19], amount of material use [20] or related environmental impacts [21], ratio of GDP to material use [3], or national monetary input-output tables expanded with environmental information [22].

When moving from the level of economies to the level of technologies or products, other life-cycle related indicators are more common. The indicators are typically not correlated, so a wide range of environmental indicators are needed [23]. For example, life cycle assessment (LCA) methodology assesses the harmful impacts of buildings in terms of global warming, ozone depletion, acidification of soil and water, eutrophication, photochemical ozone creation, and depletion of abiotic resources (elements and fossil fuels) [24,25].

The impacts from resource use, often referred to as, resource depletion, is a prominent impact category in LCA [26]. LCA methodology addresses abiotic, or non-living, resources in terms of their availability for present and future generations. The depletion of such resources can be studied from the perspective of amounts of deposits, extraction rates, future ore extractions, or exergy consumption [27].

The use of natural raw materials in building can be decreased by using lightweight structures, minimizing loss, improving durability and service life, using secondary materials and improving appropriate flexibility [28,29]. Improved space efficiency also contributes to better material efficiency when assessing it in terms of functional units (a building that fulfils the required performance).

The following equation shows how these different aspects of material efficiency relate to the wider concept of resource efficiency. Equation (1) defines the total impacts associated with the production and processing of a specific material as (adopted from [30]):

$$I = D \times M \times Y \times E \qquad (1)$$

In Equation (1), the impacts (I) are due to the demand (D) for products containing material, the average mass of material per product (M), the yield ratio of supplied material *versus* material in the final product (Y), and the average emissions per unit of material (E). The impacts of material efficiency extend to all the factors, D, M, Y, and E. In the context of buildings, the demand for new buildings is influenced by their durability, service life and flexibility. The use of lightweight structures impacts the average mass per product, and the yield ratio is affected by material losses during processes. Finally, the use of secondary materials impacts—in addition to the use of natural material resources—the average emissions, as reuse and recycling are typically significantly less energy intensive than primary production [30].

Instead of viewing material efficiency through the multiple viewpoints presented above, this research focuses on their total impacts. This research outlines the related impacts as follows: (1) depletion of natural raw materials; (2) impacts of material-related harmful emissions; (3) impacts due to material-related land use; and (4) life cycle costs due to the use of materials. The following sections discuss the importance of these different impacts, on the basis of literature.

AIM AND SCOPE

The objectives of the research were as follows:

- to outline and draw conclusions about different aspects of the material efficiency of buildings;
- to assess the significance of different building materials on the material efficiency of buildings.
- The study was founded on the premise that the importance of material efficiency is based on one or more of the following impacts:
- the depletion of raw materials and its long-term socio-economic impacts;
- land use change due to the extraction of raw materials and its environmental impacts, and impacts on the landscape and future recreational use;
- the use of energy in production processes of materials and depletion of non-renewable energy;
- harmful emissions from production processes of materials and their local and/or global environmental impacts;
- material cost impacts due to the limited availability of raw materials or a higher need for energy and/or labor in the different phases of production processes.

The different aspects of the material efficiency of buildings were outlined and analyzed with the help of a literature study. The importance of the different groups of building materials and the significance of building materials compared with the use of energy resources was studied with the help of a case study. The Abiotic Depletion Potential (ADP) was calculated in terms of ADP elements and ADP fossil, and the significance of different building materials was assessed.

With regard to the building sector, the research questions of interest are as follows: (1) "As the global availability of the main building materials is very good, should the depletion of materials be ignored in the environmental or sustainability assessment of buildings?"; (2) "Although the availability is good, are the related effects on land use, energy use and/or harmful emissions significant and should related indicators (such as GHGs) be used to indicate the material efficiency of buildings?"; (3) "Although the availability of the main building materials is very good, what proportion of buildings use scarce materials and what is the significance of these compared with the use of other materials in buildings?".

LITERATURE REVIEW

This section presents the literature review, which answers the research questions of this paper on a general level. It also points out the gaps in literature and gives reasoning for the selected case-study approach, which is presented later in this paper.

The literature review examined the impacts of material efficiency on: (1) depletion of natural raw materials; (2) impacts of material-related harmful emissions; (3) impacts due to material-related land use; and (4) life cycle costs due to the use of materials. It aimed to identify and fill potential gaps in the current knowledge and point out needs for more detailed studies.

Scarcity and Availability of Abiotic Building Materials

Material efficiency is a way to reduce the demand of abiotic building materials. Whereas the importance of material scarcity is growing in general, the issue is not as clear for building materials. Common building materials, such as metals and ceramics, are derived from ores. Some of the minerals are approaching their production peaks and some have already passed their peak [31]. There is also a continuous decrease in ore grade at which some materials are being mined [32]. The inevitability of peaking of oil is generally acknowledged, although, it is still under debate, whether or not the peak has already passed [33]. Oil is needed, for example, for production of polymer-based building materials.

The building industry uses large amounts of materials, equating to approximately 50% of European resource extraction [3], but the most common building materials are also common in nature. Aggregates, for example, are the key component of many building elements but are generally not a scarce resource [34]. However, due to their heavy and bulky nature, aggregates need to be sourced close to their markets. Viable sources may be constrained at regional and local level [35], for example in rapidly growing developing countries [36], if their viable local supply is not strategically planned [6]. Relating to these problems, approaches which account for local resources have been proposed in literature [37].

The buildings also require metallic minerals for the production of, for example, concrete reinforcements and structural steel in the building frames, roofs, façades, windows and doors of the building envelope and pipes, ducts and wirings of building systems. Despite of dependence on the import of metallic minerals in some countries [34] these resources are not considered scarce, as their global availability is good [6]. However, mining of these minerals may become critical in terms of social impacts that mining activities cause locally on land and ecosystems [38].

When buildings become more energy efficient and building systems more advanced and complex, the demand for scarcer resources may increase. Some of the components of advanced, energy-efficient building systems, such as wind turbine magnets, high-capacity batteries, energy-efficient lighting and photovoltaic cells require rare earths and critical natural resources in their production [39]. However, the exact selection and weighting of factors, which make a raw material critical or scarce, are still open research questions [40]. Raw materials may be considered critical, for instance, if they have national significance for economies and their current or future supply is at risk [39]. Other sources of criticality may rise from specific ecological, social, or political considerations [6].

Greenhouse Gases

The building sector is the single largest contributor to global greenhouse gas emissions. On the other hand, it also has a substantial emission saving potential. Material efficiency extends to all the underlying factors of resource efficiency, making it a significant contributor to resulting impacts from materials. Considering these viewpoints, material efficiency has a significant role in reducing the global GHG emissions from buildings.

The greenhouse gas emissions from buildings are related to the embodied energy of building materials and the emissions from operational energy use and the role of materials is becoming increasingly important. The research and policies have focused only on the operational energy use until recently [41,42,43]. This can be explained by the fact that, the role of embodied energy has been relatively low, at some 10%–20% [44,45], but development towards more energy efficient buildings increases the importance of materials. In low-energy buildings the role of materials can be as high as 50% [41] and ultimately, at zero-energy-level, all the energy-consumption, and related greenhouse gas emissions come from the embodied energy of building materials [42]. Due to this development, the embodied energy and related emissions cannot be omitted in life cycle assessments.

In addition to initial material consumption, the buildings also need materials for their lifetime renovations. The energy consumption of interior renovations over the lifetime of a building can account for some 20% to 30% of the initial embodied energy [46]. The need of this recurrent embodied energy can be almost halved, with the use of materials with longer service life [47].

When looking at the issue from the level of residential areas, also transport needs to be considered. Significant greenhouse gas savings can be achieved in all, embodied, operational and transport energy needs when planning residential areas [48]. From sector-level, the most important factors affecting the greenhouse gas emissions are housing size, style and location [49].

Another viewpoint to the issue is the temporal perspective of emissions from building. The initial GHG emissions emitted over a short period of time in the construction phase may compromise the greenhouse gas mitigation goals in short and medium term [50]. Therefore, the greenhouse gas emission targets cannot be achieved with energy-efficient new buildings alone.

Ruuska and Häkkinen [28] assess the total greenhouse gas emissions of a multi-storey residential building in Finland with the help of a parametric study. The results show for a concrete building case that material-related emission account for some 40% of 50-year lifetime total GHG-emissions for a passive-level building in Southern Finland. Furthermore, if soil stabilization of a building site is included in the figures, the role of materials rises to over 50% of lifetime totals.

Land Use

Construction causes irreversible land changes. Use of land means consumption of resources, in terms of changing the potential end-use and the consumption of soil materials. Buildings use land directly by occupying the land under their footprints and through their embodied land use, relating to their raw material and energy use throughout the building's value chain. An impact because of land use occurs when the land properties are modified (transformation) and also when the current man-made properties are maintained (occupation) [51]. Changes in land use can have wide-ranging environmental consequences, including biodiversity loss, changes in emissions of gases affecting climate change, changes in hydrology, and soil degradation [52].

Buildings and other construction assets cause soil sealing as land remains below constructions. Artificial sealing is generally extensive and permanent [53]. When vegetated soils are replaced with impermeable surfaces, it results in the increase of overland flow, reduction of infiltration and bypass of natural storage [54].

Although the global availability of the main building materials is good, the consideration of land use may affect the importance of material efficiency with regard to buildings. However, an LCA-based case study analysis [55] indicates that when only non-renewable material resources are considered, the land occupied by buildings is more important that the land use due to the extraction of raw materials used for buildings. However, when wood is used as a building material, the land use (in terms of occupied land area) required for the production of building materials becomes more significant than the land occupancy of the building itself.

The extraction of aggregate materials also affects the landscape and the natural geological and biological conditions. In addition to this, in Finland, the extraction of gravel affects the quality of the groundwater because the extraction increases the variety in the quality and pollution risk of the groundwater [56]. In addition to the impacts on groundwater and surface water, the production of aggregates causes local impacts, such as vibration, and noise and dust emissions.

Cost and Productivity

Material efficiency has an important effect on construction cost efficiency. The positive impacts on cost and productivity can be seen as a natural driver towards material efficiency in the building industry.

The importance of materials in relation to the investment costs of construction varies. The approximate magnitude has been estimated at 15%–40% of the investment cost (including the cost of design, interfaces, labor costs, site overheads, taxes and the contractor's profits) [29]. Minimizing the loss of materials has a direct impact on the investment costs. On the other hand, better and appropriate flexibility in the design of spaces can also have a significant impact on the life cycle costs, especially in the context of retail and office buildings.

Goodrum *et al.* [57] studied the relationship between changes in material technology and productivity in construction. The results show that changes in material technology correlate with improvements in both labor and partial factor productivity (physical output per material cost + equipment cost + labor cost). The authors found that the relationship between changes in material technology and construction productivity was weaker for labor productivity than for partial factor productivity. The strongest relationship between changes in material technology and labor productivity was also found among changes in the unit weight of materials followed by modularity, curability, and installation.

Existing Standards and Regulation

The current European regulation, as well as the work done for the development of assessment standards, reflects the stated policy targets to consider and improve the material efficiency and the overall resource efficiency of societies. However, unlike energy performance, which is defined by European Directives [58,59], material efficiency is not tightly controlled or regulated. Also, contrary to the energy efficiency of building and renovation [60,61], there are no fiscal instruments or incentives in place for improvements in material efficiency of buildings.

In Europe, the Construction Product Regulation [62] gives basic requirements for construction products. Construction works as a whole and in their separate parts must be fit for their intended use, throughout the life cycle of the works and fill the basic requirements. Sustainable use of resources is included in the requirements, and the CPR states that construction works must be designed, built and demolished in such a way that the use of natural resources is sustainable. Especially the following is highlighted: (1) re-use or recyclability of the construction works, their materials and parts after demolition; (2) durability of the construction works; and (3) use of environmentally compatible raw and secondary materials in the construction works. Even though the Construction Product Regulation emphasizes the importance of material efficiency, it does not give normative rules for it, or dictate mandatory information about material efficiency.

Assessments of resource depletion and comparisons of buildings and building products are supported by international and European standards. The current standardization and guidelines suggest using two separate impact categories for resource depletion: ADP elements for all non-renewable abiotic materials and ADP fossil fuels for all fossil resources [24,25,63]. Previously, both these items were assessed in terms of antimony equivalents [64]. However, as the two contribute towards the decrease of different resources, their ADP is characterized by different units [65]. The unit of measurement for the depletion of natural resources is the antimony equivalent (kg Sb eq) and for the depletion of natural fossil energy the resources, their net calorific value (MJ). Despite of its established status through the current standardization and guidelines, the calculation of ADP has some shortcomings. For example, the characterization factors for its calculation do not exist for many of the common building materials. The basic problem behind this is that such factors cannot be defined for many of the common building materials, such as gypsum, silica sand, construction sand, clays, limestone, and such, due to lack of data on material configurations, reserves, reserve bases, and ultimate reserves for these materials [65].

The status of ADP calculations in standardization and the identified shortcomings in the calculation method, give a basis for the case-study of this research. The literature study was unable to identify detailed ADP calculations, which would show the importance of different building materials. The case-study aims to create new knowledge on the importance of different building materials, in terms of their ADP. It also aims to compare the material-related ADP to the ADP from lifetime operational energy use. Finally, it aims to give more information on the significance of the use of different scarce materials in buildings.

QUANTIFYING THE ABIOTIC DEPLETION POTENTIAL (ADP) OF BUILDINGS

The case-study aims to add to the existing knowledge by showing the importance of different building materials, in terms of their abiotic resource depletion potential (ADP). It also studies the importance of building materials, in relation to operational energy use and the role of advanced building systems. Finally, the case-study offers new information on the current calculation method for ADP, together with its limitations. These issues were selected as the focus of the case-study, based on the gaps in the existing literature.

This section presents the case-study building, and explains the calculation method and main data sources used in the study. This case-study assesses the resource depletion of a case-building, by using impact categories of ADP elements and ADP fossil, recommended by current standardization and guidelines. The following subsections go through the calculation method, principles of the used life cycle assessment method, material quantities used in the assessment, calculation of energy consumption and, especially, calculation of ADP elements and ADP fossil.

Calculation Method

This research used life cycle assessment to determine the ADP of a case building. The calculation was carried out by using the bill of quantities (BOQ) of a real world building and assigning each of the materials with a specific characterization factor for their ADP (elements). For ADP fossil, the energy consumption associated with the materials of BOQ was completed with lifetime energy consumption information.

Life Cycle Assessment

Life cycle assessment means compiling and evaluating inputs, outputs and environmental impacts of a product system throughout its life cycle [66].

It is widely accepted as one of the best tools for environmental assessment of a variety of products and processes [67]. This research uses a process-based analysis, which is generally recognized as more accurate, but more labor, and time-intensive than, for example, input-output analysis [68,69]. The selected method and its limitations and benefits are examined in more detail in the 'discussion'-section. The life cycle assessment is limited to the abiotic depletion (ADP) of non-renewable raw materials and fossil fuels. The assessment does not aim to be exhaustive, but it aims to define the ADP of building materials with sufficient accuracy. The specific focus of this research is on the product stage, but also construction, use and end of life stages are assessed to cover the whole life cycle of the building, following the division of current standardization [24]. The assessment period of this research was 50 years.

Material Quantities

The material quantities required for the assessments of the product stage were based on the bill of quantities (BOQ) of a real-world case building, which is described in further detail later on. The BOQ was derived from the building's building information model (BIM), hence offering a high level of accuracy in material amounts. For calculation purposes, the materials of the BOQ were categorized under nine identified main material groups, namely: aluminium, concrete, copper, fossil materials, gravel, other mineral resources, steel, wood boards, and other wood-based products.

In addition to the quantities of the BOQ, the lifetime material consumption, including waste during construction stage and material requirements for use stage were also accounted for. The material loss was estimated to be 5% for all the building materials, for both construction and use phase material needs, based on literature [70]. The material needs of the use phase were assessed by estimating replacement and refurbishment needs over the lifetime of the building, for different building parts and components. The material needs of maintenance and repair were estimated to be insignificant and they were not accounted for in the assessment.

The following assumptions were made for the lifetime renovations. Firstly, the load-bearing structures were assumed to last for the whole lifetime of the building. Secondly, the roofing, building systems, windows, doors, glazing, and the surfaces of sanitary spaces were expected to be replaced (or refurbished) once over the 50-year assessment period. Thirdly, the surface finishes, fittings and furniture were expected to require replacement in every 10 years, thus, they were assumed to undergo four renewals over the assessment period.

Energy Consumption

The energy consumption of the product stage was taken into account by using life cycle inventories (LCI), which included energy consumption from raw material supply, transport and manufacturing from cradle-to-gate. Energy consumption of construction installation process and transportations were taken into account in the construction phase. For the use phase, the assessment included the energy consumption of replacement and refurbishment, transportation of materials and operational energy use of the building. The end of life phase included energy needed for deconstruction and transport of waste from site. The waste processing and disposal stages were excluded for this assessment. Data sources used for these calculation are shown in more detail in the next section.

Calculation of ADP Elements

The calculation of ADP elements had five steps. Firstly, total material needs over the lifecycle of the building were defined. This was done by combining the information from the original BOQ with the estimates on material losses during construction (5%) and assumptions on replacements and refurbishments. Each of the materials was then categorized under one of the identified nine main material groups.

The second step was to define the total abiotic material inputs for each of the main materials. This was done using the European ELCD database and its LCI data [71] for this purpose.

The third step was to derive the ADP characterization factors for each of the abiotic inputs. The characterization factors used are based on the CLM database's base reserve figures [72], as recommended in current guidelines [64].

Fourthly, after designating the ADP characterization factors for each of the abiotic inputs of the main materials, the average ADP factor for each main material was calculated.

Finally, when all the material amounts, and corresponding ADP characterization factors were defined, the ADP for each material was calculated. After this, the building-level ADP was calculated by adding together the ADPs of all the nine main materials. The results of calculations, together with references to the used data sources are presented in section "ADP Elements".

This research also considers the specific issue of soil stabilization, which may be needed in case of poor ground conditions on building site, as it has been previously found to be significant building factor impacting (GHG) emissions [28] and its main components have high embodied energy. The ADP Elements calculations follow the same methodology as described previously, the only difference being the main materials in stabilization are cement (CEMII) and quicklime (CaO) with a mixing ratio of 1:1. In addition, ADP Fossil is assessed for the soil stabilizations. The assessment results for soil stabilization, along with the data sources, are presented in section "ADP of soil stabilization".

Another specific issue studied by this research is the ADP of advanced building systems of energy-efficient buildings, because such systems typically include rare earth elements and other critical materials [39]. The components selected for study are energy-efficient lighting and PV panels. The ADP Fossil of these is not calculated due to a lack of reliable data. The calculation results and data sources are shown in section "ADP of advanced building systems".

Calculation of ADP Fossil

The ADP fossil calculations followed a similar methodology to that of the ADP elements. For material-related ADP-fossil, the calculation comprised of three stages.

Firstly, the total material needs over the lifecycle of the building were based on the total masses calculated for ADP elements.

Secondly, the non-renewable energy inputs for each of the main materials were derived from the ELCD database [71] to give a characterization factor for ADP fossil for each of the main materials.

Thirdly, when all the material amounts, and corresponding ADP Fossil characterization factors were defined, the ADP for each material was calculated. After this, the building-level ADP was calculated by adding together the ADPs of all the nine main materials. The results of calculations, together with references to the used data sources are presented in section "ADP Fossil".

In addition to the direct, material-related energy consumption, fossil energy is also consumed in material transportation. The contribution of transportations to the ADP Fossil is calculated by assuming a 50 km transport with a semi-trailer combination to the building site for all the materials and the same 50 km distance for all the materials to cover their transport off the site with earth moving lorry at the end-of-life. The construction installation process, lifetime replacement and refurbishment activities, and deconstruction of the building at the end-of-life also consume fossil energy. These are assessed using values from previous research. The assessment results for transportations, and construction, lifetime renovations and demolition are shown in section, along with the used data sources "ADP Fossil of Material Transportation and Construction Work".

To complete the ADP Fossil calculations, lifetime operational energy use is also assessed. This is done by assessing the operational energy consumption over the lifetime of a building. This research divides the operational energy consumption into three items: space heating, hot water, and electricity. The calculations are based on standard energy consumption of buildings, in terms of end use of energy. The end use of energy is then converted into non-renewable primary energy, based on country-specific energy production profile. Furthermore, as energy production is constantly developing, future energy production scenarios are used to forecast the development of the use of non-renewable primary energy over the life cycle of 50 years. All of these calculations, together with the used calculation data, are presented in section "ADP of the operational energy use".

Case Study Building

All of the ADP calculations were made for a specific case building, which was located in Southern Finland, and represented a typical Finnish contemporary building. The building under study was a six storey residential building with a basement floor. The gross floor area of the building was 3060 m² and the number of apartments was 28. The structures of the building were passive-level and the heating method was district heating. The load-bearing frame, consisting of internal and external walls, floor slabs and roof, were precast concrete structures. The bill of quantities, extracted from the building information model (BIM) of the case building was used as the basis of the calculations of this research. Material quantities of the case building are not shown here, as they are presented later in this paper, in the result tables for the ADP (Table 1 and Table 2).

RESULTS

The following subsections present the calculation results of the case-study, along with the references for the used data sources. The ADP elements and ADP fossil for the case building are shown in the first two subsections, followed by results for soil stabilization. After this, the impacts of advanced building systems are assessed, followed by the impacts of transports and construction work. The last result section shows the results for ADP from operational energy use and compares it to the material-related ADP results.

ADP Elements of Building Materials

This section shows the results for ADP Elements of building materials for the case building. The following table (Table 1) shows that the total need of building materials over a 50-year life cycle for the case building is 4960 t, or 1.62 t/m². The total material need includes the initial material needs for construction of the building (89%), recurrent material needs for replacements and refurbishments (6%), and material losses (5%). The table also shows that the production of the building materials for the case-building requires a total of 7320 t of abiotic inputs, or 2.39 t/m². According to the results, the building-level abiotic depletion potential, over the lifetime of the building, is 1.05 kg of Antimony equivalents, or 0.34 g/m².

In addition to these results, the following Table 1 also includes the ADP characterization factors used in the calculations for each of the main materials. It also shows the noteworthy information on abiotic inputs, which lack an ADP characterization factor, and are therefore not included in the calculation results.

ADP Fossil of Building Materials

This section shows the results for ADP Fossil of building materials for the case-building. The total material needs presented in the following table (Table 2) match those presented in the previous section. According to the results, the ADP Fossil of the case-building is 15,900 GJ of fossil energy inputs, or 5.2 GJ/m².

ADP of Soil Stabilization

This section studies the effect of soil stabilization on the ADP elements and ADP fossil. The total material need for stabilization is 1420 t, including material losses (5%). The following Table 3 shows that the ADP elements value of soil stabilization is 530 g, or 0.17 g/m², and that the ADP Fossil is 3500 GJ, or 1.14 GJ/m².

ADP of Advanced Building Systems

This section assesses the ADP elements of advanced building systems. The ADP Fossil is not assessed, due to lack of reliable data.

Energy-Efficient Lighting

This section shows the calculation results for ADP Elements of energy-efficient lighting. The selected lamp type is a standard T12-type fluorescent lamp with a rare earth triphosphor coating, with a coating thickness of 5 mg/cm² [73] and a total of 7 grams of phosphorous coating. In addition to this, the lamp has low-pressure mercury vapor, with an estimated amount of 25 mg per lamp [74]. Assuming a service life of 10 years for the lamps, four replacements are required over the 50-year life cycle. The case building has a total of 355 lamps.

The following Table 4 shows that ADP Elements for energy-efficient lighting is 0.12 kg of Antimony equivalents, or 0.38 g/m².

Table 1: Total mass of materials, abiotic material inputs per ton material ton, abiotic material inputs with no abiotic depletion (ADP) characterization factor, average ADP characterization factor of abiotic inputs, total ADP of materials and data source for material inputs

Material	Total mass of materials (t)	Abiotic material inputs per material ton (t/t)	Total abiotic material inputs (t)	Abiotic material inputs with no ADP$_{CF}$ (%)	ADP$_{AVG}$ of abiotic inputs (t Sb eq /t)	Total ADP of materials (kg Sb eq)	Data source for material inputs
Aluminium	29	4.8	142	87.2%	3.22×10^{-6}	0.46	[75]
Concrete	3549	1.4	5016	99.9%	8.28×10^{-6}	0.04	[76]
Copper	4	6.0	26	99.2%	1.90×10^{-5}	0.49	[77]
Fossil materials	90	2.8	256	99.9%	7.40×10^{-10}	0.00	[78]
Gravel	629	1.9	1202	100.0%	–	–	[79]
Other minerals	337	0.8	254	100.0%	2.83×10^{-10}	0.00	[80]
Steel	83	3.5	291	91.6%	1.86×10^{-7}	0.05	[81]
Wood	42	0.1	5	99.7%	4.79×10^{-9}	0.00	[82]
Wood boards	200	0.6	129	99.7%	4.84×10^{-9}	0.00	[83]
Total	4960	–	7319	99.3%	–	1.05	–

Table 2: Total mass of materials, fossil energy inputs per ton material ton, total ADP Fossil of materials and data source for material inputs

Material	Total mass of materials (t)	Fossil energy inputs per material ton (GJ/t)	Total ADP of materials (GJ)	Data source for material inputs
Aluminium	29	37.0	1088	[75]
Concrete	3549	0.8	2720	[76]
Copper	4	17.5	75	[77]
Fossil materials	90	85.6	7696	[78]
Gravel	629	0.1	38	[79]
Other minerals	337	3.7	1259	[80]
Steel	83	15.7	1297	[81]
Wood	42	0.6	27	[82]
Wood boards	200	8.7	1728	[83]
Total	4960	–	15,900	–

Table 3: Total mass of materials, fossil energy inputs per ton material ton, total ADP Fossil of materials, abiotic material inputs per ton material ton, abiotic material inputs with no ADP characterization factor, average ADP characterization factor of abiotic inputs, total ADP of materials and data source for material inputs for soil stabilization of case-building

Material	Total mass of materials (t)	Fossil energy inputs per material ton (GJ/t)	Total ADP Fossil of materials (GJ)	Abiotic material inputs per material ton (t/t)	Total abiotic material inputs (t)	Abiotic material inputs with no ADP$_{CF}$ (%)	ADP$_{AVG}$ of abiotic inputs (kg Sb eq /t)	Total ADP of materials (kg Sb eq)	Data source for material inputs
CEMII	709	3.6	2558	1.7	1199	99.6%	0.00045	0.53	[84]
CaO	709	5.4	3820	3.2	2303	100.0%	–	–	[85]
Total	1420	9	6380	5	3500	99.8%	–	0.53	–

Table 4: Total weight of selected materials, ADP characterization factors for materials and total ADP for lighting, based on a single lamp type over a 50-year life-cycle with four replacements

Material	Total weight of materials (kg)	ADP$_{CF}$ kg (Sb eq)/kg	Total ADP kg (Sb eq)	Data source for characterization factors
Mercury	0.04	2.62	0.12	[72]
Rare earth elements	12.71	0.0006	0.007	[86]
Total	12.76	–	0.12	–

Solar Panels

This section looks at the photovoltaic panels of solar panels and shows their contribution to the depletion of abiotic resources, in terms of ADP Elements. The selected panels are of two types: c-Si (Crystalline Silicone) and CIS/CIGS (Copper Indium Selenide/Copper Indium Gallium (di) Selenide). The main material is glass, which forms approximately 74% to 84% of the total mass. The remainder is aluminium (10% to 12% of the total mass) and other metals (4% to 16% of the totals), as summarized in a report to the European Commission [87]. Assuming that the lifetime of the panels is 15 years, the panels will need to be renewed three times over the lifetime.

The results in Table 5 show that the ADP for solar panels may vary from 180 to 174,000 kg of Antimony equivalents, or 60 g/m^2 to 60 kg/m^2.

Table 5: Total area of solar panels, ADP characterization factors per square meter of panel, and total ADP for solar panels over a 50-year life-cycle with three replacements.

Panel type	Total area of solar panels (m^2)	ADP$_{CF}$ Sb eq (kg/m^2)	Total ADP Sb eq (kg)
c-Si	370	0.2	180
CIS/CIGS	370	157	174,380

ADP of Material Transportations and Construction Work

This section presents the ADP fossil of material transportations and construction work. The total ADP Fossil from material transportations, and construction and demolition work is 2400 GJ, or 0.8 GJ/m^2. The results are as follows:

- Total mass (building) 4960 t;
- Fossil fuels (construction work) 0.249 GJ/t (Based on data presented in [49]);
- Fossil fuels (demolition work 0.137 GJ/t (Based on data presented in [49]);
- Fossil fuels (transportation) 0.10 t (Based on VTT LIPASTO traffic emissions [88];
- Fossil fuels (total) 0.49 GJ/t;
- ADP fossil energy (building) 2413 GJ.

ADP of the Lifetime Operational Energy Use and Comparison to the ADP of Materials

This shows the results for the ADP Fossil of the lifetime operational energy use of the case building. The calculation of the used conversion factors is also explained in this section. The case building uses a total of 3050 MWh of heating energy for spaces, 5350 MWh for hot water, and 7650 MWh of electricity over 50 years, in terms of end-use of energy (EUE). Heating and hot water is produced with district heat, whereas electricity is taken from the grid.

In order to relate the end-use of energy to the use non-renewable primary energy resources, primary energy conversion factors (PECFs) are needed. Here, these factors are based on the Finnish data [89] and those are 0.77 for district heating and 1.75 for electricity.

As the assessment period covers a 50-year timespan, these conversion factors will not remain constant due to developments in energy production. Therefore, another conversion factor is needed to translate the non-renewable energy consumption of today to match the expected average over the 50-year assessment period. The future use of non-renewable primary energy is expected to follow closely the estimated future development of Finnish GHG emissions (data prepared by the Finnish Ministry of Employment and the Economy, and presented, for example, in [28]). The conversion factors used here for translating contemporary PECFs to 50-year averages are 0.8 for district heat and 0.4 for electricity and they are named future conversion factors (FCFs).

The following table (Table 6) shows the ADP Fossil for the case building, in terms of total non-renewable primary energy use over the 50-year life cycle.

Table 6: End-use of energy, primary energy conversion factors, future conversion factors, and ADP Fossil for the case building from operational energy use over 50-year life cycle

End-use energy of energy, purpose of use	End-use of energy (EUE) MWh	End-use of energy (EUE) GJ	Primary energy conversion factor (PECF)	Future conversion factor (FCF)	ADP Fossil/Total non-renewable primary energy use, 50a (GJ)
Heating energy	3050	10,980	0.77	0.8	7063
Hot water	5350	19,260	0.77	0.8	12,389
Electricity	7650	27,540	1.75	0.4	19,278
Total	16,050	57,780	−	−	38,730

Figure 1 combines the ADP Fossil values for operational energy use from Table 6, for building materials from Table 2, for soil stabilization from Table 3, and for material-related processes from Section 5.5. The respective ADP Fossil values for the different items are as follows: heating energy 7100 GJ (2.32 GJ/m^2), hot water 12,400 GJ (4.05 GJ/m^2), electricity 19,300 GJ (6.31

GJ/m²), building materials 15,900 GJ (5.20 GJ/m²), material-related processes 2400 GJ (0.78 GJ/m²), and site construction (soil stabilization) 6400 GJ (2.09 GJ/m²).

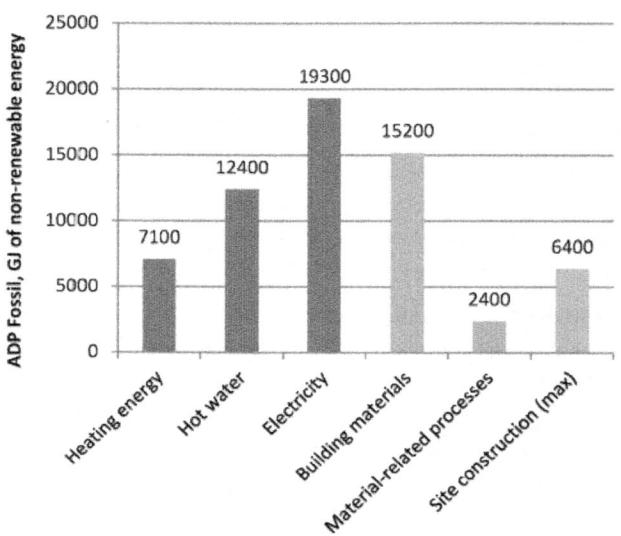

Figure 1: Fossil energy consumption (net calorific value) over the life cycle of the building.

In summary, the APD Fossil due to operational energy totals 38,700 GJ (12.65 GJ/m²) and material-related ADP Fossil is in total 17,600 GJ (5.75 GJ/m²) or 24,000 GJ (7.84 GJ/m²), depending on the stabilization needs. Therefore, the total lifetime ADP Fossil varies from 18.5 to 20.5 GJ/m². The result shows that the role of material-related non-renewable energy consumption for the case-building is at the level of 30% to 40% of lifetime total energy consumption.

DISCUSSION

The case-study of this research aimed to fill in the gaps in the current knowledge, as identified in the literature review. It looked into the depletion of natural raw materials, through an assessment of lifetime abiotic depletion potential (ADP) of a residential multi-storey case-building with concrete structures, for both ADP elements and ADP fossil, as defined in current guidelines [24,25,63]. It should be highlighted that due to the case-study approach, the generalization of the results should be done with caution, especially considering the building type and location.

The material quantities were extracted from the building information model (BIM) of a real world building, so the data accuracy for initial material consumption can be considered high. The material losses, on the other hand, were estimated to be at the level of 5% of total material consumption. Commonly used values in literature vary from 0% to 10% [70]. Also, the lifetime material needs for replacements and refurbishments were assessed through simple estimates on service lives of different building components. An analysis on the impacts of estimation errors show that a change of 25% in these factors would increase/decrease the material amounts by some 10% for the case-building.

The case-study used the European reference life cycle database, ELCD [71] to derive the abiotic material inputs and energy requirements for each of the main materials of the building. The LCIs of the database are compiled mainly by process analysis. It can be argued that this method is associated with underestimation of the impacts, as the number of processes and the order of upstream processes are limited [68], and sufficient boundaries may be difficult to cover due to the complexity of upstream processes [69]. For basic building materials, for example, the incompleteness factor, often referred to as truncation error [66] is estimated to be at least 10% [69], some estimates being as high as 60% for residential buildings [67].

It should be noted that the data sources for the ELCD-database are drawn on a wider regional level and the energy inputs for the production processes use country-level statistics and national grid-mix information and they are not pure process based analyses. This enhancement of process-based information with IO-based data can be considered to make the profiles of ELCD profiles hybrid analyses in a sense [69].

The ADP characterization factors used for the calculation of ADP elements embody significant uncertainty in them. This research used the CLM database's base reserve figures [72], as recommended in European ILCD handbook [63]. However, the current standards [24,25] do not explicitly state which reserve estimates to use, and some LCAs and EPDs may still be assessed using the ultimate reserve figures, as this has been a past recommendation [64]. The ADP characterization factor for base reserves of copper, for example, is two times bigger than that for the ultimate reserves, for iron 30 times bigger and for aluminium, 23,000 times bigger. This makes it difficult to reliably compare the results of ADP studies between each other. However, the ADP of the case building, 1.05 kg of Antimony equivalents for almost five million kilograms (4960 t) of building materials can be compared to the production of some basic metals from virgin raw materials. The production of 420 kg of copper, 41,500 kg of aluminium, or 630,000 kg of iron from virgin raw materials would produce the same ADP of 1.05 kg [72]. These comparisons suggest that the result for ADP of the building is of very low level.

Only 0.7% of the abiotic material inputs of the case building have a characterization factor in the first place, making the ADP elements assessment practically worthless. The basic issue behind this is that such factors cannot be defined for any of the common building materials, such as gypsum, silica sand, construction sand, clays, limestone, and such, due to lack of data on material configurations, reserves, reserve bases and ultimate reserves for these materials [65]. Based on the results of the case-study, the benefits and purpose of calculating ADP elements for buildings is highly questionable in its current form. Methods, which would better account for local scarcity of resources [37] or land or social impacts [38], could fit the purpose better.

The assessment of advanced building systems resulted in ADP elements of 0.12 kg. For solar panels, the figures were 180 and 180,000 kg of Antimony equivalents. The results of advanced building systems show that such systems may be of relatively high importance, compared to the building itself.

The case-study of this research also assessed the APD Fossil for the materials of the case-building. The uncertainties related to these calculations, concerning the material quantities and the used LCI database are the same, which were discussed previously for ADP elements. As ADP fossil is defined in terms of non-renewable energy, the problem of characterization factors does not have an effect on the results. The assessment results showed that the material-related ADP Fossil totalled from 17,600 GJ (5.75 GJ/m^2) to 24,000 GJ (7.84 GJ/m^2). Research on similar buildings is limited but, for example, results of two residential buildings with concrete frame and floor area of some 1200 m^2 **in Sweden, show embodied energy from 4.6 to 5.4 GJ/m^2** [90], as summarized in Ramesh *et al.* [44]. It should be pointed out that the embodied energy figures are not directly comparable to the ADP fossil figures, as the ADP fossil does not include the use of renewable energy.

The ADP fossil due to operational energy totalled to 38,700 GJ (12.65 GJ/m^2) in the case-study. The results show that the material-related non-renewable energy consumption of the case building was at the level of 30% to 40% of lifetime total non-renewable energy consumption. These results are in line with a GHG assessment of the same building, done in a previous research, showing that material-related GHG emissions accounted for 40% to 50% of lifetime total emissions [28]. The comparable result is largely explained by the fact that GHG emissions are mainly due to consumption of fossil energy resources. As discussed above, ADP fossil does not contain renewable energy. In Finland, for example, the share of renewable energy sources in energy production was 27% in the year 2010 [91].

The operational energy consumption (end-use of energy) was assessed based on standard consumption figures, stated in Finnish regulations. The energy consumption of the case building was 105 kWh/m^2. The real consumption figures may vary from this significantly, due to user behavior, as shown in previous research [92]. However, assessment of user behavior was not the focus of this study and this variation was not considered in the assessment. In order to convert the end-use of energy into non-renewable primary energy use, Finnish national-level energy production information was used [89] and, in order to take the future development towards low-emission energy production, conversion factors based on [28] were used. Whereas the present-day ratio of non-renewable primary energy to end-use of energy can be thought to be a relatively reliable figure, the future conversion factors depend on political decisions in the future and cannot be predicted accurately. For example, a decrease of 25% in these factors would impact the results significantly, indicating higher than expected share of renewable energy in the future and lower than expected share of non-renewable energy. For the case-study, such change would decrease the ADP fossil from operational energy use from 12.65 GJ/m^2 **to 9.5 GJ/m^2**. This would increase the role of material-related energy consumption from the level of 35% to 45% of lifetime totals.

The study was founded on the premise that the importance of material efficiency is based on one or more of the following impacts:

- the depletion of raw materials and its long-term socio-economic impacts;
- land use change due to the extraction of raw materials and its environmental impacts and impacts on the landscape and future recreational use;
- the use of energy in production processes of materials and depletion of non-renewable energy;
- harmful emissions from production processes of materials and their local and/or global environmental impacts;
- material cost impacts due to the limited availability of raw materials or a higher need for energy and/or labor in the different phases of production processes.

This research did a comprehensive literature study to outline and draw conclusions about different aspects of the material efficiency of buildings.

Material efficiency is a complex issue to deal with in steering because there is no widely acknowledged way to make different materials commensurable. The impacts of material efficiency extend to all the aspects of resource efficiency, as shown with Equation (1) of this paper. The demand for new buildings is influenced by their durability, service life and flexibility. The use of lightweight structures impacts the average mass per product, and the yield ratio is affected by material losses on the building site. Finally, the use of secondary materials typically reduces the emissions from production. Due to the comprehensive nature of material efficiency, the focus of policy formulation should not be on its individual components, such as yield rates, average masses per products, and such, but on the impacts caused by material efficiency. Söderholm and Tilton [14] argue similarly, that it is better to avoid policies that directly encourage specific material efficiency options, and that policies should address particular environmental problems and information externalities to enhance material efficiency in instead.

The study was founded on the premise that the importance of material efficiency is based on some of its impacts. The importance of the different impacts (indicated with indicators) can be viewed from the perspective of sustainable development. An indicator can be validated as applicable to sustainable building if it fulfils two minimum requirements: it must be related to a subject of concern for sustainable development, and buildings must have a significant impact on that issue [93].

From the perspective of sustainable development, the greenhouse gas emissions from building sector are an example of an environmental problem, on which material efficiency has a significant impact on. Greenhouse gas emissions from building sector are a significant contributor on global warming, and material efficiency has a significant impact on the issue.

The consumption of non-renewable energy resources is near analogous to the greenhouse gas emissions, as the greenhouse gas emissions are mainly the result of consumption of non-renewable fossil energy in production processes of materials. This analogy was also partly illustrated by the results of the case-study.

However, the results on material-related land use showed that the importance of material-efficiency on land-use was practically negligible, as the footprint of the building was significantly more important than the land used for the extraction of non-renewable raw materials.

From the viewpoint of costs, the results showed that the role of materials is only small, some 10% to 40% of the construction costs. This means that both savings through improved material efficiency, and additional costs through future price increases in materials, have only a limited impact on total costs.

The construction industry consumes significant amounts of raw materials globally. However, the most common building materials are also common in nature. The results suggest that the most common building materials have no significant impact on depletion natural raw materials globally, although locally this might be important. However, the case might be different for some scarcer resources, which are used in advanced building systems. The case of non-renewable energy resources is different, as discussed previously. The material efficiency has a significant impact on the consumption of non-renewable energy resources.

The impact indicators for material efficiency should be concrete and they should indicate problems, which have global significance. As such, the resource depletion indicators of the current guidelines for buildings do not fully support this. This research suggests that the material efficiency should focus on the significant global impacts of material efficiency, not on the individual factors of it. At the present-day, global warming and greenhouse gas emissions are among the biggest global problems, on which material efficiency has a direct and significant impact on. Therefore, this paper suggests that greenhouse gas emissions could be used as an indicator for material efficiency in building.

CONCLUSION

Material efficiency is emphasized as an important aspect of sustainable building, as indicated by the inclusion of the ADP aspect in EN 15804 [24] and EN 15978 [25] and the inclusion of the new basic requirement for sustainable use of resources in the Construction Product Regulation [60]. The roadmap to a resource-efficient Europe [3] addresses buildings as one of the three key sectors. However, further research is still needed to clarify and draw conclusions about the correct indicators and methods to assess the material efficiency of buildings and construction.

This research studied the different aspects of material efficiency: scarcity, land use, and environmental impacts related to the manufacturing of materials.

The preliminary results received with the help of a comprehensive case study (which was aimed at all the materials used for the case building) revealed that basic building materials have only a minor effect on the results when assessed in terms of ADP elements (as recommended by ILCD [63]). Approximately 99% of building materials have no effect on the ADP value, and, thus, approximately 1% of the materials (by weight) determine the results. The basic building materials that affect the results are the metallic materials used in buildings (steel, aluminium and copper). The result also showed that very minor material flows (in terms of weight), such as lamps and solar panels, may have a significantly bigger effect than any of the basic building materials, including all the metal used. The result raises questions of whether the ADP elements assessment method is appropriate for the assessment of buildings and construction. On the other hand, the ADP fossil fuel calculations were able to capture the material impacts more effectively. When comparing the ADP fossil values from material-related sources with the values from operational energy use, the share of materials accounted for approximately 30% to 40% of the lifetime totals.

Despite the relatively low impact on the depletion of abiotic resources, the building materials still have local impacts on the landscape and natural environment. The impacts of the extraction of gravel on ground water may also be substantial on local level. The impact of land use of abiotic materials is small compared with the footprint of the building. The land use of the building itself dominates the results (unless the land used for wood used for heating energy production is taken into account). If the use of wood is taken into account, its impact dominates in terms of land use and but also with regard to biodiversity impacts.

The greenhouse gas emissions from building sector are examples of environmental problems, on which material efficiency has a significant impact on. Greenhouse gas emissions from buildings are affected by all the aspects of material efficiency, and improvements in material efficiency can have significant impacts on the amount of emissions. This paper suggests that greenhouse gas emissions could be used as an indicator for material efficiency in building.

ACKNOWLEDGMENTS

This study was a part of the Sustainability and performance assessment and benchmarking of buildings (SuPerBuildings) project (FP7 EU Project 2010-2012) and the Ownership in sustainable building (OKRA) project (2011-2014) funded by TEKES – the Finnish Funding Agency for technology and Innovation.

AUTHOR CONTRIBUTIONS

The article was done in collaboration of the authors. The case study was done by Antti Ruuska, the study of literature was done equally by both authors. The second author, Tarja Häkkinen was the supervisor while the main writing was done by Antti Ruuska.

REFERENCES

1.   The Federation of Finnish Technology Industries (Teknologiateollisuus) 2013. Kilpailukykyä ja uutta liiketoimintaa materiaalitehokkuudesta. (In Finnish). Available online: http://www.teknologiateollisuus.fi/file/15592/ Materiaalitehokkuusjulkaisu2013.pdf.html (accessed on 13 February 2014).

2. European Commission. A resource-efficient Europe—Flagship initiative under the Europe 2020 Strategy. COM (2011) 21. Brussels, 26.1.2011. Available online: http://ec.europa.eu/resource-efficient-europe/pdf/resource_efficient_europe_en.pdf (accessed 23 June 2014).

3. European Commission. Roadmap to a Resource Efficient Europe. COM (2011) 571 final. Brussels, 20.9.2011. Available online: http://ec.europa.eu/environment/resource_efficiency/pdf/com2011_571.pdf (accessed 23 June 2014).

4. Berge, B. *The Ecology of Building Materials*, 2nd ed.; Elsevier: Italy, 2009; p. 427.

5. Azapagic, A. Developing a framework for sustainable development indicators for the mining and minerals industry. *J. Clean. Prod.* **2004,** *12*, 639–662.

6. European Commission. Critical raw materials for the EU. Technical Report, June 2010. Available online: http://ec.europa.eu/enterprise/policies/raw-materials/files/docs/report-b_en.pdf (accessed on 21 January 2014).

7. Brent, A.C.; Hietkamp, S. The Impact of Mineral Resource Depletion. *Int. J. Life Cycle Assess.* **2006,** *11*, 361–362.

8. Steen, B.A. Abiotic resource depletion. Different perceptions of the problem with mineral deposits. *Int. J. Life Cycle Assess.* **2006,** *1*, 49–54.

9. Strauss, K.; Brent, A.; Hietkamp, S. Characterisation and normalisation factors for life cycle impact assessment mined abiotic resources categories in south Africa. The manufacturing of catalytic converter exhaust systems as a case study. *Int. J. Life Cycle Assess.* **2006,** *11*, 162–171.

10. Meadows, D.H.; Randers, J.; Meadows, D.L. *Limits to Growth. The 30-Year Update*; Earthscan: Oxford, UK, 2005; p. 338.

11. Jolliet, O.; Müller-Wenk, R.; Bare, J.; Brent, A.; Goedkoop, M.; Heijnungs, R.; Itsubo, N.; Peña, C.; Pennington, D.; Potting, J.; *et al.* The LCIA Midpoint-damage Framework of the UNEP/SETAC Life Cycle Initiative. *Int. J. Life Cycle Manag.* **2004,** *9*, 394–404.

12. Yellishetty, M.; Ranjith, P.G.; Tharumarajah, A.; Bhosale, S.A. Life cycle assessment in the minerals and metals sector: A critical review of selected issues and challenges. *Int. J. Life Cycle Assess.* **2009,** *14*, 257–267.

13. Finnveden, G. The Resource Debate Needs to Continue. *Int. J. Life Cycle Assess.* **2005,** *10*, 372–372.

14. Söderholm, P.; Tilton, J.E. Material efficiency: An economic perspective. *Resour. Conserv. Recycl.* **2012,** *61*, 75–82.

15. Vieira, M.D.M.; Ponsioen, T.C.; Goedkoop, M.J.; Huijbregts, M.A.J. Surplus cost as a life cycle impact indicator for mineral resource scarcity. Available online: http://www.lc-impact.eu/userfiles/D_1_4_mineral_and_fossil_resource_use.pdf (accessed on 10 December 2013).

16. Ponsionen, T.C.; Vieira, M.D.M.; Goedkoop, M.J. Surplus cost as a life cycle impact indicator for fossil resource depletion. Available online: http://www.lc-impact.eu/userfiles/D_1_4_mineral_and_fossil_resource_use.pdf (accessed on 10 December 2013).

17. Rees, W.; Wackernagel, M. Urban ecological footprints: Why cities cannot be sustainable—And why they are a key to sustainability. *Environ. Impact Assess. Rev.* **1996,** *16*, 223–248.

18. Haberl, H.; Wackernagel, M.; Krausmann, F.; Erb, K.-H.; Monfreda, C. Ecological footprints and human appropriation of net primary production: A comparison. *Land Use Policy* **2004,** *21*, 279–288.

19. European Environment Agency. Land accounts for Europe 1990–2000, towards integrated land and ecosystem accounting, EEA Report No 11/2006. Available online: http://www.eea.europa.eu/publications/eea_report_2006_11(accessed 23 June 2014).

20. Eurostat. *Economy-wide Material Flow Accounts and Derived Indicators—A Methodological Guide*; Office for Official Publications of the European Communities: Luxembourg, Luxembourg, 2001. Available online: http://epp.eurostat.ec.europa.eu/portal/page/portal/environmental_accounts/documents/3.pdf (accessed 23 June 2014).

21. Van der Voet, E.; van Oers, L.; Moll, S.; Schütz, H.; Bringezu, S.; de Bruyn, S.; Sevenster, M.; Warringa, G. *Policy Review on Decoupling: Development of Indicators to Assess Decoupling of Economic Development and Environmental Pressure in the EU-25 and AC-3 Countries*; Department Industrial Ecology, Leiden University: Leiden, The Netherlands, 2004.

22. Tukker, A.; Huppes, G.; van Oers, L.; Heijungs, R. *Environmentally Extended Input-Output Tables and Models for Europe*; Eder, P., Delgado, L., Neuwahl, F., Eds.; EC, JRC, IPTS Technical Report Series; EUR 22194 EN. Seville, Spain, 2006. Available online: http://ftp.jrc.es/EURdoc/eur22194en.pdf (accessed on 23 June 2014).

23. Berger, M.; Finkbeiner, M. Correlation analysis of life cycle impact assessment indicators measuring resource use. *Int. J. Life Cycle Assess.* **2011,** *16*, 75–81.

24. EN 15804:2012 Sustainability of construction works—Environmental product declarations—Core rules for the product category of construction

products. 2012. Available online: http://standards.cen.eu/dyn/www/f?p=204:110:0::::FSP_PROJECT:40703&cs=1C696AB3A6B08F09003DC00E3E3B2DA17 (accessed 23 June 2014).

25. EN 15978:2011. Sustainability of construction works—Assessment of environmental performance of buildings—Calculation method. 2011. Available online: http://standards.cen.eu/dyn/www/f?p=204:110:0::::FSP_PROJECT:31325&cs=16BA44316931 8FC086C 4652D797E50C47 (accessed 23 June 2014).

26. Stewart, M.; Weidema, B.P. A Consistent Framework for Assessing the Impacts from Resource Use—A focus on resource functionality. *Int. J. Life Cycle Assess.* **2005,** *10,* 240–247.

27. Pennington, D.W.; Potting, J.; Finnveden, G.; Lindeijer, E.; Jolliet, O.; Rydberg, T.; Rebitzer, G. Life cycle assessment Part 2: Current impact assessment practice. *Environ. Int.* **2004,** *30,* 721–739.

28. Ruuska, A.; Häkkinen, T.; Vares, S.; Korhonen, M.-R.; Myllymaa, T. Environmental Impacts of Building Materials (Rakennusmateriaalien ympäristövaikutukset), Finnish Ministry of Environment, 8/2013 (In Finnish). Available online: http://www.ym.fi/download/noname/%7B1FAF46B2-2649-41ED-B3AA-5EA789C9512F%7D/37571 (accessed on 31 January 2014).

29. Salmi, O.; Haapalehto, T.; Harlin, A.; Häkkinen, T.; Kangas, H.; Mroueh, U.-M.; Qvintus, P. *The Development of Material Efficiency in the Finnish Industries*; Technical Report for The Ministry of employment and the economy (In Finnish): Helsinki, Finland, 2013; p. 46.

30. Allwood, J.M.; Ashby, M.F.; Gutowski, T.G.; Worrell, E. Material efficiency: A white paper. *Resour. Conserv. Recycl.*2011, *55,* 362–381.

31. Prior, T.; Giurco, D.; Mudd, D.; Mason, L.; Behrich, J. Resource depletion, peak minerals and the implications for sustainable resource management. *Glob. Environ. Chang.* **2012,** *22,* 577–587.

32. Wouters, H.; Bol, D. *Material Scarcity*; Materials Innovation Institute: Delft, The Netherlands, 2009.

33. De Almeida, P.; Silva, P.D. The peak of oil production—Timings and market recognition. *Energy Policy* **2009,** *37,* 1267–1276.

34. European Commission. The raw materials initiative—Meeting our critical needs for growth and jobs in Europe. COM 699. Brussels, 4.11.2008. Available online: http://ec.europa.eu/enterprise/sectors/metals-minerals/files/com699_en.pdf(accessed 23 June 2014).

35. Ahmed, M.S.; Vidyadhara, H.S. Experimental study on strength behaviour

of recycled aggregate concrete. *Int. J. Eng. Res. Technol.* **2013,** *2*, 76–82.

36. Lohani, T.K.; Padhi, M.; Dash, K.P.; Jena, S. Optimum utilization of quarry dust as partial replacement of sand in concrete. *Int. J. Appl. Sci. Eng. Res.* **2012,** *1*, 391–404.

37. Habert, G.; Bouzidi, Y.; Chen, C.; Jullien, A. Development of a depletion indicator for natural resources used in concrete. *Resour. Conserv. Recycl.* **2010,** *54*, 364–376.

38. Yellishetty, M.; Mudd, G.M.; Ranjith, P.G. The steel industry, abiotic resource depletion and life cycle assessment: A real or perceived issue? *J. Clean. Prod.* **2011,** *19*, 78–90.

39. U.S. Department of Energy, Critical Materials Strategy Summary. 2011. Available online: http://energy.gov/sites/prod/files/DOE_CMS2011_FINAL_Full.pdf (accessed on 7 January 2014).

40. Gleich, B.; Achzet, B.; Mayer, H.; Rathgeber, A. An empirical approach to determine specific weights of driving factors for the price of commodities—A contribution to the measurement of the economic scarcity of minerals and metals. *Resour. Policy* **2013,** *38*, 350–362.

41. Sartori, I.; Hestnes, A.G. Energy use in the life cycle of conventional and low-energy buildings: A review article.*Energy Build.* **2007,** *39*, 249–257.

42. Hernandez, P.; Kenny, P. Development of a methodology for life cycle building energy ratings. *Energy Policy* **2011,** *39*, 2779–3788.

43. Stephan, A.; Crawford, R.H.; de Myttenaere, K. Towards a more holistic approach to reducing the energy demand of dwellings. In Proceedings of the 2011 International Conference on Green Buildings and Sustainable Cities, Procedia Engineering, Bologna, Italy, 15–16 September; 2011; pp. 1033–1041.

44. Ramesh, T.; Prakash, R.; Shukla, K.K. Life cycle energy analysis of buildings: An overview. *Energy Build.* **2010,** *42*, 1592–1600.

45. Yung, P.; Lam, K.C.; Yu, C. An audit of life cycle energy analyses of buildings. *Habitat Int.* **2013,** *39*, 43–54.

46. Aktas, C.B.; Bilec, M.M. Impact of lifetime on US residential building LCA results. *Int. J. Life Cycle Assess.* **2011,** *17*, 337–349.

47. Rauf, A.; Crawford, R.H. The relationship between material service life and the life cycle energy of contemporary residential buildings in Australia. *Archit. Sci. Rev.* **2013,** *56*, 252–261.

48. Stephan, A.; Crawford, R.H.; de Myttenaere, K. Multi-scale life cycle energy analysis of a low-density suburban neighbourhood in Melbourne, Australia. *Build. Environ.* **2013,** *68*, 35–49.

49. Fuller, R.J.; Crawford, R.H. Impact of past and future residential housing development patterns on energy demand and related emissions. *J. Hous. Build. Environ.* **2011**, *26*, 165–183.

50. Säynäjoki, A.; Heinonen, J.; Junnila, S. A scenario analysis of the life cycle greenhouse gas emissions of a new residential area. *Environ. Res. Lett.* **2012**, *7*, 034037:1–034037:10.

51. Milà i Canals, L.; Bauer, C.; Depestele, J.; Dubreuil, A.; Freiermuth Knuchel, R.; Gaillard, G.; Michelsen, O.; Müller-Wenk, R.; Rydgren, B. Key elements in a framework for land use impact assessment within LCA. *Int. J. Cycle Assess.* 2007, *12*, 5–15.

52. Marshall, E.; Shortle, J. Urban Development Impacts on Ecosystems. In *Land Use Problems and Conflicts: Causes Consequences and Solutions*; Goetz, S., Shortle, J., Bergstrom, J., Eds.; Routledge Publishing: New York, NY, USA, 2005.

53. Scalenghe, R.; Marsan, F.A. The anthropogenic sealing of soils in urban areas. *Landsc. Urban Plan.* **2009**, *90*, 1–10.

54. Wheater, H.; Evans, E. Land use, water management and future flood risk. *Land Use Policy* **2009**, *26*, 251–264.

55. Häkkinen, T.; Helin, T.; Antuña, C.; Supper, S.; Schiopu, N.; Nibel, S. Land Use as an Aspect of Sustainable Building.*Int. J. Sustain. Land Use Urban Plan.* **2013**, *1*, 21–41.

56. Finnish Ministry of Environment. Sustainable use of soil materials. Guidelines of the Ministry of environment. Available online: http://www. ymparisto.fi/download.asp?contentid=101195&lan=fi (accessed on 30 August 2013).

57. Goodrum, P.M.; Zhai, D.; Yasin, M. Relationship between Changes in Material Technology and Construction Productivity. *J. Constr. Eng. Manag.* **2009**, *135*, 278–287.

58. European Parliament, European Council, Directive 2002/91/EC of the European Parliament and of the Council of 16 December 2002 on the energy performance of buildings, 2002. Available online: http://eur-lex. europa.eu/legal-content/EN/TXT/PDF/?uri=CELEX:32002L0091&from=EN (accessed 23 June 2014).

59. European Parliament, European Council, Directive 2010/31/EU of the European Parliament and of the Council of 19 May 2010 on the energy performance of buildings, 2010. Available online: http://eur-lex.europa. eu/LexUriServ/LexUriServ.do?uri=OJ:L:2010:153:0013:0035:EN:PDF (accessed 23 June 2014).

60. Cansino, J.M.; Pablo-Romero, M.P.; Román, R.; Yñiguez, R. Promoting renewable energy sources for heating and cooling in EU-27 countries. Assessment of the sustainable building steering mechanisms in selected EU member states. *Energy Policy* **2011**, *39*, 3803–3812.

61. Tuominen, P.; Klobut, K.; Tolman, A.; Adjei, A.; de Best-Waldhober, M. Energy savings potential in buildings and overcoming market barriers in member states of the European Union. *Energy Build.* **2012**, *51*, 48–55.

62. European Union. The Construction Products Regulation (EU) No 305/2011 (CPR). 2011. Available online: http://eur-lex.europa.eu/LexUriServ/LexUriServ.do?uri=OJ:L:2011:088:0005:0043:EN:PDF (accessed 23 June 2014).

63. European Commission Joint Research Centre (JRC). *ILCD Handbook: Recommendations for Life Cycle Impact Assessment in the European Context*; Publication Office of the European Union: Luxembourg, Luxembourg, 2011. Available online: http://publications.jrc.ec.europa.eu/repository/bitstream/111111111/26229/1/jrc61049_ilcd%20handbook%20final.pdf(accessed on 4 February 2014).

64. Guinée, J.B.; Gorrée, M.; Heijungs, R.; Huppes, G.; Kleijn, R.; Koning, A.; de Oers, L.; van Wegener Sleeswijk, A.; Suh, S.; de Haes, H.A.; *et al. Handbook on Life Cycle Assessment. Operational Guide to the ISO Standards. I: LCA in Perspective. IIa: Guide. IIb: Operational annex. III: Scientific background*; Kluwer Academic Publishers: Dordrecht, The Netherlands, 2002; p. 692.

65. Van Oers, L.; de Koning, A.; Guinée, J.B.; Huppes, G. *Abiotic Resource Depletion in LCA, Improving Characterization Factors for Abiotic Resource Depletion as Recommended in the New Dutch LCA Handbook*; Road and hydraulic engineering institute: Amsterdam, The Netherlands, 2002.

66. EN ISO 14040:2006. In *Environmental Management. Life Cycle Assessment. Principles and Framework*; The International Organization for standardization: London, UK, 2006.

67. Crawford, R.H. Validation of a hybrid life-cycle inventory analysis method. *J. Environ. Manag.* **2008**, *88*, 496–506.

68. Suh, S.; Lenzen, M.; Treloar, G.J.; Hondo, H.; Horvath, A.; Huppes, G.; Jolliet, O.; Klann, U.; Krewitt, W.; Moriguchi, Y.; Munksgaard, J.; Norris, G. Critical review: System Boundary Selection in Life-Cycle Inventories Using Hybrid Approaches. *Environ. Sci. Technol.* **2004**, *38*, 657–664.

69. Treloar, G.J. Extracting Embodied Energy Paths from Input-Output Tables: Towards an Input-Output-based Hybrid Energy Analysis

Method. *Econ. Syst. Res.* **1997**, *9*, 375–391.

70. Dixit, M.K.; Culp, C.H.; Férnandez-Solís, J.L. System boundary for embodied energy in buildings: A conceptual model for definition. *Renew. Sustain. Energy Rev.* **2013**, *21*, 153–164.

71. European Commission Joint Research Centre (JRC). European reference Life Cycle Database, ELCD, European Platform on Life Cycle Assessment. Available online: http://eplca.jrc.ec.europa.eu/?page_id=126 (accessed on 23 June 2014).

72. University of Leiden, CML-IA Characterisation Factors, CML, 2013. Available online: http://www.leidenuniv.nl/cml/ssp/databases/cmlia/cmlia.zip (accessed on 26 January 2014).

73. Soules, T.F.; Whitman, P.K.; Chirayath, D.R. Fluorescent lamp with phosphor coating of multiple layers. European Patent Specification EP 0807958B1, 30 October 2012.

74. U.S. Environmental Protection Agency. *Office of Solid Waste, Mercury Emissions from Disposal of Fluorescent Lamps*; Office of Solid Waste U.S. Environmental Protection Agency: Washington, DC, USA; 30; June; 1997.

75. European aluminium association (EAA). Aluminium extrusion profile, LCI data set, European reference Life Cycle Database, ELCD. 2013; Permanent dataset URI. Available online: http://lca.jrc.ec.europa.eu/lcainfohub/datasets/elcd/processes/09215eb0-5fc9-11dd-ad8b-0800200c9a66.xml (accessed on 23 June 2014).

76. PE International, Pre-cast concrete, LCI data set. European reference Life Cycle Database, ELCD. 2013; Permanent dataset URI. Available online: http://lca.jrc.ec.europa.eu/lcainfohub/datasets/elcd/processes/898618b0-3306-11dd-bd11-0800200c9a66.xml (accessed on 23 June 2014).

77. European Copper Institute, Copper tube, LCI data set. European reference Life Cycle Database, ELCD. 2013; Permanent dataset URI. Available online: http://lca.jrc.ec.europa.eu/lcainfohub/datasets/elcd/contacts/42a11490-573c-11dd-ae16-0800200c9a66.xml (accessed on 23 June 2014).

78. PE International, Polypropylene fibres (PP), LCI data set. European reference Life Cycle Database, ELCD. 2013; Permanent dataset URI. Available online: http://lca.jrc.ec.europa.eu/lcainfohub/datasets/elcd/processes/db00901b-338f-11dd-bd11-0800200c9a66.xml (accessed on 23 June 2014).

79. PE International, Gravel 2/32, LCI data set. European reference Life Cycle Database, ELCD. 2013; Permanent dataset URI. Available online: http://lca.jrc.ec.europa.eu/lcainfohub/datasets/elcd/processes/898618b2-3306-11dd-bd11-0800200c9a66.xml (accessed on 23 June 2014).

80. Eurogypsum, Gypsum Plasterboard, LCI data set. European reference Life Cycle Database, ELCD. 2013. Available online: http://lca.jrc.ec.europa.eu/lcainfohub/datasets/elcd/processes/cc39e70e-4a40-42b6-89e3-7305f0b95dc4.xml(accessed on 23 June 2014).

81. Steel sections, LCI data set. Worldsteel, European reference Life Cycle Database, ELCD. 2013; Permanent dataset URI. Available online: http://elcd.jrc.ec.europa.eu/ELCD3/resource/processes/09d61948-238a-40e7-8e1f-afdc0c98f902?format=html&version=03.00.000 (accessed on 23 June 2014).

82. PE-International, Pine wood, European reference Life Cycle Database, ELCD. 2013; Permanent dataset URI. Available online: http://lca.jrc.ec.europa.eu/lcainfohub/datasets/elcd/processes/621e64d0-f471-4023-9ebc-a52cd8ee573f.xml(accessed on 23 June 2014).

83. PE-International, Particle board, European reference Life Cycle Database, ELCD. 2013; Permanent dataset URI. Available online: http://lca.jrc.ec.europa.eu/lcainfohub/datasets/elcd/processes/bd7fdac9-40d5-4613-9374-6969803269d9.xml (accessed on 23 June 2014).

84. Cembureau, Portland cement, European reference Life Cycle Database, ELCD. 2013; Permanent dataset URI. Available online: http://eplca.jrc.ec.europa.eu/ELCD3/resource/processes/600573dd-dfa5-44e5-b458-8727e793ffd7.xml(accessed on 23 June 2014).

85. European Lime Association, Quicklime, European reference Life Cycle Database, ELCD. 2013; Permanent dataset URI. Available online: http://eplca.jrc.ec.europa.eu/ELCD3/resource/sources/7983f4c6-a355-4250-aaa8-5780a72cc1df.xml(accessed on 23 June 2014).

86. European Commission Joint Research Centre (JRC). *Characterisation Factors of the ILCD, Recommended Life Cycle Impact Assessment Methods, Database and Supporting Information, JRC Technical Notes*; European Union, Publications office of the European Union: Luxembourg, Luxembourg, 2012.

87. Bio Intelligence Service. *Study on Photovoltaic Panels Supplementing the Impact Sssessment for a Recast of the WEEE Directive*, Final report to European Commission DG ENV. 14 April 2011. Available online: http://ec.europa.eu/environment/waste/weee/pdf/Study%20on%20PVs%20Bio%20final.pdf (accessed on 23 June 2014).

88. LIPASTO—A calculation system for traffic exhaust emissions and energy consumption in Finland, VTT Technical Research Centre of Finland. Available online: http://lipasto.vtt.fi/indexe.htm (accessed on 21 January 2014).

89. Keto, M. *Energy Factors, General Principles and Factors for Realized Production of Electricity and District Heating*; Technical Report for The Ministry of Environment; Aalto University: Espoo, Finland; November; 2010.

90. Adalberth, K.; Almgren, A.; Holleris, P.E. Life-Cycle assessment of four multi-family buildings. *Int. J. Low Energy Sustain. Build.* **2001,** *2,* 1–21.

91. Statistics Finland, Energy Statistics Year book 2011. Available online: http://www.stat.fi/tup/julkaisut/tiedostot/julkaisuluettelo/yene_enev_201100_2012_6164_net.pdf (accessed on 23 June 2014).

92. Blom, I.; Itard, L.; Meijer, A. Environmental impact of building-related and user-related energy consumption in dwellings. *Build. Environ.* **2011,** *46,* 1657–1669.

93. Häkkinen, T. *Sustainable Refurbishment of Exterior Walls and Building Facades*; VTT: Espoo, Finland, 2012. Available online: http://www.vtt.fi/inf/pdf/technology/2012/T30.pdf (accessed on 23 June 2014).

CITATION

CHAPTER 1

Schijndel, A. (2015) Evaluation of Inverse Modeling Techniques for Pinpointing Water Leakages at Building Constructions. Modeling and Numerical Simulation of Material Science, 5, 15-25. doi: 10.4236/mnsms.2015.51002.

CHAPTER 2

Chiara Scrosati and Fabio Scamoni, Managing Measurement Uncertainty in Building Acoustics, doi: 10.3390/buildings5041389.

CHAPTER 3

María Dolores Andújar-Montoya, Virgilio Gilart-Iglesias, Andrés Montoyo and Diego Marcos-Jorquera, A Construction Management Framework for Mass Customisation in Traditional Construction, doi:10.3390/su7055182.

CHAPTER 4

Ian Skelton, Peter Demian, Jacqui Glass, Dino Bouchlaghem and Chimay Anumba, Lifting Wing in Constructing Tall Buildings—Aerodynamic Testing, doi:10.3390/buildings4020245.

CHAPTER 5

Junli Yang, Ibuchim Cyril B. Ogunkah, A Multi-Criteria Decision Support System for the Selection of Low-Cost Green Building Materials and Components, http://dx.doi.org/10.4236/jbcpr.2013.14013

CHAPTER 6

Roger Birchmore, Andy Pivac, and Robert Tait, Impacts of an Innovative Residential Construction Method on Internal Conditions, doi:10.3390/buildings5010179.

CHAPTER 7

Agnieszka Zalejska-Jonsson, Hans Lind, and Staffan Hintze, Energy-Efficient Technologies and the Building's Saleable Floor Area: Bust or Boost for Highly-Efficient Green Construction?, doi:10.3390/buildings3030570.

CHAPTER 8

Hyo Seon Park, Sewook Son, Se Woon Choi and Yousok Kim, Wireless Laser Range Finder System for Vertical Displacement Monitoring of Mega-Trusses during Construction, doi:10.3390/s130505796.

CHAPTER 9

Ma, J. (2015) Influence Analysis of a New Building to the Bridge Pile Foundation Construction. Open Journal of Civil Engineering, 5, 109-117. doi: 10.4236/ojce.2015.51011.

CHAPTER 10

Antti Ruuska and Tarja Häkkinen, Material Efficiency of Building Construction, doi:10.3390/buildings4030266.

INDEX